01/2021

APOLLO'S ARROW

ALSO BY NICHOLAS A. CHRISTAKIS

Death Foretold

Connected (with James H. Fowler)

Blueprint

APOLLO'S ARROW

THE PROFOUND AND ENDURING IMPACT OF CORONAVIRUS ON THE WAY WE LIVE

NICHOLAS A. CHRISTAKIS

Little, Brown Spark
New York Boston London

Little, Brown Spark
Hachette Book Group
1290 Avenue of the Americas, New York, NY 10104
littlebrownspark.com

First Edition: October 2020

Little, Brown Spark is an imprint of Little, Brown and Company, a division of Hachette Book Group, Inc. The Little, Brown Spark name and logo are trademarks of Hachette Book Group, Inc.

The publisher is not responsible for websites (or their content) that are not owned by the publisher.

The Hachette Speakers Bureau provides a wide range of authors for speaking events. To find out more, go to hachettespeakersbureau.com or call (866) 376-6591.

Illustrations designed by Cavan Huang

ISBN 978-0-316-62821-1
LCCN 2020942642

1 2020

LSC-C

Printed in the United States of America

For my teacher and dear friend
Renée C. Fox,
who survived and studied epidemics, who deeply
understands how illness and society intersect, and
who has influenced generations of fortunate students

And for my many other teachers
across a lifetime, including
Paul V. Piazza, Tom S. Reese, Leopold J. Pospisil,
John B. Mulliken, Allan M. Brandt,
Arthur M. Kleinman, Paul D. Allison,
Sankey V. Williams, and Arthur H. Rubenstein

Contents

Preface   xii

1. An Infinitesimal Thing   3

2. An Old Enemy Returns   34

3. Pulling Apart   85

4. Grief, Fear, and Lies   137

5. Us and Them   171

6. Banding Together   206

7. Things Change   247

8. How Plagues End   295

Acknowledgments   325

Notes   327

Index   357

And [Apollo] descended from the summits of Olympus, enraged in heart, having upon his shoulders his bow and quiver covered on all sides. But as he moved, the shafts rattled forth with upon the shoulder of him, enraged; but he went along like unto the night. Then he sat down apart from the ships, and sent among them an arrow, and terrible arose the clang of his silver bow. First, he attacked the mules, and the swift dogs. But afterward dispatching a pointed arrow against [the Greeks] themselves, he smote them, and frequent funeral piles of the dead were continually burning. Nine days through the army went the arrows of the god; but on the tenth, Achilles called the people to an assembly; for to his mind the white-armed goddess [Hera] had suggested it; for she was anxious concerning the Greeks, because she saw them perishing.

—Homer, *The Iliad*

Acronyms

CDC: Centers for Disease Control and Prevention, the leading government agency charged with epidemic control, based in Atlanta, Georgia.

COVID-19: The clinical disease caused by SARS-2, involving a range of symptoms and severities; also used to refer to the pandemic itself.

NIAID: National Institute of Allergy and Infectious Diseases, the leading government agency charged with scientific research into infectious disease, based in Bethesda, Maryland.

NPI: A nonpharmaceutical intervention, such as quarantining, used instead of or in addition to drugs to combat an epidemic.

PPE: Personal protective equipment, such as masks, face shields, gloves, etc., worn by health-care personnel and others to avoid contracting an infection.

SARS: Severe acute respiratory syndrome, a serious clinical illness involving shortness of breath that can result from infection with various pathogens or from other injuries to the lungs; also used as the name of a condition caused by the SARS-1 virus.

SARS-1: Virus from the coronavirus family that emerged in 2003 and caused a small pandemic.

SARS-2: Virus, also known as SARS-CoV-2, from the coronavirus family that emerged in 2019 and caused a large pandemic.

Preface

The gods of Greek mythology were ever present in my childhood. They were constant companions of my imagination, the subjects of my immigrant parents' bedtime stories, and even the names of children I played with when we visited our cousins in Greece. I was fascinated by the gods' duality: immortality and power contrasted with frailty and vice. The god Apollo, for example, was both a healer and the bringer of disease. During the Trojan War, with his silver bow and quiver of arrows, he rained a plague down on the Greeks to punish them for kidnapping and enslaving Chryseis, the daughter of one of his favored priests.

I found myself thinking again about Apollo and his vengeance as I contemplated our own twenty-first-century barrage more than three thousand years after the events described in *The Iliad*. It seemed to me that the novel coronavirus was a threat that was both wholly new and deeply ancient. This catastrophe called on us to confront our adversary in a modern way while also relying on wisdom from the past.

Despite the advances we have made in medicine, sanitation, communication, technology, and science, this pandemic is nearly as ruinous as any in the past century. Lonely deaths. Families unable to say goodbye to loved ones or perform proper funerals and acts of mourning. Destroyed livelihoods and stunted educations. Bread lines. Denial. Fear and sadness and pain. As I write, on August 1, 2020, over 155,000 Americans and over 680,000 people worldwide have died, and many more are still uncounted. A second wave of the pandemic is imminent, whether or not the hopes for a rapid vaccine are realized.

However, even in the midst of the onslaught, many people believe that the efforts to contain the virus have been excessive. Some Americans feel that our response has been overblown, yet another reflection of this nation's modern inability to accept hard realities. But I believe this thinking is wrong on two counts. First, it has required extraordinary force, including all our twenty-first-century wealth and know-how, to contain the virus to "only" this many deaths. I share the view of many good scientists that vastly more Americans would have died—perhaps a million—had we failed to deploy the resources we marshaled, belatedly, in the spring of 2020 to cope with the first wave of the pandemic. To compare this COVID-19 pandemic without mitigation efforts (or even with mitigation efforts!) to a typical flu season, as some have done, is a misreading of reality. Second, it is a misreading of history to think that in our time we would somehow be spared the burden of having to deal with a pandemic or that other people in other times have not faced the same fear and loneliness, the same polarization, the same fights over masks and business closures, the same call to neighborliness and cooperation. They have.

In late January 2020, as the virus was gathering force, I shifted the work of the many talented young scientists and staff in my research group at Yale to focus on it. First, working with Chinese colleagues, we published a study that used the mobile-phone data of millions of people in China to track the spread of the virus in January and February 2020. Then my lab began to plan studies of the biology and impact of the virus in the isolated region of Copan, Honduras, where we had a long-term field site and close relationships with thirty thousand residents in one hundred seventy-six villages. We also started exploring how mass gatherings, like elections and protests, might intersect with the spread of the virus throughout the United States. And in May 2020, we developed and released Hunala, an app based on network science and machine-learning techniques that people could use to assess their risk of infection.

The atmosphere in the whole scientific community in early 2020 was charged with urgency and probity. Colleagues all around the world pivoted to work on the coronavirus and broke down barriers to research, collaboration, and publishing. But very quickly it also became clear that there was an emerging vacuum of public information and few effective ways to communicate the problem that was unfolding. Along with a broad range of scientists, including epidemiologists, virologists, physicians, sociologists, and economists, I turned to Twitter to post tutorial threads on coronavirus-related topics such as the mortality rate in children and the elderly, the reasons we had to "flatten the curve," the nature of immunity after infection with the virus, and the extraordinary approach China had used to deal with the outbreak.

This book is another way I hope to help our society cope with the threat before us. In the middle of March 2020, Yale University closed down—though many laboratories, including my own, continued to work remotely. I wrote this book between March and August 2020 while in isolation with my wife, Erika, and our ten-year-old son in our home in Vermont. Our adult children intermittently sheltered with us as well, as they too were cut off from the lives they led before the disease struck.

I hope to help others understand what we are confronting, both biologically and socially, to outline how humans have faced similar threats in the past, and to explain how we will get to the other side of this, which we will, albeit after tremendous sorrow. The ability to understand a contagious and deadly disease builds directly on my years of teaching about public health, implementing global health interventions, serving as a hospice physician caring for the dying and bereaved, analyzing contagions using network science, and working as an academic sociologist studying social phenomena.

The COVID-19 pandemic is still a moving target, however. As of this moment, there is much that is unknown—biologically, clinically, epidemiologically, socially, economically, and politically. In part, the reason is that our actions are changing the outcome of the story. It's hard to know for sure what will happen. And there is much that only the passage of time will reveal, including the long-term health effects of the infection and the long-term consequences of our response to the contagion (such as how our physical and social distancing might affect the mental health and education of our children and the economic prospects of a generation of young people presently entering adulthood). We also do not know whether or when a vaccine will be available, how risky it will be, and how long the immunity it confers might last. Despite these uncertainties, we must all, as individuals and as a society, make the best decisions we can at the moment, informed by the broadest consideration of views and the best understanding of scientific facts.

The plague Apollo unleashed at Troy did eventually end due to the intercession of Achilles and of Hera, the queen of the gods. After ten days and many deaths, Apollo's terrible arrows ceased, and he put down his bow. Epidemics end. But how we get to that point defines us and our own moment facing down this ancient threat.

APOLLO'S
ARROW

1.

An Infinitesimal Thing

Humanity has but three great enemies: fever, famine, and war; of these by far the greatest, by far the most terrible, is fever.
—Sir William Osler, "The Study of the Fevers of the South" (1896)

In the late fall of 2019, an invisible virus that had been quietly evolving in bats for decades leaped in an instant to a human being in Wuhan, China. It was a chance event whose most subtle details we will probably never know. Neither the person to whom the virus gravitated nor anyone else was fully aware of what had transpired. It was a tiny, imperceptible change.

Scientists later came to suspect that this initial move by the virus might have happened at the Huanan Seafood Wholesale Market in Wuhan, because many of the first recorded patients were vendors or visitors there. But the picture was confusing. Huanan is known as a wet market because, as at many other markets throughout the world, one can buy fresh produce, fish, meat, and live animals there, and sometimes even wildlife (such as hedgehogs, badgers, snakes, and turtledoves). Some of these animals are butchered in the market, on the spot. Unlike the antiseptic supermarkets many of us are accustomed to, the pavements in such places are hosed down during the day to keep them clean. Hence, the markets are "wet."[1]

As far as we know, bats were not for sale at Huanan, though bats are consumed in China.[2] In a prescient article published a year before the virus slipped unseen into our species, scientists suggested that "bat-animal and bat-human interactions, such as the presence of live bats in wildlife wet markets and restaurants in Southern China, may lead to devastating global outbreaks."[3]

The first person with a confirmed case of the disease that would come to be known as COVID-19 developed symptoms of severe acute respiratory syndrome (SARS) on December 1, 2019. There may have been other patients earlier; we do not know. However, this patient (and a few other early cases) did not have contact with bats or wildlife or the Huanan market. This has led to concerns that perhaps the virus initially leaped to humans in some other way, such as through researchers in Wuhan who collected samples of the virus directly from wild bats and analyzed them in laboratories with inadequate protective procedures.[4] The Wuhan Center for Disease Control and Prevention, which does research with bat viruses, is just a few blocks from the Huanan market, and the Wuhan Institute of Virology is also a few miles away from it. However, Chinese authorities have claimed that there was no chance the virus leaked from these facilities.[5]

Notwithstanding the mysterious origin of the virus, 66 percent of the first forty-one people to contract the disease, during the month of December, did indeed have a direct connection to the Huanan market as shoppers, traders, or visitors.[6] If the market was not the place where the virus first found its way to humans, it was the place where it first became easy for us to detect. The market, with its densely packed stalls and large number of people, provided a fertile environment for the virus to spread rapidly and easily, generate a localized cluster of cases, and therefore come to our attention.[7]

One of the first doctors to sound the alarm about the disease was Dr. Jixian Zhang of the Hubei Provincial Hospital of Integrated Chinese and Western Medicine. On December 26, 2019, she noticed seven cases of atypical pneumonia; three patients were in the same family, and four were from the Huanan market and knew each other. She reported them to the Wuhan Center for Disease Control the next day.[8] Eventually, as part of an effort to cover up their initial inaction as the pandemic took root, the authorities gave her a merit award for reporting the cases.[9] But later investigation revealed that there had been other cases of atypical pneumonia earlier in December, above the threshold for notifying the central Chinese Center for Disease Control in Beijing, that had gone unreported. Precious time to contain the outbreak was lost. In fact, a later analysis documented that there were 104 cases and 15 deaths during the month of December.[10]

The authorities began to realize what was happening and shut down the market on January 1, 2020.[11] By then, the initial patients, dispersed to various hospitals, were being collected and transferred to a specially designated facility, Jinyintan Hospital.[12] On January 27, 2020, analyses released by the Chinese CDC (and later regarded by some as possible misinformation) noted that 33 of 585 environmental samples (such as swabs of surfaces) collected at Huanan from January 1 to January 12 contained the RNA of a novel coronavirus, later named SARS-CoV-2. The positive samples were highly concentrated on surfaces in the western part of the market, where the wild animals were sold.[13]

On December 30, 2019, two days before the market was shut down, a thirty-three-year-old ophthalmologist, Dr. Wenliang Li, became aware of the emerging cluster of cases after reading an alarming report by one of his colleagues. Dr. Ai Fen, the head of the emergency department at Wuhan Central Hospital, had received a lab report for a patient with atypical pneumonia indicating that the patient had SARS.[14] On a private WeChat group with a few medical-school classmates, Li spread the alarm. "There are seven confirmed cases of SARS at Huanan Seafood Market," he said. "The latest news is, it has been confirmed that they are coronavirus infections, but the exact virus strain is being subtyped. Protect yourselves from infection and inform your family members to be on the alert."[15]

By January 3, 2020, local authorities caught wind of Li's communications. There was a Chinese Communist Party meeting scheduled for later in the month, on January 12, and news of a local outbreak, much less a serious one, was not welcome. Indeed, until at least January 11, the public was wrongly assured that no new cases had been observed in Wuhan.[16] Li was called to meet with the police and accused of "rumor-mongering" and "making false statements on the internet." He was forced to retract what he had said and sign a letter promising that he would not engage in "illegal activities."[17] This was not the last time that the truth about COVID-19 would be suppressed or ignored as the pathogen spread around the world.

Of course, Dr. Li was completely correct. Later, the authorities would publicly apologize, and he would become a hero to ordinary Chinese people tired of constraints on free expression and disillusioned by misinformation from their leaders.[18] Alas, as eventually happened to many other health-care workers in China (and in many other countries), Li died of COVID-19, on February 7.[19] He had contracted the disease on January 8 while taking care of a glaucoma patient. That patient was a shopkeeper at the Huanan market.

The Chinese became aware fairly quickly that the disease could spread from person to person and was not independently and repeatedly acquired from a fixed animal reservoir. This worrisome fact was confirmed in a report about the first forty-one known cases published online in the British medical journal *The Lancet* on January 24.[20] The Chinese were also aware that the disease was serious. Of these first patients, six (15 percent) died. The article concluded that the virus "still needs to be deeply studied in case it becomes a global health threat."

The virus spread—first slowly, then quickly—through Wuhan and then through all of Hubei Province, home to fifty-eight million people. By January, while the overall percentage of infected people in Wuhan was still tiny, it was high enough that when large numbers of people left the city, some of them carried the pathogen with them.

The virus had announced itself with extremely unfortunate timing, right at the start of the annual *chunyun* (春运) migration in China that was taking place in the run-up to the Lunar New Year festival on January 25, 2020. During this period, over three billion trips are typically made, a mass movement that puts the annual Thanksgiving travel in the United States to shame.[21] To make matters worse, Wuhan is a central transportation hub for China. Nearly twelve million trips were taken through Wuhan in January (as research in my lab, in collaboration with Chinese scientists, later documented), thus carrying the virus throughout China by the middle of February.[22] The more people from Wuhan who went to a particular destination, as shown in figure 1, the worse the SARS-2 outbreak at that destination would later be. The initial "imported" cases set off local outbreaks via cascades of what epidemiologists term *community transmission*.

Authorities initially silenced voices like Li's, but later they abruptly yielded to reality and changed course—as other politicians in dozens of other countries would also eventually do. China scrambled to contain the outbreak, and more honest reporting was now encouraged. As Chinese president Xi Jinping said in his first public statements regarding the situation, on January 20: "It's necessary to release epidemic information in a timely manner and deepen international cooperation."[23] The Communist Party's central political and legal affairs commission, a group not known for encouraging transparency, offered its own stern warning on a popular social media site in China: "Whoever deliberately delays and conceals reports will forever be nailed to history's pillar of shame." The post was later deleted.[24]

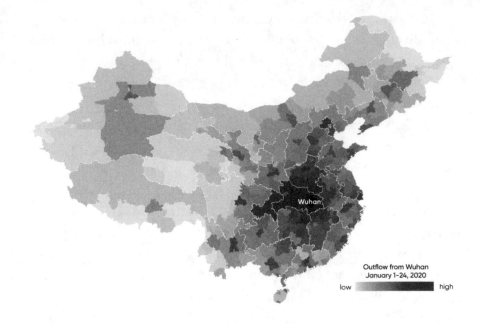

Figure 1: Population outflow from Wuhan in January 2020 carried the SARS-2 virus.

On January 17, nine days after Dr. Wenliang Li contracted SARS-2, seventy-two-year-old Dr. Lanjuan Li, a well-known physician and epidemiologist at Zhejiang Medical University in Hangzhou, one of China's oldest medical schools, learned from private communications that some medical personnel in Wuhan had fallen ill with a new kind of pneumonia.[25] That day, she contacted the National Health Commission in Beijing seeking permission to go to Wuhan, and the next day, China sent her there as part of a six-member team. Also on the team was Dr. Nanshan Zhong, an eighty-three-year-old pulmonologist renowned for his role in identifying the nature and severity of the prior SARS viral outbreak in 2003. Both Li and Zhong enjoyed tremendous respect in China and around the world. Dr. George Fu Gao, the head of the Chinese CDC in Beijing, had been alarmed at what was happening in Wuhan (since hearing of informal reports in late December, he had been prodding local authorities to be more forthcoming), and he also joined the mission.[26]

On January 19, the team visited hospitals, the Wuhan Center for Disease Control and Prevention, and the Huanan market. The city's health-care system was already inundated. In a few days, China would begin construction of a 645,000-square-foot field hospital with thirty intensive care units and a thousand beds to supplement the existing infrastructure in Wuhan. Construction would be completed in ten days.[27] On the evening of January 19, the team returned to Beijing and briefed the National Health Commission. Their report was alarming. At eight thirty a.m. the next day, January 20, the six experts took part in a cabinet meeting in Zhongnanhai, the Chinese leadership compound adjacent to the Forbidden City. Because the disease could spread from person to person, the team advised the government to implement stronger control measures, and they recommended closing off Wuhan. The Wuhan government announced at two a.m. on January 23 that it would impose a lockdown at ten o'clock that morning. A lockdown of the whole surrounding province of Hubei followed almost immediately.[28]

By January 25, nearly all of China was shut down.[29] According to an analysis conducted by one of my Chinese students soon afterward, 934 million people lived in provinces that were subject to new rules, described as "closed-off management" (封闭管理). The scale of the practices, reminiscent to some extent of the degree of social control under Chairman Mao, was breathtaking. It was the largest imposition of public health measures in human history.

"Closed-off management" involved many features.[30] People were required to shelter in their homes and were given permission to leave only once or twice a week for essentials. Shoppers waited in lines and kept six feet of separation between themselves and others—a development that stunned both local and foreign observers familiar with the usual press of bodies in China. And simply everyone wore a mask in public. Movement of people and vehicles was checked with special exit-entrance permits in every area, often down to the neighborhood level. Collectivist slogans made a reappearance everywhere, from little notations on these permits ("It is every-one's responsibility to fight the virus") to huge red banners in the streets. Every person's temperature was checked at the entrance to every community. Schools were moved online for millions of pupils. Vehicles and public places were regularly disinfected. Food and other essentials were carefully delivered on an enormous scale. The Chinese authorities encouraged delivery companies to distribute goods, and the companies vouched, via the ubiquitous apps used to place orders, that their drivers were wearing masks and did not have fevers.

The rules were enforced by block captains, local officials, and Communist Party members.[31] This was made easier by the authoritarian government and collectivist norms in China, and enforcement of this new regime was not just top-down. For instance, rural residents set up crude roadblocks of felled trees to keep outsiders out, and they interrogated visitors in local dialects in order to detect interlopers.[32]

This control sometimes came with modern twists. In February, a state-run military electronics company released an app that allowed citizens to enter their names and ID numbers and be informed of whether they might have come into contact with a carrier of the virus while using planes, trains, or buses. This technology struck many people in countries around the world as creepy, yet similar ideas would soon strike them as desirable, even normal.[33]

The Chinese government began to gingerly lift some of these restrictions in some parts of the country in late March, but the Chinese continued to implement many other procedures on a large scale.[34] For instance, people in elevators used disposable toothpicks, provided on wall-mounted pincushions, to push the buttons. Elevators in many cities allowed only four people at a time, their positions marked by tape newly placed on the floor. Signs in the elevators said PLEASE BE PATIENT AND WAIT FOR THE NEXT ELEVATOR. LET'S UNITE TOGETHER TO FIGHT THE VIRUS IN THIS SPECIAL PERIOD. As workers returned to offices and factories, the restaurants and cafeterias that served them were modified. Customers were separated by cardboard or Plexiglas dividers and instructed to eat quickly. Only one person was allowed per table, and there was no talking and no socializing. Gallows humor emerged, as with other aspects of the lockdowns, and many workers observed, "This one-table-per-person experience reminds me of my old school days taking exams."

In its approach, China had essentially detonated a social nuclear weapon. And so it was able to stop the spread of the virus. By late March, the number of new reported cases in the nation dropped from thousands per day to less than fifty per day.[35] By April, the daily case count hit zero, and this in a country of 1.4 billion people. There has been some criticism of the Chinese standards for reporting cases (for instance, initially they did not include asymptomatic cases of infection in their counts) and of the honesty of the reporting (certainly, information about the earliest cases in Wuhan was suppressed).[36] But the enormous reduction in cases once China mobilized to control the epidemic was an astonishing achievement from a public health point of view, even if some of the Chinese numbers were fuzzy.

To be clear, China, and other countries that subsequently implemented their own lockdowns, had not eradicated the virus; it had merely temporarily stopped its spread. When the lockdowns were lifted, the virus would come back.[37]

My personal involvement with COVID-19 research began the day after Wuhan initiated its lockdown. On January 24, I was contacted by some Chinese colleagues with whom I had been collaborating for several years, analyzing mobile-phone data from China. Previously, we had been looking at how high-speed rail lines and earthquakes reshaped how people interacted with one another to form social networks, a topic of interest to me since 2001. Maybe, we thought in late January 2020, we could use similar data to study the burgeoning epidemic. As a result, I began to concentrate on what was happening in China. And I became increasingly alarmed. I realized that COVID-19 was not going to be an epidemic solely in China. It would be a serious pandemic of historic proportions.

As I was studying all these things happening in China, I began to realize that the inundated hospitals, the lockdowns, the home-schooling, the Plexiglas partitions, even the toothpicks would all be coming to the United States before long. I could not think of a reason they would not. But when I tried to sound the alarm in my own household in early February, my wife, who usually takes me reasonably seriously, thought I was having prepper fantasies.

>———→

By the time the outbreak in China was brought under control, SARS-2 was well into its spread all over the world. In fact, it had made it to at least one person in America by the middle of January. The first case to come to public attention was a thirty-five-year-old man who was diagnosed in Snohomish, Washington. This information was announced by the CDC in a press release on January 21. The patient had returned to Washington from Wuhan on January 15.[38] Genetic analysis found he had a variant of the virus, recorded as USA/WA1/2020, or WA1 for short, that was closely related to variants seen in Fujian, Hangzhou, and Guangdong Provinces in China.[39] By pure chance, one of the forty-one initial patients in Wuhan or some intermediate person had infected this man. By the time the case was announced, the United States had started to do some cursory checks on incoming passengers from Wuhan, but only at certain airports, such as New York, Los Angeles, and San Francisco, and only beginning on January 17, two days after this man had arrived. This patchy approach illustrated what later became abundantly clear: border closings usually have a very limited effect on a pandemic like COVID-19.

This same CDC press release noted "growing indications that limited person-to-person spread is happening." And the published clinical report about this first detected case would provide additional evidence of this: the patient had not visited the Huanan market or any health-care facilities and had not had contact with anyone he knew to be sick. He had acquired the disease from a person who had almost certainly been asymptomatic. This shortly proved to be one of the most bedeviling aspects of the infection—as the pandemic spread around the globe, asymptomatic transmission made the disease much harder to track and control. We could not rely on people's symptoms to know who had it.

That the patient was diagnosed at all was a stroke of luck. He had seen a CDC alert about the virus, and when he developed a slight temperature and a cough four days after returning from Wuhan, on January 19, he sought treatment at an urgent-care clinic north of Seattle. The clinic staff knew to take a specimen and send it to the CDC on an overnight flight. The patient was discharged and told to self-isolate at home, which he did. On the afternoon of January 20, his test came back positive. And by eleven o'clock that night, he was in a plastic-enclosed isolation gurney heading to a biocontainment ward previously set up to handle Ebola patients at Providence Regional Medical Center in Everett, Washington. He had become— to use a term that should not imply the man had any personal responsibility for his predicament—our first known, test-confirmed case, our "Patient Zero."[40]

He worsened and developed pneumonia. While he was in the hospital, he was cared for by staff wearing protective equipment, including face masks, and a robot was used to take his vital signs. He often used a video link to communicate with his doctors and nurses so that they could keep their distance and avoid contracting a disease that might kill them as it had Wenliang Li and many health-care workers in Wuhan. This impersonal, isolated medical care forecast the sort of care many other hospitalized patients would later receive. By January 30, Patient Zero was much better, and he was discharged soon after. By February 21, he was deemed no longer infectious, and he was allowed to leave home isolation.

Contact tracing—the gumshoe work of the public health system whereby one goes backward from a known case to see who the patient has been in contact with—revealed that at least sixty people had been exposed to Patient Zero. Amazingly, none of them got sick. Later genetic analyses confirmed that this patient was very likely not responsible for the epidemic taking hold in Seattle. The existence of such dead ends in viral transmission is another important, if perplexing, feature of this pandemic. Based on such analyses, which are discussed in more depth below, it seems that some other unknown person, possibly an American citizen with ties to China, arrived from Hubei Province around February 13 and seeded the outbreak in Washington with a different variant of the virus.[41]

It was this later variant of the virus that wound up in the Life Care Center nursing home in the nearby city of Kirkland. The large number of intrinsically vulnerable elderly people provided fertile ground for the virus to spread; it caused a localized cluster of cases that soon drew attention. In February, paramedics noticed that they were making much more frequent emergency visits to this facility— there had been seven visits in January but roughly thirty in February. First responders were getting sick themselves. The fire department declared Life Care a hot zone and required ambulance personnel to wear full protective equipment to enter. And the Life Care staff was sometimes asked to wheel out the patients, in masks, and leave them at the curb for the paramedics to pick up. It became clear that the deaths were due to the novel coronavirus only when positive test results came back on February 28. Two days afterward, on March 1, a man in his seventies became the first Life Care resident to die of the disease.[42] A CDC report published later, on March 27, revealed a total of 167 cases linked to the facility: 101 residents (more than two-thirds of the facility's population), 50 health-care personnel, and 16 visitors; at least 35 of these people eventually died.[43] As of

March 8 in Seattle, there were only 118 documented infections and 18 deaths—and almost all of the deaths were from Life Care.

The clustering of vulnerable elderly people in nursing homes provided a kind of petri dish for the virus around the country. Soon, the small morgues typically present in such facilities proved hopelessly inadequate to the rapid pace of death. By April, disturbing newspaper headlines appeared: "After Anonymous Tip, 17 Bodies Found at Nursing Home Hit by Virus" and "Almost Every Day Has Brought a New Death from Coronavirus at the Soldiers' Home in Holyoke; 67 Have Died So Far." In the latter case, this meant that one-third of all the residents at the facility had died.[44] Nursing homes had become the inadvertent "pesthouses"—shelters once used for plague victims—of the early twenty-first century. Other old people living at home alone would sometimes die of COVID-19 so fast that statisticians later had to revise their estimates of the lethality of the virus upward to take into account these previously unnoticed deaths.

Given that the Seattle area was home to Patient Zero, the first reported cluster of infections, and the first deaths of the epidemic, it seemed like it must have been where the virus first took hold in the United States. However, later studies identified patients who became ill even earlier in other parts of the West Coast. In the United States, the bodies of those who have died unattended or whose deaths are deemed suspicious are subject to autopsy by the local medical examiner. This is how the medical examiner in Santa Clara, California, came to perform an autopsy on fifty-seven-year-old Patricia Dowd, who had fallen ill with flu-like symptoms in late January. She stayed home from work and told her family she could not make it to a reunion in nearby Stockton. At eight o'clock on the morning of February 8, she was in touch with a colleague from work, but she was found dead two hours later.[45] Initially, her death was thought to be due to a heart attack, but subsequent testing revealed the presence of SARS-2. Since the time from infection to death from

COVID-19 is typically about three weeks, the virus likely arrived in the Bay Area in the middle of January, roughly when Patient Zero arrived in Seattle. And since Dowd had not been to China herself, this meant that community transmission had already begun.

The first documented case of person-to-person transmission, excluding Dowd, since she got the pathogen from an unknown person, was between a married couple in Illinois.[46] The wife had returned to the United States from Wuhan on January 13, 2020, and she infected her husband. They were both hospitalized with serious illnesses, and they both recovered. Interestingly, however, like Patient Zero, this couple infected no one else. State public health officials traced 372 people they had come in contact with, including 195 health-care workers. The virus reached this couple and went no further.

Back in Seattle, however, after the other new importation noted above, the virus kept going. Dr. Helen Chu, an infectious disease expert, had worried about the first Seattle case when she heard about it in late January. And she was in a position to do something about her concerns. She was spearheading an ongoing effort, begun in 2018, called the Seattle Flu Study (supported by philanthropist Bill Gates), and its staff had been collecting nasal swabs from people with respiratory symptoms as part of a surveillance project in the Puget Sound area. She realized that she could test some of the more recent samples (from January and February) to determine if and when the coronavirus had begun to spread.

Unable to get permission from state and federal officials and increasingly frantic that the disease was spreading, Chu and her colleagues began to analyze their specimens without final approval on February 25. They immediately discovered that a fifteen-year-old teenager with no travel history to China (or anywhere else) had caught SARS-2 in the preceding weeks. He had sought medical care for an upper respiratory infection on February 24. Although he lived just fifteen miles from Patient Zero, the variant of the virus he had was different, meaning he did not contract it from Patient

Zero.[47] When this diagnosis was made, local health-care workers, including a doctor I know very well, raced to track down the boy. They found him at school that day. And why shouldn't he have been at school? He had recovered from what seemed like a routine illness and gone about his business. The boy was yanked from the premises the moment he was reached, and his school was soon shut down.

Upon discovery of this case, Chu realized with dread that the disease was "just everywhere already."[48] The Seattle Flu Study researchers went on to test more previously collected specimens and find more cases. Like the teenager, these patients were then informed of their infections. In fact, at this point, COVID-19 had already been responsible for the deaths of two people in the Seattle area. Both were older. This, too, rapidly became a familiar feature of the illness: the young largely seemed spared from its worst effects.

The fact that this teenager had contracted COVID-19 without leaving the country was additional clear evidence that community transmission was well under way in the United States. However, because of the scarcity of tests, the CDC initially recommended that people with respiratory complaints be tested only if they had a travel history to China or an exposure to a known COVID-19 case, guidance that would persist until February 27. As a result, in the six weeks after the identification of Patient Zero, only fifty-nine other cases were detected in the entire country.[49] Rules restricting access to tests were widespread, not for any clinical reason but simply because there were not enough of them. My wife, Erika, was seriously sick with flu symptoms in early March, but she was denied a test at our local hospital, a major medical center, on the grounds that she had "too many symptoms." Astoundingly, the inability to perform an adequate number of tests persisted, nationwide, through the summer.

Americans had put on blindfolds when they should have put on masks. The lack of testing was a huge blunder, and it drastically slowed the response to the early infections. Experts suspected then what we all know now: the disease was indeed everywhere. By March 25, Washington State alone, by virtue of more testing, confirmed 2,580 cases and 132 deaths. In the United States as a whole on that date, these numbers stood at 68,673 and 1,028.[50]

The infections in Seattle seem to have seeded the outbreak on the *Grand Princess* cruise ship, one of many instances where cruise ships (and, eventually, even a U.S. nuclear aircraft carrier, the USS *Theodore Roosevelt*) became hot zones. In many cases, passengers died on these ships. Incredibly, ships were not allowed to make landfall and were kept offshore by authorities as the epidemic raged on board, taking more lives due to the lack of medical care and the close quarters.[51]

On February 11, with over 2,400 passengers and 1,111 crew members, the *Grand Princess* left San Francisco for a cruise to Mexico; it returned to port on February 21. Most of the crew and 68 passengers remained on board, and on February 21, the ship set sail for Hawaii with 2,460 mostly new passengers. On March 4, a case of COVID-19 was diagnosed in one of the passengers who had completed the first journey, and the ship diverted course in the Pacific and began to return to port. By then, as feared, COVID-19 had broken out on board, with two passengers and nineteen crew testing positive.[52] The ship made landfall on March 8, and passengers and crew were taken to military bases for quarantine.[53] By March 21, seventy-eight people had tested positive. The CDC would release a rather acerbic advisory shortly thereafter: "All persons should defer all cruise travel worldwide during the COVID-19 pandemic."[54]

Another ship, the *Diamond Princess*, was quarantined in Yokohama, Japan, on February 3, and it would play a crucial role in the epidemic, providing scientists with a kind of grim natural experiment. Despite how critical experiments are to scientific knowledge, there are many situations in which scientists *cannot* do experiments for practical or ethical reasons. For example, we cannot experimentally assess whether the loss of a spouse increases a person's risk of death (due to what is known as "broken-heart syndrome") because we cannot randomly kill or otherwise remove people's spouses!

But scientists can sometimes take advantage of natural experiments, situations where the "treatment" has been assigned to subjects by chance—such as observing the effects of people put in close physical contact with one another who are therefore at risk of becoming infected with a deadly germ. Of course, natural experiments do not have the careful controls of designed experiments; we cannot always be sure that the treatments really are allocated truly by chance, among other limitations. For instance, with the *Diamond Princess*, scientists must take into account that the people who go on cruises are older and richer and possibly more sociable than other people.

But the ship still offered observable evidence, and in the confusing early days of the pandemic, scientists pored over the data in dozens of papers, searching for any kind of signal in the noise. This defined and contained population of 3,711 people who were not allowed to disembark allowed epidemiologists to ascertain what fraction of a population SARS-2 could infect and how many of those people, once afflicted, would die.[55] Analysis showed that at least 712 of the (relatively elderly) passengers contracted the virus, and at least 12 of them (or 1.7 percent) eventually died.[56] Both numbers alarmed experts. Indeed, the number of cases on the ship was so large that in the international league tables of COVID-19 cases maintained at the time by the WHO, the ship was listed just below China and Italy—as if it were its own country.

By the middle of March, the United States was jolted awake to the danger posed by COVID-19. The cluster of deaths at Life Care forced West Coast leaders to recognize that something had to be done. Beginning on March 5, heads of large Seattle technology firms, like Amazon and Microsoft, encouraged their employees to work from home if they could (a few days earlier, an Amazon employee had tested positive for the virus and had been quarantined).[57] Later analyses of the decline in restaurant bookings, among other data sets, showed that ordinary residents of the city, having read about the local events, stopped going out as much without having to be told. On March 17, the governor of Washington, Jay Inslee, issued orders for all bars, dine-in restaurants, and entertainment and recreation facilities to close. On March 19, California governor Gavin Newsom issued a statewide order for people to stay at home except for essential activities.[58] On March 23, Inslee followed suit in Washington.

»———→

Modern genetic techniques were instrumental for understanding the fundamentals of the virus and ascertaining where it spread. The first step was mapping the virus's genome—a more manageable task for a virus than for more complex organisms. The genome of every virus contains instructions for just a handful of proteins, and since viruses work by taking over our genetic machinery to reproduce in our bodies, they do not need the equipment to do that on their own. The coronavirus has a genetic code that is just 29,903 letters long. Its genome was rapidly sequenced (from a sample taken from a Huanan market vendor) by Chinese scientists led by Dr. Yong-Zhen Zhang at Fudan University in Shanghai and publicly released on January 11, 2020, in order to pave the way for the development of diagnostic tests.[59] The next day, the Chinese government, in an absurd move reflecting the desire to control scientific information, shut down the lab that had done this important work for "rectification."[60]

SARS-2 is in the family of viruses knowns as coronaviruses. Some species of coronaviruses cause the common cold in humans; others afflict some of our domesticated animals, like pigs, cats, and chickens. Genetic sequencing showed that SARS-2 is 96.2 percent identical to a coronavirus found years ago in a bat in a cave in Yunnan, China (that virus is known as RaTG13). This confirms that SARS-2 originated in bats, where it probably circulated unnoticed for decades, but the virus also might have spent some time in pango-lins before coming to our own species, in a confusing trajectory we may never fully unravel.[61]

Bats have been the origin for many other epidemics, such as the deadly Ebola and Marburg agents and the rare Hendra, Nipah, and St. Louis encephalitis viruses. It's not known exactly why bats are such a prolific source of human pathogens, but they have haunted our species in other ways for a very long time as objects of mythology associated with death; they are found in folklore from Nigeria to Oaxaca to Europe (as in the stories regarding Dracula). One theory suggests that the bat immune system is strangely similar to our own and that pathogens that adapt to bats can more easily afflict us. Another theory posits that because bats are the only mammals that can fly, it is easier for them to more widely disseminate the viruses they harbor to other mammals, including us.

An important consequence of mapping the virus's genome is that it allows us to reliably identify different variants of the virus and therefore track its spread across the world.[62] Over time, the viral genome undergoes tiny mutations—slight changes in its genetic code that usually have no effect on the virus's function. These changes occur at fairly regular intervals, like a molecular clock—one tiny mutation every two weeks, on average. Since those mutations happen at random places in the code, the genome of a virus in one part of the world will be slightly different than it is in other parts. By studying these cumulative, haphazard mutations collected from many thousands of specimens around the globe, scientists can reconstruct the movements of the virus. These mutations act like stamps in a passport, recording where the virus has been and when it crossed our borders. For instance, it's thanks to this technique that we were quickly able to confirm that the outbreak on the *Grand Princess* leaving San Francisco was connected to the earlier one in Seattle, which was, in turn, connected to the original one in Wuhan.

Sequencing of the Washington outbreak began in the middle of February in the laboratory of Trevor Bedford, an infectious disease specialist at the University of Washington who was also part of the Seattle Flu Study.[63] At the beginning of the Seattle outbreak, Bedford's team tried to track down the sources of the virus by analyzing the genomes of the viruses taken from different infected individuals throughout the region. One possibility was that SARS-2 was introduced into the Seattle area by Patient Zero on January 15, 2020, and that it then spread cryptically for a while before and after this patient came to medical attention, ultimately causing the outbreak at Life Care and also afflicting the teenager detected by the Seattle Flu Study. The second possibility was that there was a second, parallel importation of the virus, or even several separate importations, leading to other local outbreaks.

Distinguishing between these possibilities is important in getting a sense of scale—it tells researchers how many fronts we are fighting the virus on. Such data are also very useful for assessing the infectiousness and course of the virus. In this way, it was possible to deduce that the variant of the virus responsible for the community transmission in Seattle did not originate with Patient Zero but rather with some later importation. By late February, this latter variant of the virus was responsible for 85 percent of the confirmed infections in the region, though other variants had also begun arriving in the area from still other travelers, all starting their own family trees.

Distinguishing among such alternatives is also difficult because it is not always possible to be absolutely certain. A key reason for this is that the rapid rate at which people transmit SARS-2 (the average interval between one person contracting it and then transmitting it to someone else is about a week) is faster than the rate at which the virus has a mutation that allows us to discern if it is unique (about once every two weeks). This is like visiting a new country every week but getting your passport stamped only every other week; it makes it hard to know exactly where you have been.

Figure 2 shows what these sorts of genetic trees look like. It's taken from work by the lab of Michael Worobey, an evolutionary biologist at the University of Arizona, in a collaboration with other labs. Each dot is a variant of the virus (as ascertained by sequencing its genome). The variant afflicting Patient Zero, WA1 (on the bottom left), was a dead end, spawning no subsequent variants or infected individuals. Other variants (in the upper right) seeded the outbreak in Washington sometime around February 13. And from there, they spread to California, New York, and elsewhere.[64]

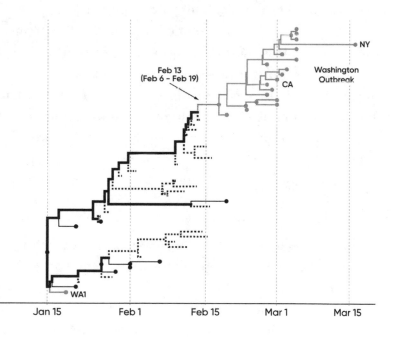

Figure 2: Genetic mapping of variants of SARS-2 with estimated times of divergence makes tracing the path of the virus possible.

Bedford's team sequenced the initial community-transmission case (in the teenager) and, like Dr. Zhang had done in China a month before, immediately posted the results. Bedford put the information on Twitter, which would become a major means for scientists to rapidly share information, on February 29. An international consortium began to share viral genome sequences from around the world on the online Nextstrain platform, which functions somewhat like the online tools that ancestry buffs use to define family trees in humans.

In March, after SARS-2 showed up in Connecticut, similar methods were used by researchers trying to trace whether cases stemmed from a domestic source or an international one.[65] Scientists sequenced the genomes of nine specimens of the virus and examined data regarding air travel at nearby airports. The United States had placed broad travel restrictions on China on January 31, on Iran on February 29, and on Europe on March 11. But it became clear through these genetic analyses that the greatest risk to Americans was from *domestic* importations from other U.S. states rather than from foreign arrivals. The SARS-2 variants found in Connecticut at that time came from several other locations, mostly Washington, and none of the patients examined at that point had been abroad. Since there are numerically so many more domestic than international travelers, it should not be surprising that the domestic risk is greater. These scientists concluded that the imposition of restrictions on international travel had a limited effect on the virus's spread.

However, on March 1, 2020, the first confirmed case of coronavirus in New York State was reported—in a patient who had indeed traveled internationally and returned to New York City from Iran on February 25, just ahead of the travel ban.[66] Of course, the disease had been circulating for some time in New York City, and by late March, hospitals were inundated with patients. By early April, nearly one thousand patients were dying of the virus *every day* in the city, a statistic that dropped only a few weeks after widespread efforts to engage in physical distancing began. By March 19, the virus had been detected in all fifty U.S. states. Figure 3 shows the estimated dates of key early arrivals of the virus that successfully seeded the epidemic in our country.[67]

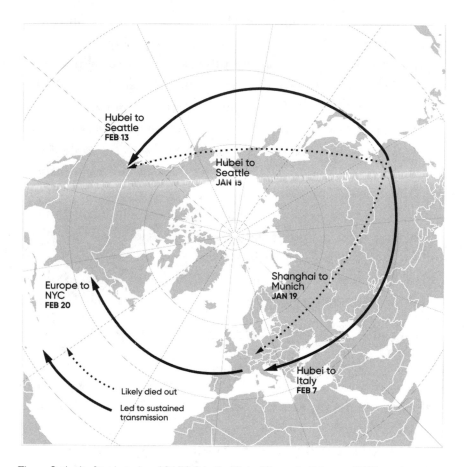

Figure 3: Arrival trajectories of SARS-2 to the United States in February 2020 can be inferred from genetic analysis of viral variants.

Sandwiched between China's outbreak in January and America's in March were major deadly outbreaks in Italy and Iran, both of which were devastating.[68] Spain, France, and many other European and some Asian and Latin American countries were also hit hard. By early April, for example, Ecuador was ravaged; dead bodies accumulated so fast that they were shrouded in sheets of plastic, covered in a few rocks, and abandoned on the sidewalks.[69] My lab has a public health field site involving over thirty thousand people in the rural state of Copan in western Honduras. We ceased our activity in March in order to avoid any risk of contributing to a similar calamity.

On March 11, roughly four months after the virus slipped into our species, the World Health Organization declared COVID-19 to be a pandemic. This was a formality, of course, since the virus's pervasiveness was already evident to all sophisticated observers and to countless residents of beleaguered cities around the globe. As of April 1, 2020, there were 219,622 known cases and 5,114 deaths in the United States and 936,851 cases and 47,210 deaths worldwide. By May 1, COVID-19 had become the leading daily killer in the United States, far eclipsing the deaths caused by the seasonal flu and even surpassing cancer and heart disease. By July 1, there were 130,761 recorded deaths in the United States and 518,135 deaths worldwide—and no end in sight.

$$\gg\!\!\longrightarrow$$

The class of virus to which SARS-2 belongs gets its name from its appearance under an electron microscope. When this type of virus was first visualized, in 1968, it was seen to have a crown-like feature on the outside, hence the name *coronavirus* (*corona* comes from an old Greek word denoting a wreath worn as a crown).[70] This crown is actually composed of the *spike proteins* of the virus, and these proteins turn out to be crucial to its ability to harm us. The spike protein binds to a protein on the surface of human cells (known as ACE2) and thereby initiates the process that results in the virus being internalized into a cell. The virus then releases its RNA and uses our genetic machinery to reproduce itself, which releases more virus into our bodies.

As respiratory infections go, COVID-19 (which is the *disease* caused by the *virus* SARS-2) is particularly protean, encompassing a great variety of symptoms, from fever to cough to muscle pain to the loss of the sense of smell (anosmia). The symptoms that a patient experiences depend in part on which cells in the body the virus infects and how the body responds immunologically. The primary form of the disease is respiratory and involves cough and fever, the two most common symptoms, and shortness of breath. Two less common manifestations of the disease are musculoskeletal symptoms (muscle pain, joint pain, and fatigue) and gastrointestinal symptoms (abdominal pain, vomiting, and diarrhea). But many other symptoms—such as rash, headache, dizziness, and so on— can occur. Anosmia, while uncommon, is more indicative of SARS-2 than of some other respiratory pathogen.[71]

The disease also displays a great range of severity across patients. Perhaps half of those infected are entirely asymptomatic. For the remainder, the range of outcomes stretches from a mild illness (in most cases) to hospitalization (in perhaps 20 percent of cases) to death (in perhaps as many as 1 percent of cases). Among those who fall ill, some recover quickly and others have lingering health problems.

In mild cases, symptoms may never intensify past a sore throat, muscle aches, or a low-grade fever. Some people assume they have the flu or a long cold; some have even attributed the symptoms to jet lag.[72] The only symptoms one sixty-seven-year-old man experienced during his quarantine on the *Diamond Princess* were a brief fever, mild shortness of breath for three days, and a cough.[73] "If I were at home with similar symptoms, I probably would've gone to work as usual," he said.[74] Many in this category suffered most from dizziness and fatigue. People describe taking more naps and note that climbing up the stairs and showering become Herculean tasks. Another man, age seventy-three, explained, "It was just a loss of all energy and drive....My brain wasn't working very well. I was calling it 'the corona fog.'"[75]

Patients with severe cases report their symptoms with intense language. "It feels like an alien has taken over your body" is how one forty-one-year-old resident of Chicago described her weeks-long battle with COVID-19.[76] Like many others who got sick, she experienced a sudden onset and swift progression of the classic coronavirus symptoms: a heaving dry cough, loss of taste and smell, head and body aches, and extreme exhaustion: "It's not the flu.... It doesn't feel like anything you've ever had."

One forty-three-year-old patient started her week with back pain and a cough and ended it with an ambulance ride to the emergency department, where she was rapidly put in a medically induced coma and intubated: "I felt so beat up, like I had been in a boxing ring with Mike Tyson.... The cough rattled through my whole body. You know how a car sounds when the engine is sputtering? That is what it sounded like."[77] Health-care workers describe the suddenness with which patients with COVID-19 can "crump," appearing fine one moment, then horribly short of breath, and then rapidly in need of intubation. Some patients get so sick so fast that they die at home before anyone realizes what is happening.

Even young adults who are fit and healthy can be left gasping for breath when trying to speak or walk across a room. A nineteen-year-old patient described feeling like "an elephant was lying on my chest." Another patient described the sensation of breathing as being "stabbed in the chest with an ice pick," while another said it felt like there were "cotton balls" in her lungs that could not be coughed up.[78]

Once in the lungs, the virus kills the cells that line the alveoli, little globular sacs that are responsible for oxygen exchange. Blood and fluid leak from the injured lung tissue into the sacs, which makes patients short of breath. They drown in their own fluids. The virus can also infect other tissues in the body, which is why it may give rise to symptoms outside the respiratory system (e.g., by affecting intestines and causing diarrhea). To make matters worse, sometimes the immune system overreacts to the invader, setting in motion something known as a "cytokine storm" that worsens, rather than improves, the situation. In the process, the body releases substances intended to help coordinate the defense against the invader, but the substances wind up harming cells in the lungs (and elsewhere), irritating them and causing damage that worsens oxygen exchange and leads to shortness of breath.[79]

Finally, even when patients feel like they are recovering, the virus's erratic nature may suddenly inflict upon them a new symptom, extreme fatigue, or another fever spike. For those who return from the hospital after severe cases of COVID-19, the coughing fits and weakness can last long after the fever has subsided. The recovery time can range from two weeks for a mild case to six weeks or more for a severe one.[80] Exhaustion from the physical toll of the virus is intensified by its emotional toll, given the contagious nature of the disease. Many patients felt guilty about having unknowingly transmitted the virus to others when they were still asymptomatic. Doctors also think that many patients may have long-term consequences of infection with SARS-2 in many organ systems, in what is called "post-COVID syndrome." Such patients may have permanently scarred or injured lungs or kidneys or hearts, for example, or even, in rare cases, neurological deficits. It will likely be some years before scientists truly understand the lasting implications of the disease for patients, including children, who, though uncommonly symptomatic, may indeed have rare complications.

»————————➤

Most people have not had personal experience with deadly epidemics. But plagues have always afflicted human beings, at least since we started living in large enough groups in cities about three thousand years ago. There was the plague of Athens in 430 BCE. The plague of Justinian in 541 CE. The Black Death in 1347. The Spanish flu in 1918. There were gods of plagues in ancient times—not only the Greek god Apollo, but the Vedic god Rudra and the Chinese deity Shi Wenye. Plague is an old, familiar enemy. And so, in 2020, a plague once again appeared.

How do humans in the twenty-first century respond, personally and collectively, to this reappearance? The challenges and responses, both good and bad, are timeless. Plagues reshape our familiar social order, require us to disperse and live apart, wreck economies, replace trust with fear and suspicion, invite some to blame others for their predicament, embolden liars, and cause grief. But plagues also elicit kindness, cooperation, sacrifice, and ingenuity.

The world is quite different now than it was during prior plagues; today we have exceedingly dense cities, electronic technology, modern medicine, better material circumstances, and the ability to know what is happening in real time. Scientists can track the outbreaks from space—as they watch cities shut down. And from the ground—as they observe people's mobile phones ceasing their translocation. And at a molecular level—as they use genetic techniques to analyze mutations, capturing the spread of the virus.

But from the point of view of the virus, the climate is ripe and things are as simple as ever. It is having a field day. In terms of evolutionary biology, the virus has had what is known as an "ecological release." This refers to the expansion of range and the population explosion that occurs when a species is freed from constraints it previously faced. The typical example of this is invasive species introduced by humans such as the cane toads that overwhelmed Australia, the rats that overwhelmed New Zealand (nearly wiping out the dinosaur-like tuataras that had occupied the island for millions of years, until 1250 CE), and the kudzu plants in the southeastern United States. The new arrivals suddenly find wide-open terrain for them to exploit. Our species has no natural immunity to the virus. We have never seen this particular pathogen before. It is a "virgin soil epidemic."[81] And so the coronavirus swept through humanity like a wave.

There is some debate among experts about whether viruses are living things. But this virus is surely acting like any other living thing would: it found available and untouched habitat, and it seized it.

And the virus will keep infecting humans until we develop immunity or invent a vaccine. Even then, SARS-2 will most likely become like other viruses that circulate in our species, such as influenza, measles, and the common cold. No matter what, humans will have to reach a modus vivendi with this virus. But it will kill many of us before we do. A new pathogen has been introduced into our species, and in some form, it will now circulate among us forever.

>———→

When I was fifteen, my father, who had trained as a nuclear physicist, told me a story about a butterfly flapping its wings in China and causing a hurricane offshore from where we lived, in Washington, DC. I did not believe that could happen.

This image had first been conjured by an MIT professor of meteorology, Edward Lorenz, on December 29, 1972, at the 139th meeting of the American Association for the Advancement of Science. The details were not exactly the same; Lorenz had used the image of a butterfly flapping its wings in Brazil setting off a tornado in Texas.[82] But the image was so powerful that, with a puff, it gave rise to countless expressions about where exactly the butterfly was flapping its wings (China, Brazil), what exactly it caused (storms, tsunamis, people falling off skyscrapers, stock market crashes), and where exactly it had these effects (Japan, London, New York City).

This idea also changed diverse branches of mathematics and physics. Lorenz was proposing that minuscule disturbances and seemingly irrelevant modifications to the starting conditions in a complex system could, over time, result in dramatically different final outcomes. Some of us remember Jeff Goldblum's character describing this idea in the 1993 film *Jurassic Park*—he placed a drop of water on Laura Dern's knuckle and explained how tiny perturbations could affect which way the water flowed.

Lorenz had chanced on his observation because of a small jostle in his own life. In the winter of 1961, he was using a primitive computer to model the weather and predict weather patterns. At one point, he decided to rerun a program in order to take a closer look at what was happening. The previous run had yielded 0.506127 as an output. Lorenz rounded it to 0.506 and resumed the calculations. He went down the hall to get a cup of coffee, and when he came back, he discovered that the computer had generated forecasts that were completely at odds with the ones it had generated before. This tiny numerical alteration had drastically changed the two months of simulated weather that his model was predicting.[83]

And so was born the idea that infinitesimal things can have huge effects. Some systems are exquisitely sensitive to their initial conditions, and to the extent this is true, predicting the future can be nearly impossible. In 1963, Lorenz wrote a paper about this titled "Deterministic Nonperiodic Flow."[84] The paper was initially ignored, but it eventually became a classic. Ultimately, it had an impact far outside meteorology, and by the 1970s and 1980s, it was recognized as a foundational effort in the emerging field of chaos theory.

A colleague of Lorenz observed that, if his theory was correct, the flap of a seagull's wings "would be enough to alter the course of the weather forever." Lorenz later noted, "The controversy has not yet been settled, but the most recent evidence seems to favor the gulls."[85] Eventually, Lorenz shifted to the more poetic metaphor of a butterfly, and he expressed a bit more hesitancy. For the rest of his life, he struggled with the answer to the question he had raised. "Even today, I am unsure of the proper answer," he said in a lecture in 2008, over forty years after the rounding of 0.506127 to 0.506 had sent his life in a radically new direction.[86]

One of the reasons the butterfly metaphor has engaged so many people and entered popular culture is, I think, that it is so disturbing. It upends our belief that the world should be predictable, ordered, or even comprehensible. It threatens the idea that things happen for reasons and that we might, by using science, be able to discern those reasons, however obscure they might be. It threatens the idea that we can make rational predictions and plans.

And although I have devoted much of my career to understanding the inertia of social systems—their unchanging reality, their fundamental evolutionary origins, and the ways in which they are stable and fixed—it has now been impressed upon me that social systems, in ways that had not previously engaged me, can indeed be extraordinarily unstable, like the weather.

As the virus was leaping into our species in late 2019, I was, like most people, still making plans that I thought would come to fruition. This idea of unpredictability was nowhere in my mind. My family was planning to travel to Greece to see my elderly father. We were looking forward to our daughter's college graduation. My university was deciding which faculty members to recruit and what conferences to hold. I was assigning new projects in my lab, and we were doing community-health research in Honduras and India. People throughout the world had no inkling that their jobs and livelihoods might soon evaporate, that their loved ones might be separated from them, that the next couple of months would be so utterly and unfathomably different from the past few. Who could have predicted that the most innocuous acts—a handshake, brushing hair across one's face, singing in a choir—would suddenly seem unthinkable, even repellent? In November of 2019, staffers on political campaigns were planning their strategies for the spring. Owners of small businesses were ordering inventory. Farmers were choosing crops. And economists were forecasting ongoing growth.

None of these events would happen—because of a tiny thing that we cannot see that made a move we could not observe. A butterfly flapping its wings in China could cause a hurricane in Washington, DC.

2

An Old Enemy Returns

> Everybody knows that pestilences have a way of
> recurring in the world; yet somehow we find it
> hard to believe in ones that crash down on our
> heads from a blue sky. There have been as many
> plagues as wars in history, yet always plagues
> and wars take people equally by surprise.
>
> —Albert Camus, *The Plague* (1947)

In 1918, six-year-old Marilee Harris fell sick with the Spanish flu.
At her age, she faced a roughly 1 percent chance of death. As she
described in her autobiography—published in 2015 when she was
102—she knew she had recovered when she finally walked down the
stairs at her home and saw her father eating breakfast. After her
isolation, rejoining her family for a meal was a powerful memory. (As
it turned out, Marilee had avoided death once before, because her
mother, "a proper Victorian woman," did not want to "have children
in her forties" and had tried to abort her by drinking castor oil.)[1]

This was not her last brush with mortality during a pandemic.
In 2020, Marilee—now Marilee Shapiro Asher and an accomplished
artist—contracted COVID-19 at Chevy Chase House, a senior-living

community in Washington, DC. This time, at the age of 107, she faced a risk of death surely over 50 percent. When she stopped eating, on April 18, she was taken to the hospital and her family was notified that she might die within hours. She survived. After five days, and without having been put on a ventilator, she recovered enough to go home, and she planned to continue making art.[2]

There were over ninety thousand centenarians in the United States in 2020, and many of them might have survived a childhood case of the 1918 flu. But only a small number of people are both sufficiently old and mentally healthy enough to remember it. Since major pandemics come along so rarely, very few people have actually experienced them. But the fact that they are not in people's living memory does not mean that we can neglect them. They inevitably recur. We should pay attention.

>────────→

While the coronavirus causing the 2020 pandemic is new, we have lived with other coronaviruses for a long time. In our species, four types of coronavirus are endemic, meaning we have reached a détente with these pathogens. They circulate among us and afflict us in a steady manner at some baseline level, and we have become biologically and socially accustomed to them. These four types of coronavirus cause nothing more than the common cold, accounting for between 15 to 30 percent of cases of that nuisance condition (more than two hundred different virus species, including corona-viruses, cause colds).[3]

But human beings have had run-ins with more serious coronavirus species that have done much more damage than wintertime sniffles. In 2003, we faced the first pandemic of a coronavirus that closely resembles the one that afflicted us in 2020. But reviewing how it *differed* from SARS-2 helps shed light on why and how the pandemic of 2020 came to be so very serious, and this also provides a way for us to understand the epidemiology of such pathogens. Like COVID-19, the pandemic of 2003 began in China; the first known case was a farmer from the Shunde district in Guangdong Province who developed symptoms on November 16, 2002.[4] This virus also came from bats and was characterized by fever, dry cough, shortness of breath, muscle pains, and sometimes a deadly pneumonia. And it too spread far and wide, reaching the United States and twenty-nine other countries.[5]

But what we now call SARS-1 had certain intrinsic epidemiological qualities that kept it from overwhelming us. Efforts to respond to SARS-1 were ultimately so successful that the World Health Organization declared the pandemic "contained" on July 5, 2003, just eight months after it had begun.[6] As of August 1, 2003, a total of just 8,422 people had been infected worldwide, roughly the same number of COVID-19 cases detected just in the state of Idaho as of July 1, 2020, seven months into the pandemic.[7] Examining the difference between the two viruses helps us appreciate why SARS-2 overwhelmed the world while its cousin SARS-1 did not.

The virus made its appearance in November 2002, but the Chinese government did not notify the World Health Organization about the outbreak of disease until February 12, 2003, when it reported 305 total cases, including 105 health-care workers, and five deaths. The cause was unknown at the time. Later, China was much criticized for this reporting delay, which also involved an initial cover-up. The early stages of the outbreak were very slow, but the case count accelerated starting on January 31 when Zuofeng Zhou, a fishmonger, was admitted to the Sun Yat-sen Memorial Hospital in Guangzhou. During Zhou's hospital stay, thirty nurses and doctors were infected. As with the 2020 pandemic, this tight cluster of cases and the fact that the virus spread to health-care personnel sounded the alarm, prompting officials to take action and indicating both the severity of the disease and its contagiousness.

One of the doctors who got infected by Zhou was sixty-four-year-old

Jianlun Liu. He was feeling well enough on February 21 that he embarked on a three-hour bus ride south to Hong Kong for his nephew's wedding. Liu felt a bit sick during the ride, but when he and his wife arrived, they were easily able to go out for lunch and see his relatives.[8] At five o'clock p.m., they checked in to room 911 of the three-star Metropole Hotel in the Kowloon region of the city.

By the next morning, Liu was sicker. He walked five blocks down Waterloo Road to the Kwong Wah Hospital to seek treatment. Having cared for patients with SARS himself, he warned the personnel that he should be put in isolation (although he might not have needed to warn them, because they were already aware of the outbreak).[9] The next morning, February 23, Liu had to be sedated and intubated. One doctor and five nurses caring for him fell ill, but having been properly warned, they had all been wearing N95 masks, gloves, and gowns. This probably reduced the amount of virus they were exposed to, which tends to make such illnesses milder, and all of them recovered. Alas, Liu did not recover; he died on March 4.

While Liu was unconscious, on March 1, his brother-in-law was also admitted to the same hospital for the same condition. He, too, deteriorated and died, on March 19, but not before doctors biopsied his lungs. From that sample, a team at Hong Kong University was able to grow the virus, and on March 21, they announced that they had identified it: under a microscope, it had the characteristic surface spikes of a coronavirus.[10]

Like the fishmonger before him, Liu proved to be a super-spreader—twenty-three guests of the Metropole also developed SARS, including seven from the ninth floor, where he had stayed. These guests went on to seed the epidemic throughout the world. Later, the World Health Organization reported that nearly half of all the cases seen worldwide from this pandemic could be traced back to Liu's twenty-four-hour stay at the Metropole Hotel.

Maps of this floor of the hotel, shown in figure 4, became famous among epidemiologists. Just by looking at the map, one could see that being closer to room 911 was a big risk for contracting the disease. Ultimately, perhaps 80 percent of the 1,755 patients who were infected with the virus in Hong Kong could be traced back to Liu.[11] A leading theory (never definitively proven) of why so many people got sick was that Liu had vomited on the carpet in the hallway outside his room. Cleaning up the mess involved a vacuum cleaner, which may have aerosolized viral particles, spreading them widely through the hall and possibly sending them into the ventilation system. Amazingly, however, none of the hotel's three hundred employees fell ill, a mystery that was never explained.

The outward spread of this pathogen from the point source of this hotel was prodigious. For instance, a twenty-six-year-old airport technician visited a friend at the Metropole several times from February 15 through February 23. On March 4, the same day that Liu died, the man was admitted to ward 8A of the Prince of Wales Hospital in Hong Kong.[12] While there, he was given a nebulizer treatment to ease his breathing. By design, this device produces a fine mist, and it appears to have accidentally spread the virus around the ward.[13] Eventually, at least ninety-nine hospital workers who came into contact with him became infected.

The beds in the hospital started to fill up with the hospital's own staff. Dr. Joseph Sung, the head of the medical faculty, observed, "There were two dozen of my colleagues sitting in the same room; everybody was shaking and running a high fever; many were coughing.... That was the beginning of the nightmare, because from that day on, every day we saw more and more people developing the same illness."[14] Sung divided the hospital staff into two teams. One would take care of all non-SARS patients, and the other—the "dirty team," as they called it—would risk infection and care for the SARS patients. Those with young children were exempted from service on the dirty team. But those who were single or whose children were grown were encouraged to volunteer. Sung would later describe their predicament: "I needed a continuous supply of manpower to go in. And I was very touched by the fact that after we exhausted everybody in the medical department, surgeons, orthopedic people, gynecologists, even ophthalmologists came to help us."

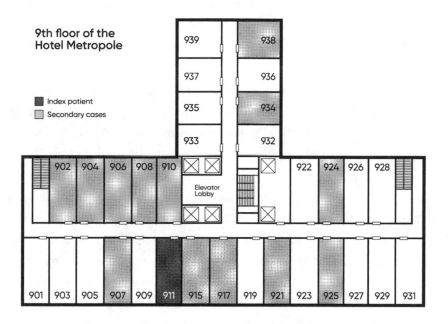

Figure 4: The ninth floor of the Metropole Hotel in Hong Kong was a key location of spread of SARS-1 in 2003. Dr. Liu, the index patient who would be a super-spreader, stayed in room 911.

A thirty-three-year-old patient with kidney disease who was hospitalized for some blood tests on ward 8A on March 13 contracted the infection while he was there. The next day and then again on March 19, he visited his older brother who lived on the seventh floor of the Amoy Gardens, a densely packed high-rise housing complex in Hong Kong. During his visit, the patient was sick with diarrhea and had to use the bathroom. In a bit of detective work that would become famous among epidemiologists, it was eventually determined that many of the later cases in the complex were related to dried-out sewage pipes in the apartment. Running exhaust fans and flushing toilets with buckets disturbed desiccated sewage and released virus-laden aerosols (termed *gaseous plumes*) into multiple bathrooms.[15] Within days, a large outbreak involving 321 patients had started at the Amoy Gardens.

This airborne mode of transmission is far more alarming to public health workers than the more common droplet transmission, whereby sick patients expel virus-laden droplets when they cough, sneeze, or, possibly, just talk forcefully. Such droplets are heavy and typically fall to the ground within six feet of the person expelling them—a distance that has become familiar due to physical-distancing guidelines during the 2020 pandemic (though, to be clear, six feet is not always enough). But with airborne transmission, tiny, light-weight viral particles can float, possibly quite far, through the air. This seems to have happened at the Amoy Gardens, and from a fecal source, no less.[16]

Across the hall from Liu at the Metropole, in room 910, was Johnny Chen, a forty-seven-year-old Chinese-American garment merchandiser from Shanghai.[17] He flew on to Vietnam, where he was admitted to the French Hospital of Hanoi on February 26. Chen died on March 13 (after having been evacuated back to Hong Kong on March 5), but not before infecting thirty-eight members of the hospital staff in Hanoi. The personnel took the extraordinary step of locking themselves in the hospital in order to protect the outside world.

Among the doctors in the Hanoi hospital was Carlo Urbani, an infectious disease specialist at the World Health Organization, who was in town. He pitched in and worked tirelessly for several weeks. Like the other sharp-eyed doctors who noted the transmission to health-care workers as evidence of human-to-human spread, Urbani recognized that a serious new infectious disease had emerged. On February 28, just two days after Chen had been admitted to the hospital in Hanoi, just a week after Liu had been admitted to the hospital back in Hong Kong, and just sixteen days after the notification China had given to the WHO, Urbani informed the WHO of what he had observed.[18] On March 11, he traveled from Vietnam to Thailand, but he fell ill on the flight. When he arrived in Bangkok, he warned a friend who had been waiting for him not to touch him and asked for an ambulance to take him immediately to a hospital.[19] On March 29, Urbani died. He remains much revered in public health circles.

Still another hotel guest to become infected was seventy-eight-year-old Sui-Chu Kwan. She had chosen the Metropole Hotel because an airline had given her a free voucher to stay there. She checked out on February 23 and returned to Toronto, Canada, where she transmitted the disease to her forty-four-year-old son and four other members of her family. She died on March 5 and her son died on March 13, but not before they had spread the disease to health-care workers at Scarborough Grace Hospital.[20] The hospital was closed to new patients and visitors on March 25, and thousands of residents of Toronto were told to self-isolate at home. As had occurred in Hong Kong earlier, local public health authorities scrambled to contain the epidemic by tracing exposed people and isolating cases. Ultimately, Canada would have 241 cases (108 of them health-care workers) and 41 deaths due to several importations from Asia.

Still other guests and visitors from the Metropole Hotel would travel to other countries around the world, including Singapore and Taiwan, starting epidemics there too. The outbreak in Singapore appears to have been the first seed for the outbreaks that occurred in the United States. The first case in the country was detected on March 15 in a fifty-three-year-old man who had traveled to Singapore and then fallen ill on March 10. Subsequent U.S. clusters were identified based on importations from Guangdong, Vietnam, and Hong Kong.[21] And three patients who started U.S. clusters had stayed at the Metropole. But in the end, our entire nation had only thirty-three cases and no deaths. The last case in the United States was recorded in July of 2003.

Back in China, at the origin of the pandemic, the disease continued to spread, ultimately resulting in 5,327 cases and 349 deaths. Dr. Nanshan Zhong (who, years later, would participate in the six-person expert team that went to Wuhan in January 2020) was at that time the director of the Guangzhou Institute of Respiratory Diseases in the capital of Guangdong Province. The first patient with SARS arrived at the institute early in the outbreak, on December 20, 2002, and as cases accumulated in the coming month, Zhong discerned what was happening. Opposing the official line about the etiology, origin, and prevalence of the disease, Zhong held a press conference on February 11, 2003, the day before China officially notified the WHO, at the Guangdong Department of Health, and he explained the cause, prevention, and management of the new disease.[22] On April 20, the mayor of Beijing and the minister of health, both of whom had, unlike Zhong, downplayed the threat, were removed from their Communist Party posts. Zhong was lionized.

The last probable case of SARS-1 in China was reported on June 25, 2003. By then, Chinese scientists had identified the likely source. In May of 2003, viruses similar to the one causing the disease in humans were isolated from animals that were eaten as delicacies. Twenty-five animals from eight species (including masked palm civets, raccoon dogs, badgers, beavers, and hares), all of which were obtained from the Dongmen Market in Shenzhen in southern Guangdong, were tested. All six of the civets and the raccoon dog tested positive for the coronavirus. It was logical to look at these animals. From the outset, epidemiological data had suggested that SARS originated with them. Animal handlers and chefs who prepared these animals for consumption were overrepresented among the initial Chinese cases of SARS. And later studies found high levels of antibodies to SARS-1 in animal vendors in several markets in Guangdong in 2003, in keeping with their having been exposed to the virus. The animal traders, for instance, had much higher levels of the antibodies than did the vegetable traders from the same markets who served as a control group.[23] Genetic analyses revealed that the ultimate source of the virus was bats, probably via their excretions.[24]

Commercial airlines in the first SARS pandemic were like the cruise ships in the 2020 pandemic. Five international flights were associated with the transmission of the disease, though SARS patients are known to have boarded many more. In the worst-affected flight, twenty-two of one hundred twenty people on board became ill from the index case. Sitting close to an infected patient increased the risk, unsurprisingly, but it was not essential for catching the virus. On one flight, two people who were on the other side of the plane, seven rows and a passageway apart from the patient, still got infected.[25]

Because of transmission on planes, a previously unused technique debuted during this epidemic: thermal screening at airports. I noted that it was still in use on a trip I took to Hong Kong in 2005 and it struck me as something out of a science fiction movie. I did not like seeing the heat of my own body—or, for that matter, of the bodies of people near me—exposed on video monitors that were surveilled by stony-faced security personnel. The humiliation hardly seemed worth the effort because in the end, only a minuscule number of patients were identified using these scans. That may be why, when I returned to Hong Kong in 2016, they were gone. In fact, one analysis that combined results from Canada, China, Taiwan, and Hong Kong showed that not a single case of SARS was detected by thermal scanning of over thirty-five million international travelers—a result that was pertinent as airports in the United States began to implement thermal scans in the summer of 2020.[26] The history of pandemics is unfortunately littered with such examples of attempted technological fixes, often expensive or overreaching, that can divert attention away from more effective measures. But hindsight is clear, and it's hard to fault those acting in real time to stem the spread of an invisible invader.

On March 12, the day after Dr. Urbani was hospitalized in Thailand and two weeks after he had notified the World Health

Organization of the outbreak there, the WHO issued a global alert about a new infectious disease causing atypical pneumonia.[27] By April 16, the virus causing the disease was definitively identified.[28] By May 13, outbreaks at all the initial countries had been contained due to a mix of contact tracing, quarantine, mask usage, and public education.[29] The travel advisories that had been issued for countries like Vietnam started to be lifted that month. It is likely that, in the Northern Hemisphere, the arrival of summer—which often attenuates respiratory disease outbreaks—played a role in the evanescence of this pandemic.[30]

The 2003 SARS epidemic was the first to be addressed in certain ways with modern advances in genetics, which allowed its genome to be entirely sequenced nearly in real time and which identified variants of the virus and their geographic distribution—tools that were later put to good use in the SARS-2 pandemic, as we saw in chapter 1. Rapid efforts were made to create a vaccine for the virus, and these got as far as animal testing but were then abandoned due to lack of commercial justification when the epidemic sputtered out. Of the 8,422 people who had been infected worldwide, 916 died; health-care workers accounted for 20 percent of these deaths.[31]

———»———

And so, the 2003 SARS pandemic ended almost as abruptly as it had started. It was not even officially declared a pandemic by the WHO—though in my view, it clearly met the definitional criteria, namely, "the worldwide spread of a new disease."

Why did the SARS-1 outbreak die out after its extraordinary opening act—featuring rapid global spread, many super-spreader events, and rising alarm—whereas the SARS-2 outbreak did not? It was not just that there was somehow a more efficient public health response in 2003. After all, SARS-1 spread to many countries and it did so in many places, from markets to hotels to hospitals to apartment buildings to airplanes. The reason SARS-1 petered out is that the virus itself was different from SARS-2 in subtle but important ways that made it harder to spread and thus made the pandemic easier to control. The divergent features of the two pathogens shed light on exactly why SARS-2 has become so destructive.

One of these features, paradoxically, was that SARS-1 was actually *too deadly.* Epidemiologists quantify the lethality of pathogens in two primary ways. The *infection fatality rate* (**IFR**) is the probability a person will die if he or she gets infected. The *case fatality rate* (CFR) is the probability a person will die after being diagnosed with the condition by a health-care provider. Sometimes an alternative for the CFR is used, the *symptomatic case fatality rate* (sCFR), which is the probability that a person will die if he or she simply shows symptoms of the infection. This can be a better metric, given the vagaries involved with whether people seek or get medical care.

Because the SARS-1 pandemic is well behind us, it is very easy to compute its CFR by simply dividing all observed deaths by all observed cases. Since the disease killed 916 out of 8,472 people that came to medical attention worldwide, that's a crude CFR of 10.9 percent. But for some populations, such as elderly people in Hong Kong, as many as 50 percent of infected patients died.[32] When a disease is very deadly, it kills its victims so rapidly that the pathogen does not have much time to spread. This is why the super-deadly Ebola epidemics that kindle in Africa every few years tend to burn out. The CFR for Ebola and its even deadlier cousin the Marburg virus can reach the terrifying range of 80 to 90 percent in some outbreaks.[33] Even though SARS-1 was milder in this regard than Ebola, the virus still killed its victims too fast for it to spread effectively.

In comparison, how deadly is SARS-2? Getting a fix on the IFR and CFR for SARS-2 was very hard in the first few months of the epidemic. This difficulty presents itself every time there is a new epidemic, for many reasons. Discerning the true denominator—of either the number of people who are infected or the number who have symptoms—is hard. For one thing, a lack of tests, especially early on, makes it hard to know who is infected, particularly if many are asymptomatic. Beyond that, many people who are sick do not seek medical care if their symptoms are mild. Or, conversely, they may not be diagnosed if they die too quickly to get to a doctor.

The numerator can also be hard to compute. Even in the *Diamond Princess*, where the denominator was known with certainty (since we knew how many people were on the ship and no one could leave it), the calculated CFR kept getting worse with the passage of time because more patients who were sick but alive when the calculations were first computed ultimately died; this makes the CFR a moving target until well after the outbreak. Another problem affecting the numerator is that people who were thought to have died of a different cause, such as a heart attack, were later discovered to have died of COVID-19 (as in the case of Patricia Dowd, whose death in San Francisco was initially misidentified) and so need to be added to the toll of mortality.

So, early in an epidemic, it is hard to know the CFR. Still, most authorities, using a broad array of data, samples, and methods from countries around the world, concluded that the overall CFR for COVID-19 was in the range of 0.5 to 1.2 percent. Since roughly half (or more) of patients with SARS-2 are asymptomatic, this meant that the IFR was half as much as the CFR, so in the range of 0.25 to 0.6 percent. A CFR of about 0.5 to 1.2 percent is at least ten times *less* deadly than SARS-1. Compare that also to the usual seasonal flu, which has an overall CFR of about 0.1 percent. To summarize, SARS-1 was ten times deadlier than SARS-2, which in turn is ten times deadlier than the ordinary flu.

But there is a further wrinkle. While SARS-2 is less deadly than SARS-1 in a given single case, that does not mean that it's less dangerous overall. Imagine a population of one thousand people and a pathogen that infects twenty of them; this pathogen makes those people seriously ill and kills two of them. That yields a CFR of 10 percent. Now imagine another population of one thousand people and another pathogen that does the same thing—it makes twenty people seriously ill and kills two of them—but this pathogen also infects another one hundred eighty people, making them mildly or moderately ill but not killing them. The CFR for the latter case is two deaths out of two hundred patients, or 1.0 percent, making the second disease appear much milder. But in reality, it's actually a worse disease overall. No one would prefer to be in the latter group of a thousand people rather than the former.

Some scientists think that, given the large array of symptoms and severities we saw in chapter 1, SARS-2 is like the latter case. It can infect the lungs, making people severely ill with pneumonia, but it can also infect the upper airway, giving people mild symptoms.[34] What we have learned already about SARS-2 biology supports this idea.

Only seven types of the coronavirus are known to infect humans. Four of them cause the common cold. Two of those four, OC43 and HKU1, originally came from rodents, and the other two, 229E and NL 63, came from bats. The other three types that afflict human beings are SARS-1, SARS-2, and Middle East respiratory syndrome, known as MERS. Of course, there are more types of coronavirus, some still undiscovered, but those affect bats and other animals. The viruses causing the common cold in humans bind more easily to cells in the upper airway (above the larynx), whereas the viruses causing more serious disease bind more easily to cells in the lungs, which helps explain why they are so much more deadly. SARS-2 is special because it can bind to cells in both places, which contributes to its wide range of symptoms and also makes it more transmissible. In other words, SARS-2 appears to have the ease of transmission of a common cold but the lethality of SARS-1.

Another crucial feature that made SARS-1 easier to control than SARS-2 is that it was generally not transmissible before a patient was symptomatic. That was why a large percentage of SARS-1 infections showed up in medical professionals—they were exposed to SARS-1 patients who were already quite sick. It was precisely when these patients went to the hospital, often near death, that they were most infectious. But SARS-2 is transmissible before symptoms show up. Back in February and even March 2020, in the United States, schools and workplaces were still advising people to stay home only if they showed obvious signs of illness. Churchgoers still shared physical signs of peace at Sunday services—and even drank Communion wine from a shared chalice—as long as they were not feeling feverish or having coughing fits in their pews. These milquetoast advisories were often made in spite of the fact that scientists were learning that asymptomatic carriers were a real problem (the head of the Centers for Disease Control had publicly announced this in mid-February).[35] Astoundingly, the governor of Georgia (a state that has been home to the CDC since its founding in 1946) claimed he had not realized that asymptomatic infection was possible until April 1, despite the fact that it was noted in the scientific literature in January.[36]

The time between when a person is infected and when he or she shows signs or symptoms is called the *incubation period*. This ranges from two to fourteen days in SARS-2 (hence the recommended fourteen days of isolation) and is typically about six to seven days. For SARS-1, the incubation period was shorter, ranging from two to seven days. But there is another important interval, the *latent period*, which is the time between when a person is infected and when he or she becomes *infectious*—that is, able to spread the disease to others.

The incubation period and the latent period are not always the same. The latent period is often shorter than the incubation period in SARS-2, but that was generally not the case in SARS-1. The difference between the latent period and the incubation period is sometimes known as the *mismatch period;* it's calculated by measuring the incubation period and subtracting the latent period. The difference between these two periods can be positive or negative.

In a disease in which the difference is positive, such as HIV, the incubation period is longer than the latent period, and asymptomatic patients can be infectious for some time (in this situation, the mismatch period is called the *period of subclinical infectiousness*), as shown in figure 5. In a disease in which the difference is negative, such as smallpox, the latent period is longer than (or the same duration as) the incubation period, and patients must be symptomatic before they are infectious.

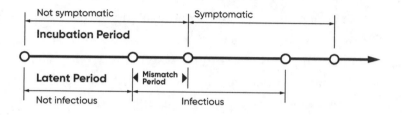

Figure 5: Important periods in the course of an infection, with illustrative durations, show that if the latent period is shorter than the incubation period, as it is in SARS-2, there can be asymptomatic transmission (a "quarantine loophole") during the mismatch period.

In the case of SARS-1, the latent period was the same as or possibly slightly longer than the incubation period, so patients generally were not infectious until they had symptoms. But in SARS-2, this is not the case. While COVID-19 patients on average take about seven days from exposure to show symptoms, a meaningful percentage of carriers can spread the disease for two to four days *before* they are symptomatic. The announcement of this finding was unpleasant news, news that many politicians and others had hoped to deny and that many doctors, myself included, had hoped, at the beginning of the pandemic, would not be the case. But from the outset, published evidence from China supported this finding. A careful early study of 468 infector-infected pairs (for instance, within families who had traveled together) found that 12 percent involved transmission *before* the infector was symptomatic.[37] And an examination of 124 Wuhan patients with a clear contact history found that 73 percent of secondary cases were infected *before* the onset of symptoms in the first cases.[38] Subsequent studies from different countries confirmed these observations.[39] In fact, in many cases, it seemed that the one to two days before a person manifested symptoms was when COVID-19 was possibly most contagious.

Asymptomatic transmission can present an obvious challenge to public health management since blithely unaware patients can spread the disease. If most transmission occurs before disease is apparent (as happens in HIV), *reactive* control measures (where public health workers wait for cases to appear in order take measures like contact tracing and quarantine) will be ineffective. Conversely, successful disease control (as happened with SARS-1) is facilitated by low transmission by asymptomatic people, since people who are clearly sick are easier to identify and quarantine before they can spread the disease too much. That is, even if we were to quarantine all people with symptoms of SARS-2, other infected people, lacking symptoms, would escape the dragnet and would be able to transmit the disease. To be clear, the existence of asymptomatic transmission of SARS-2 does *not* mean doctors should not bother to evaluate patients for symptoms such as fever and ask them to self-isolate. Isolating carriers and patients is still essential. But such an action alone cannot stop the epidemic. This is the reason that *testing* asymptomatic or minimally symptomatic people is so important and why the lag in implementing widespread testing for COVID-19 in the United States (compared to, say, South Korea) was so unfortunate. Testing allows patients to know they are infected even when they do not have symptoms, and it allows the authorities to suggest (or enforce) home isolation in such cases.

In addition to the CFR and the mismatch period, another key parameter of SARS-2 was investigated in early 2020: How many new cases does each case give rise to? For each person who becomes infected, how many other people, on average, does that individual infect? This number is known as the *effective reproduction number,* denoted R_e (sometimes also known as the *effective reproductive rate*).

The R_e is different from a more basic parameter, the R_0 (pronounced "R naught"), which was what Kate Winslet's character scribbled on a whiteboard for an incredulous media team in the movie *Contagion*. R_0 is the average expected number of secondary infections for each primary case *in a naive and wholly susceptible population with no prior history of the disease.* The R_0 captures the capacity of a pathogen to *start* an outbreak, and it reflects the degree to which it is infectious in the *absence* of any measures to control it. R_e, however, reflects the real-time spread of the epidemic later in its course, when the population is no longer "naive." The R_e is susceptible to human responses.

Some pathogens are intrinsically more transmissible than others and spread very easily from one person to another while other pathogens are much less infectious. For instance, measles is one of the most infectious diseases known, with an R_0 estimated to be 12 to 18 (that is, a single infected person typically can infect somewhere between twelve and eighteen other people). Chicken pox is 10 to 12. Smallpox is 3.5 to 6. Ebola is 1.5 to 1.9. Seasonal influenza ranges from 0.9 to 2.1.[40]

When the reproduction number (whether the R_0 or R_e) is above 1, this means that the number of cases will grow. This is the very definition of an infectious epidemic. When the reproduction number is below 1, the case counts will fall with time (since each existing case cannot even replace itself), and the epidemic is contained. When the reproduction number is approximately 1 (either naturally or because of the actions society takes), the disease is holding steady— which is part of what is meant when people say that an infectious disease has become endemic in a population.

In other words, it's not just the intrinsic infectiousness of the pathogen that matters. Aspects of the *host* and the *environment* also matter, and that is why the effective reproduction number, R_e, is important. For instance, imagine that we, the unfortunate hosts of SARS-2, adopted the extreme behavior of each of us living alone in a mountaintop hut. In that case, people who were initially infected simply could not spread the pathogen to anyone else. Or imagine that, at some point late in the epidemic, most people became immune to the pathogen. This would surely affect its transmissibility. In either case, as the epidemic progressed, a sick person would be less and less likely to encounter susceptible people to infect. The R_e declines naturally as an epidemic proceeds because susceptible people become infected and either die or survive and acquire immunity (to a greater or lesser extent). Thus, behaviors and attributes of the pathogen's host matter. How does the environment matter? Imagine that a pathogen is very sensitive to heat and so spreads easily in cold weather but not in hot weather. As a consequence of such factors related to both host and environment, the same pathogen would fare very differently if it infected hermits on Mount Athos in the summer or partygoers in New York City in the winter.

For SARS-1, the R_0 was computed to be in the range of 2.2 to 3.6, and it's probably between 2.6 and 3.0.[41] But the R_e could vary, depending on our collective response to the disease. For instance, at the outset of the SARS-1 outbreak in Singapore, the R_e was perhaps 7 during the first week, but Singapore responded promptly to stop the epidemic by implementing procedures such as quarantine, and it fell to 1.6 by the second week, then below 1 by the third week.

As with the incubation and latent periods, these descriptions of the reproduction number have so far related to the epidemic situation *on average*. With an average R_0 of 3.0, a single case of SARS-1 created three new ones. But there was variation. Some patients infected no one else or just one other person, whereas other patients infected dozens of other people, as we saw with Dr. Liu (albeit not necessarily through any fault of their own). Such people are known as super-spreaders. Figure 6 illustrates how some chains of transmission might dead-end and others might involve many victims. The circles indicate people and the lines indicate social interactions and transmission opportunities. The dark circles are infected people and the open circles are uninfected people. The index case, on the far left in black, infects four people who in turn have variable numbers of connections, and they go on to infect no people, three other people, or many people in a super-spreading event.

This *variation* in R_0 across individuals in a population can be quantified, and this quantity can have subtle but important effects on the course of an epidemic. The higher this variation (or dispersion), the more likely an epidemic will feature *both* super-spreading events *and* dead-end transmission chains. That is, an epidemic involving a population of people for whom the R_0 is a steady 3 for every person can have a very different course than an epidemic involving a population for whom the R_0 ranges from 0 to 10, even if the average R_0 is still 3. If this variation in R_0 is large, the risk of an outbreak starting from any given person falls substantially because there will be many more people who cannot spread the germ than people who can. In such a case, many importations of infected people from one place to another are required to seed the epidemic in the new location. We saw an example of this in Seattle in chapter 1, where the chain of transmission actually ended with Patient Zero. Other importations into Seattle were needed to set off the epidemic there.

Consider the following simplified illustration of this idea. Say there is a group of one hundred people who are infected with a virus. One is a super-spreader who can spread disease to three hundred people, and ninety-nine are not infectious at all. The average R_0 in this hypothetical population of one hundred is 3.0, but there is a very large variation in infectiousness. If just one person from such a population is chosen at random to travel to another place, that means that, ninety-nine out of a hundred times, there will be *no epidemic* as a result. By comparison, if there is a group of one hundred people who can each spread the disease to three people, the average R_0 is still 3.0, but now there is no variation in infectiousness. Picking

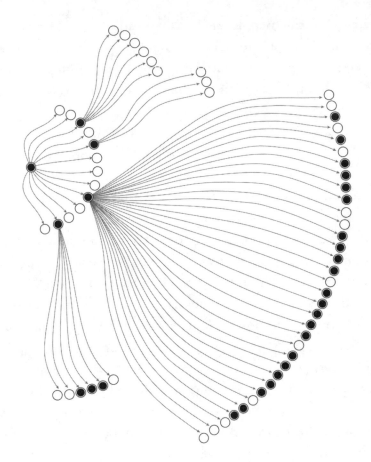

Figure 6: Variation in the spread of a pathogen in a set of social ties can result in a super-spreading event and in dead-end transmission chains (filled circles are infected people, and open circles are uninfected people).

one member of this second group to travel to another place means that *surely* an epidemic will start there. Although in both groups the pathogen has the same average R_0, the fact that the *variation* of the R_0 is *smaller* for the latter group means that the germ is much more likely to seed new infections in other communities.

An epidemic with large variation in individual R_0 manifests itself with many super-spreaders and super-spreading events. This is what happened with SARS-1.[42] For SARS-1, it was estimated that four importations were necessary for one transmission chain to be initiated (the other three importations would fail to start epidemics and die out). However, when an outbreak did occur, it was likely to be explosive. For SARS-2, it looks like the variation in R_0 is somewhat lower than for SARS-1, so while super-spreading events do occur, they are less common than the more frequent, humdrum chains of transmission.[43]

The origin of super-spreading and of the variation in reproduction number, whether quantified as R_0 or R_e, may relate, once again, not only to the pathogen but also to the host and the environment. Individual differences in hosts are important. For reasons that are not well understood, some people may shed more virus or may shed it for a longer period, setting them up to create more secondary cases. Or they might have variation in their propensity to cough that makes them more likely to spread the virus. Individual differences in behavior can also matter; people who like to spend more time in large groups socializing or who fail to wash their hands often may be more likely to become super-spreaders.

Some cases of super-spreading may relate simply to the fact that some people come into contact with a great number of other people in the routine course of their lives or they may have more social connections. To be fair, popular people are more likely to become infected themselves as well as more likely to infect numerous others (illustrated by the many politicians and actors who were infected early in the COVID-19 pandemic). For instance, consider the social network diagram in figure 7, which shows 105 real people. Every circle is a person, and the lines indicate friendships between pairs of people. Person A has four connections and B has six connections. B will be able to spread the germ more easily than A. But now look at C compared to B. Both have six connections. But C has another property: C's friends have more connections than B's friends. This makes C more central in the network, which is apparent in the image itself. And this means that C can, in general, spread the pathogen more rapidly to more people than either A or B.

In many real-world social networks, most people have very few contacts and a small minority have many connections. This small minority are the ones who often go on to become super-spreaders. So SARS-2 is more likely to reach these well-connected people, and they are more likely to spread it to a large number of people.[44] In fact, mathematical models of a disease spreading over such networks with super-spreaders closely mirror the observed trajectory of real cases of COVID-19.[45] However, just having a large number of contacts, however defined or ascertained, does not mean someone is *necessarily* a super-spreader—remember that when health officials traced the contacts of the first lab-confirmed case of person-to-person transmission of SARS-2, the married couple in Illinois, they found that *none* of their 372 contacts had become infected.[46]

The existence of natural variation in the number of friendships and connections that people have has a further implication. It means that it may actually not be necessary to have quite so many people exposed to a virus before a population reaches the important threshold known as herd immunity. Herd immunity is the idea that a group of people can be collectively immune to an infectious disease even if not everyone in the population is individually immune. The term has veterinary origins, but it applies equally well to human beings. The concept is that, if a sufficiently large number of people have acquired immunity to a disease (either by getting it and surviving or by being vaccinated), then any individual in this population who somehow contracts the illness is unlikely to encounter another person to whom he or she can transmit it. Hence, even if the chain of transmission somehow got started, it would die out.

But this is where the structure of social networks comes in again. Since people vary in how many social connections they have, popular people with many connections (like person C) tend to be infected earlier in the course of an epidemic than random people

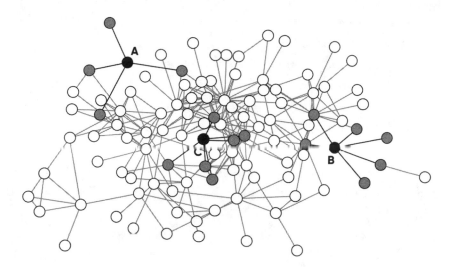

Figure 7: This is a real social network of 105 people and the connections among friends. Three people (also known as nodes in a network) with different numbers of connections are highlighted. Node A has four friends. Nodes C and B have six friends, but C is more likely to get sick from, and to spread, a pathogen like SARS-2 because C is more central.

chosen from the same population. Because of their many social interactions, popular people have an increased risk of exposure. For example, when my lab analyzed an outbreak during the 2009 H1N1 influenza pandemic, we found that, for every extra person who identified an individual as a friend, that well-liked individual was prone to get the flu eight days earlier in the course of the epidemic. This means that, in figure 7, B would get infected roughly sixteen days earlier than A because B has two more friends than A.[47]

But this also means that popular people are more likely to become immune early on in the course of an epidemic. And if all the popular people became immune early, relatively more paths for the virus to spread through society would be cut off. Unpopular people are less concerning from an epidemic-control point of view because they intrinsically infect fewer other people, so their immunity matters less. This also means, incidentally, that vaccinating people with many connections is more helpful than vaccinating people with few connections.

Finally, the *environment* may facilitate super-spreading events, such as a SARS-2 outbreak in a choir in Washington State in March 2020. The act of singing involves the expulsion of air at high force, which likely contributes to spread. In this case, an infected person was part of a group of sixty-one singers who were in a densely packed area for two and a half hours.[48] As many as fifty-two of them caught the virus from the index case; three of them were hospitalized, and two of them died. An analysis in South Korea showed that outbreaks were more common in Zumba classes than Pilates classes for a similar reason.[49] Heavy, rapid, deep breathing or shouting may be a risk factor for transmission, whereas slow, gentle breathing is not. But being indoors itself plays an important role. Other super-spreading events have involved the assembly of large numbers of people in indoor environments. At a funeral in Georgia on February 29, 2020, someone spread the virus to two hundred other people, which accelerated the broader epidemic in that state.[50] Something similar happened at a February meeting in Boston involving executives of a biotech firm; dozens of people became infected.[51] And there have been super-spreading outbreaks in prisons, nursing homes, hospitals, factories, and other indoor places where people are densely packed. A study of 318 outbreaks of three or more cases in China in the early days of the pandemic found that all but one of them took place indoors.[52]

One of the positive implications of the existence of super-spreading events is that, if people or environments prone to super-spreading can be identified, infection control efforts can be targeted, possibly leading to large gains in both efficacy and efficiency. We do not necessarily need to stop all social interactions; we can focus on tightly packed conventions and nightclubs, for instance. And we can move religious or other events outside or online.

For one more comparison, let's very briefly consider the second deadly coronavirus epidemic prior to this one: MERS, which stands for Middle East respiratory syndrome. MERS first appeared in 2012 and has symptoms that are very similar to the two SARS pandemics. As of 2020, there have been only about twenty-five hundred cases worldwide. Most of the cases (80 percent) have occurred in Saudi Arabia, though the disease has cropped up in twenty-six other countries, almost always in people who had traveled to the Middle East.[53] This disease also originated in bats, but the intermediate host is camels, who apparently acquired the virus some decades ago. Contact with an infected camel (or camel products) can lead to infection, as can substantial contact with an infected person.

The estimated CFR for MERS is 35 percent, which makes it about three times as deadly as SARS-1. But the R_0 is low. Human-to-human transmission is not easy. Most cases are among intimate household contacts or health-care workers caring for an infected patient. Several studies have suggested that the R_0 of MERS is actually approximately 1.0, or even below (though other studies estimated the R_0 to be on the order of 2.0 to 2.8).[54] As we already learned, such a low value would be below the "epidemic threshold," since an R_0 of 1 means that each patient transmits the virus to only one new person, effectively making it impossible for an epidemic to grow in size.

From this review of basic epidemiological parameters, we can see that, compared to SARS-1 in 2003, SARS-2 in 2020 has intrinsic features that make it much more of a threat.

Most estimates place the R_0 of SARS-2 as roughly the same as that of SARS-1, about 3.0. This is actually rather worrisomely high for a pathogen; compare it to the R_0 of ordinary influenza, which is 0.9 to 2.1.[55] But SARS-2 has a smaller dispersion in R_e, meaning that transmission chains are somewhat less likely to be dead ends, which makes it easier to reliably spread SARS-2 than SARS-1. SARS-2 is also less deadly than SARS-1, with a CFR of less than 1 percent (compared to around 10 percent for SARS-2). As we saw, this makes

SARS-2 paradoxically *more* concerning, because larger numbers of people survive, and move about for longer, to transmit it. Finally, and perhaps most important, unlike SARS-1, SARS-2 has a positive mismatch period; this allows for asymptomatic transmission and makes traditional public health responses based on detection and quarantine very difficult.

The upshot of all these factors is that there is a very high probability that SARS-2 will infect a large percentage of the population before the pandemic has run its course—an epidemiological parameter known as the *attack rate* (the total number of infected cases in a population at the end of an epidemic divided by the total population). For SARS-1, the attack rate was infinitesimal, with only a tiny fraction of the planet infected by the end of the pandemic, since 8,422 people out of 6.314 billion is just 0.00013 percent. For SARS-2, however, probably at least 40 percent of the human population worldwide will be infected in the end, and perhaps as much as 60 percent.

»———→

Just as these epidemiological parameters help us compare SARS-2 to other coronaviruses, they can also help us learn from respiratory disease pandemics caused by other species of virus, such as the influenza virus.

For nearly thirty years, I have been teaching about the decline in mortality in the United States during the twentieth century. This decline had many causes, including the rise in wealth in our society, changes in behaviors and public health practices (like sanitation and vaccination), and the emergence of modern medicine (though, as we will see in chapter 3, medical interventions have played a smaller role than one might think). Regardless of the reason for it, however, when we look at the curve of overall death rates in our society in the twentieth century, mortality drops and drops and drops across time, with only occasional small upward blips resisting our progress. But on the graph in figure 8, there is an enormous spike of death in 1918. I always highlighted this outlier spike to my students—this is the so-called Spanish flu pandemic. My students treated it as an odd curiosity, and I spoke of it as a relic safely in the past, the way a parent might talk about how an adult child had survived leukemia as a toddler. But it was foolish to imagine that such a thing could not happen again.

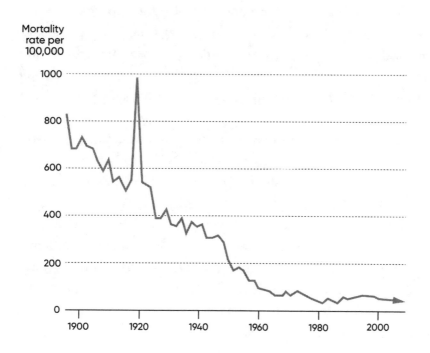

Figure 8: Mortality in the United States since 1900 has generally declined, but the spike in mortality during the 1918 influenza pandemic stands out.

It's not surprising, of course, that few people other than historians and epidemiologists remember the details of the 1918 influenza pandemic, the deadliest plague of the past century. But most people I have spoken with since the SARS-2 pandemic began, including those old enough to have been around, do not remember the influenza pandemics of 1957 or 1968 either. I was six in 1968, so that is my excuse. I do, however, remember the 1976 swine flu pandemic that never materialized, despite its being widely anticipated. Mostly I remember it because of the fact that the vaccine that was released to deal with it caused a serious but reversible neurological condition (Guillain-Barré syndrome) in one out of every one hundred thousand people, and one of my teachers at the Harvard School of Public Health wrote a book about the topic that I read in 1987.[56] I also well remember the H1N1 pandemic of 2009, another swine flu, this one originating in Mexico, since by that time I was a professor at Harvard Medical School, and I conducted research regarding the outbreak. Even though the 2009 pandemic is recent, most people overlook it simply because it was so mild and not many people died. So it gets no respect either.

As someone who has studied all of the flu outbreaks that occurred in the past century, I naturally used them as points of reference as COVID-19 began to gain steam. In February 2020, as preliminary estimates about the CFR and R_0 for SARS-2 started streaming in from China, I became increasingly convinced that the pandemic that COVID-19 would most resemble was the 1957 influenza pandemic. I knew it would not be mild, like the 2009 H1N1 pandemic, and I feared (but doubted) that it would be like 1918.[57] At the time, I was teaching a public health class at Yale. At the beginning of the semester, in January—when I was not yet paying attention to the butterfly flapping its wings in China—I had given my usual lecture to the students about the "aberrancy" that the 1918 pandemic had caused in the curve plotting the steady decline in mortality during the twentieth century. I did not expect that within two months, we would face a real risk of another such spike.

On March 5, 2020, I gave a lecture about COVID-19 and its likely severity. The students were leaving for spring break the next day, and as I looked out over the sea of inquisitive and happy faces, I tried to be both hopeful and matter-of-fact. I feared they might not be allowed to return to campus, as the pandemic was likely to get worse during their break. And I was right.

In order to benchmark myself, I used a classic approach. I assembled a sample of respiratory pandemics from the twentieth century and graphed them on a chart using two numbers: the CFR (which indexed the severity of the disease) and the R_0 (which indexed the transmissibility), as shown in figure 9.[58] Doing this really focused my attention. I used estimates of these parameters that had emerged from China in February, and it seemed to me that SARS-2 had intermediate lethality and intermediate transmissibility, making it at least as bad as the influenza pandemic of 1957. Ultimately, 115,700 excess deaths occurred in the United States over three years due to the 1957 pandemic (that year, the population size of the country was 172 million, and cancer killed 255,000 people).[59] This would be equivalent to roughly 300,000 Americans dying from COVID-19 by the end of the pandemic, a figure we are sure to surpass despite our extensive shutdowns.

To be clear, the influenza virus that caused the 1957 pandemic is totally different from the coronavirus that causes COVID-19. Both are riboviruses, meaning that they use RNA rather than DNA for their genetic code, but that is a very broad classification; it's sort of like saying that dolphins and elephants are both mammals. The two types of viruses come from very different branches of the genetic tree, from different phyla. There are four broad types of influenza virus, labeled A, B, C, and D. Influenza pandemics result from the emergence of viral strains that are novel, often from genetic recombination in animal reservoirs (usually birds or pigs). The 1957 pandemic was caused by influenza A, and it was of the H2N2 subtype.

Influenza A viruses are further divided into subtypes based on two proteins on their surfaces: hemagglutinin (H) and neuraminidase (N). There are eighteen different kinds of hemagglutinin (labeled H1 to H18) and eleven different kinds of neuraminidase (N1 to

N11). While this in principle potentially yields one hundred ninety-eight different combinations, only one hundred thirty-one subtypes have been detected naturally. The types of influenza that currently routinely circulate in our species and cause the familiar seasonal influenza include the A (H3N2) and A (H5N1) subtypes. Virus species causing outbreaks can be genetically different but still express the same protein types on their surface.

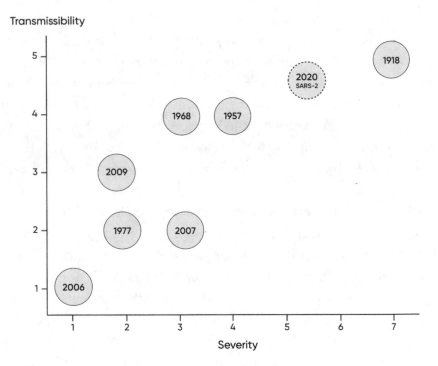

Figure 9: Graphing a century of respiratory pandemics in terms of their transmissibility and severity (lethality) showed that the basic epidemiology of SARS-2 made it a serious threat.

The 1957 pandemic likely started in central China, and it became known to the rest of the world in April of 1957.[60] Globally, it killed 1.1 million people, but there was great regional variation; for instance, there were 0.3 deaths per 10,000 people in Egypt and 9.8

deaths per 10,000 people in Chile. This variation across locations—which is common in respiratory pandemics—is sometimes perplexing. In 1957, the wealth of a country and its latitude explained 43 percent of the variation in excess mortality from the flu pandemic.[61] The way people in a particular area live (from population density to household size to the timing of school openings) may also play a role. It's possible that prior exposures to similar pathogens may confer some immunity in certain areas, and the chance occurrence of an early super-spreading event may affect outcomes. But overall, annoyingly and often, we are unable to account for why some parts of the world or some parts of a country are more stricken than other parts. A lot of the time, it's just a matter of luck.

The first rumblings of the 1957 pandemic in the United States were in a tiny article, fewer than one hundred fifty words, in the *New York Times* on April 17, 1957: "Hong Kong Battles Influenza Epidemic." It noted that two hundred fifty thousand people in Hong Kong were afflicted (10 percent of the city's population). The pandemic touched down in the United States six weeks later, on June 2, in a group of military personnel aboard destroyers in Newport, Rhode Island. But other outbreaks, unconnected to Rhode Island, soon occurred in California. The first began on June 20 at a camp for teenage girls in Davis. Over the next few weeks, there were more than fifteen similar outbreaks in children's camps throughout the state. And then, in late June, there was a convention involving eighteen hundred young people from forty-three states and several foreign countries held in Grinnell, Iowa. Several of the people attending this event had been in contact with prior cases in California. About two hundred people got sick at the conference, and another fifty cases emerged after attendees had returned home. And yet, the International Boy Scout Jamboree was held in Valley Forge, Pennsylvania, from July 10 to July 24 with hardly any problems. Only a few sporadic cases were noted, despite more than fifty-three thousand boys and personnel from all parts of the world gathering there.[62]

All of this was carefully traced later on by a team of epidemi-ologists who published a paper in 1961 with a hand-drawn map showing the movements of infected people in bold black arrows; it looked like a map of military operations from World War II.[63] The pandemic waned a bit later in the summer, but it recurred when schools opened in the fall. By September, it was everywhere in the United States and there was a peak in excess death in October. By December 16, 1957, the first wave had almost entirely subsided. However, there were subsequent waves of the pandemic, as is typical, with another peak in excess death in February of 1958 and then again in April of 1959.

Waves of cases and peaks of deaths have to do with many factors. Respiratory diseases are influenced by the weather. For example, the regular (non-pandemic) flu has a baseline seasonality—roughly 6 percent of all deaths are caused by pneumonia and influenza in the summer and 8 percent of all deaths are caused by them in the win-ter.[64] Typically, these respiratory viruses travel between the Northern and Southern Hemispheres, affecting whichever is in winter.

In the summer, the relative heat and humidity may affect how long the virus survives outside our bodies (or, possibly, how well our bodies can resist it).[65] But in addition to temperature, the rhythm of work and school life changes seasonally. In the winter, people are indoors and are more densely packed. All of these reasons can cause the effective reproduction number of a pathogen to decline in the summer to some extent. When this happens, in a process known as seasonal forcing, there can be an accumulation of susceptible individuals during the period of reduced transmission.[66] People who would have gotten sick do not, and they pile up for the winter. Like a dam, the weather and the host behavior hold the epidemic back. But eventually, the dam breaks. This results in recurrent outbreaks (and, ultimately, a seasonal pattern) after the initial introduction of the pathogen.

The pattern of school-age children and working adults being

infected but recovering and very young and very old people dying is a general phenomenon. The virus spreads among people who are out and about at school and work; it is then brought home, where it kills the age extremes—infants and the elderly—who are at the end of the transmission chains. This is also the reason, incidentally, that immunizing the elderly, while it will reduce their deaths, does not have much effect on the actual course of the epidemic. Immunizing working-age people helps break chains of transmission through social networks and can be much more effective in preventing deaths on a population level (an idea that resembles what we discussed above with respect to targeting socially connected people for immunization).

It's important to note that, by the end of the 1957 pandemic, *not* everyone had been infected. The overall attack rate in the United States appears to have been about 24 percent, though this varied from place to place (it was as high as 41 percent in part of Louisiana).[67] The 1957 pandemic ended after three years, when enough people had become immune that the population acquired herd immunity. In part, this was achieved through widespread vaccination (flu vaccines were invented in 1945).[68] Possibly the virus also became less virulent over time, which is another typical feature of infectious diseases.

»——————→

The 1918 flu pandemic (incorrectly labeled as originating in Spain) affected and killed many more people than the 1957 pandemic. Perhaps thirty-nine million people died worldwide, which was 2.1 percent of the world population, and some experts put the worldwide toll as high as one hundred million (given possible misidentification of deaths and poor reporting).[69] Life expectancy was still low in the United States at the time (for men in 1915, it was 52.5 years; for women, 56.8), and infectious disease was still a major cause of death. But even so, the 1918 flu had a powerful adverse impact. The United States had a cumulative death rate of 0.52 percent of the population, or 550,000 people, including one in every one hundred men between the ages of twenty-five and thirty-four (again, some estimates are higher).[70] For comparison, in the United States in 2020, that would translate to 1,721,000 deaths. The impact of the pandemic was so substantial that, for a while, it even had an effect on the total average life expectancy in the United States, shaving off ten years. As we saw in figure 7, it really interrupted the trajectory of improving mortality in the nation.

The fatality rate of the 1918 flu varied from country to country and place to place. For example, in Bristol Bay, Alaska, it was as high as 40 percent, but on the lower end it ranged from 1.6 percent in Rio de Janeiro, Brazil, to 3 percent in Zamora, Spain, to 6 percent in Gujarat, India, to 10 percent in Ciskei, South Africa.[71] Some of this had to do with differing public health responses within and across countries. For instance, Korea was an outlier in 1918 (as it is now); one policeman in Tokyo wistfully observed that the authorities in Korea had banned all mass gatherings "but we can't do this in Japan."[72]

Given how deadly this pandemic was, it's surprising that it's not more salient in our collective memory outside of public health circles. Everyone learns facts about World War I, but few learn much about the pandemic, which was much deadlier. Perhaps that's because dying of an illness at home seems less dramatic than dying in the trenches. But death from influenza in 1918 was totally horrific. People gasped for breath as their lungs filled with bloody fluid. Symptoms often started with two telltale mahogany spots on the cheeks that very quickly darkened the whole face (known as *heliotrope cyanosis*). The illness progressed from "dusky, reddish plum" coloring on the skin to blue coloring and, finally, to black, spreading from feet and fingers to the whole body. The patient's chest and abdomen became distended. Some of the patients died rapidly and directly from the virus, but more often, it was the secondary bacterial pneumonia that killed them. In short, it was nothing like the ridiculously glamourous death from the Spanish flu shown in an episode of *Downton Abbey*.

The origin of the 1918 pandemic is not clear, but we know it was an H1N1 type of influenza. Reports from the time indicated that there were often simultaneous outbreaks in humans and pigs, but it's unclear which species gave it to which. The natural reservoir for influenza viruses is thought to be wild waterfowl, but humans are not often susceptible to catching the disease directly from birds. Since pigs can be infected with both bird and human strains of flu, it is suspected that they are often integral to the process whereby a virus becomes capable of infecting humans—which is why we intermittently call flu pandemics "bird flu" or "swine flu." From time to time, there is a mixture of genetic material from these avian strains with existing human or pig strains, and, in a Frankenstein fashion, this leads to the creation of a new variety of the virus that can then cause major or, depending on the mutations, minor outbreaks.

Over the past twenty years, a kind of genetic archaeology has allowed us to learn more about what happened in 1918. Preserved tissue specimens from the lungs of two American soldiers who died in September 1918 (one in New York and one in South Carolina) and a very precious sample from an Inuit woman who had been buried in the permafrost since November 1918 have allowed us to reconstruct the genetic sequence of the virus and even bring it back to life (in a secure laboratory at the CDC) to better understand these mutations and their physiological and epidemiological implications.[73]

The first wave of the pandemic occurred in the spring and summer of 1918. The virus was highly contagious, but it caused relatively few deaths. For reasons that are not entirely clear, in August of that year, a much more lethal and infectious form of the disease emerged and spread worldwide. In the United States, the main impact was felt between September and November 1918, and in some American cities, many thousands of people per week died. There were three primary waves of the pandemic: a peak in the spring of 1918, a rapid second wave in the fall of 1918 (peaking in October), and a third wave a year later in the winter of 1919. It finally subsided that summer, as shown in figure 10.

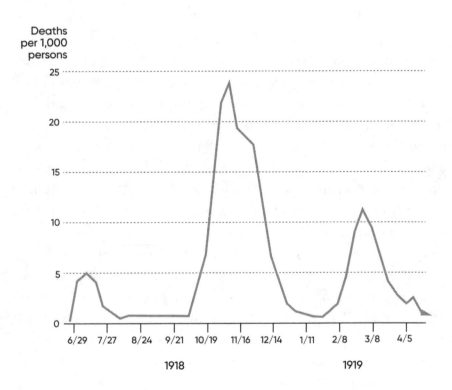

Figure 10: There were three waves of deaths during the 1918 Spanish influenza pandemic; the second wave was four times deadlier than the first.

There is some evidence that the disease first crossed from pigs to humans in Kansas, afflicting a mess cook at Camp Funston, a U.S. Army training facility, on March 4, 1918.[74] From there, it spread with troop movements to the East Coast and then to France by April. This resulted in the pandemic causing significant military disruption in Europe. While morbidity was highest in the U.S. Army (26 percent, with over one million men sickened in 1918), the German army reported 700,000 cases, and the British Expeditionary Forces reported 313,000 cases in France (perhaps half of the British forces were afflicted).[75]

In August 1918, the flu erupted again in three locations: Boston (arriving there from European ships); Brest, France (via troop movements); and Freetown, Sierra Leone (via a British naval vessel). From these points and others, it spread throughout the United States and the whole world. Most of the deaths occurred in the period between September and December 1918. The disease was devastating. For example, in early September 1918, Dr. Victor Vaughan, acting surgeon general of the army, received urgent orders to go to Camp Devens near Boston. When he arrived, he was stunned: "I saw hundreds of young stalwart men in uniform coming into the wards of the hospital. Every bed was full, yet others crowded in. The faces wore a bluish cast; a cough brought up the blood-stained sputum. In the morning, the dead bodies are stacked about the morgue like cordwood."[76] On the day he arrived, sixty-three men died from influenza.

A detailed study from Norway, where partial data on the socioeconomic status of victims were available, showed that the first wave hit the poor but the second wave hit the rich.[77] It afflicted world leaders, including President Woodrow Wilson, French prime minister Georges Clemenceau, and UK prime minister David Lloyd George. Some authorities believe that the flu even hastened the end of World War I. This second wave in the United States ended very abruptly; for example, after peaking in New York City on October 20, 1918, the flu all but disappeared from the metropolis by November 24.[78]

There are many theories about why the second wave was worse. One is that it struck an already weakened population even more worn-out by war and famine. It also may have been related to how the particular circumstances of the war favored the evolution of a deadlier strain of the virus. Ordinarily, pathogens evolve to be less deadly, since it does not suit their interests to kill their hosts. A dead host cannot easily spread the germ to others, so causing milder illness is "better" for the pathogen from a Darwinian point of view. People who are very sick stay home in bed or die, and those who are only mildly sick continue with their lives, preferentially spreading milder strains of the pathogen. But in the trenches of World War I, the process was reversed. Soldiers with a mild strain stayed put on the battlefield or wherever they were, often dying of other causes, while patients who were seriously ill or near death were transported to crowded hospitals, often on crowded trains. Thus it's possible the deadlier strain of the virus could have been advantaged, spreading to more people.

Philadelphia was among the hardest-hit American cities, along with other densely populated industrial cities such as Pittsburgh, Lowell, and Chicago. The initial wave of the flu came to Philadelphia on a British merchant ship. The health director of the city, a gynecologist named Wilmer Krusen, had to decide whether to shut down the city entirely or take more limited measures. The problem was that a parade in support of the war effort was planned for September 28, 1918 (not dissimilar to the large meeting of the Chinese Communist Party in Wuhan in 2020). It was not canceled, and an estimated two hundred thousand people—over one-tenth of the population of the city—attended. The Philadelphia Liberty Loans Parade was two miles long and even had a marching band led by John Philip Sousa. Within two days, hospital capacity in the city was exceeded. Shortly thereafter, by October 3, the epidemic took off like wildfire. People died so fast that caskets piled up in the streets and volunteers had to dig mass graves.[79] It has been called the "deadliest parade in American history."

In New York, the epidemic also arrived by sea. On August 11, 1918, the Norwegian vessel *Bergensfjord* radioed ahead to say that ten of its passengers and eleven of its crew were ill. The ship was met in Brooklyn the day it arrived; the patients were brought to nearby hospitals and the ship was placed under quarantine.[80] But the epidemic still took hold in New York City and spread fast; at its peak, on October 20, 1918, over eight hundred people died (from influenza and pneumonia combined). This death count at the peak of the second wave was not dissimilar to the first wave of COVID-19 in New York City in 2020, but the city was less than a third the size back then. More than thirty thousand people died in New York City by the end of the 1918 pandemic.

In response to the influenza pandemic, New York City took advantage of what was a very sophisticated public health system for its time, honed in part by years of experience dealing with other infectious diseases, such as tuberculosis, yellow fever, and diphtheria. The city staggered business hours to avoid rush-hour crowding and established over one hundred fifty health centers to manage disease surveillance and coordinate care for patients in (mostly) voluntary home-based isolation.

Then, as now, officials struggled with whether to close the schools, theaters, and subways. The health commissioner, Royal S. Copeland, a homeopathic physician and later a U.S. senator, generally favored keeping things open, but the sick were strictly quarantined at their homes or at special institutions. Most patients were not infectious prior to the onset of symptoms, which meant that such isolation could be employed as a means of disease control. Masks were encouraged. Sharing cups at drinking fountains was discouraged. And there was a massive public health–education campaign; thousands of placards and a million leaflets written in English, Italian, German, and Yiddish told citizens to cover their mouths and noses and stop spitting in public. An anti-spitting campaign had been started twenty years earlier, but it was stepped up during the pandemic. Signs throughout the city carried messages such as SPIT SPREADS DEATH. Public funerals were banned; only spouses were allowed.

Copeland famously did not close the schools on the grounds that homes were more dangerous. His rationale was "New York is a great cosmopolitan city and in some homes there is careless disregard for modern sanitation.... In the schools, the children are under the constant guardianship of the medical inspectors. This work is part of our system of disease control. If the schools were closed, at least 1,000,000 would be sent to their homes and become 1,000,000 possibilities for the disease. Furthermore, there would be nobody to take special notice of their condition."[81] Dr. S. Josephine Baker, the progressive head of child hygiene, argued for keeping children in school, where they could minimize exposure to unhygienic home situations and receive food and care if sick.[82] The schools were also viewed as a conduit to get health information to families. Copeland endured enormous criticism, but the strategy was vindicated by the results. Public health decisions always involve difficult, utilitarian trade-offs between benefits and costs to different people.

New York City ended up keeping quarantine measures in place for seventy-three days, and it began to quarantine relatively early compared to other cities, such as Philadelphia (roughly two weeks earlier in the course of its epidemic). Consequently, in the end, New York City had approximately half the excess deaths that Philadelphia had.[83]

Quantifying excess deaths is a statistical technique often employed by modern scholars studying epidemics, but it was first proposed by a founder of epidemiology, William Farr, in London in 1847. Farr defined this quantity as the number of deaths observed during an epidemic in excess of those expected under normal circumstances.[84] This approach is especially appropriate when quantifying historical epidemics, which may have had different definitions of disease or for which few records are available.

The technique again found use in 2020 for the COVID-19 pandemic, given the fog of war in knowing accurately how many people are actually dying of the disease. By tracking excess deaths overall or for a specific category of causes of death, such as "pneumonia and influenza," it is possible to detect a spike in cases in the United States in late February and early March, even before COVID-19 was widely recognized as a problem; this helps quantify possible underreporting due to misdiagnosed or overlooked cases. For instance, an analysis of national data from early January 2020 through March 28, 2020, by scientists at Yale found that California reported 101 deaths due to COVID-19 during this period, but there were actually 399 excess deaths due to pneumonia and influenza. Many or most of those deaths were surely from COVID-19. Looking at all causes of death combined in New York and New Jersey in this same period, the investigators found that there was as much as a threefold increase in total deaths as the pandemic swept through those two states during its first wave, regardless of how many cases of COVID-19 were actually diagnosed by health-care personnel.[85] And looking at the entire United States through the end of May 2020, researchers concluded that the reported number of COVID-19 deaths was likely 22 percent lower than the true number of deaths.[86]

Finally, the quantification of excess deaths allows us to summarize the overall impact of the pandemic on people's health. The virus kills some people directly, by infecting them, and others indirectly, by, for example, prompting people to delay going to the hospital for other conditions and thus needlessly dying, or by increasing suicides as a result of depression due to job loss or social isolation. But the pandemic also saved some lives too. For instance, motor-vehicle fatalities fell during the winter and spring of 2020, as fewer people were on the road; there were fewer deaths due to complications from noncritical medical procedures, as hospitals had canceled elective procedures; fewer babies were born premature (possibly because their homebound mothers were under less physical stress or were less exposed to all pathogens); and fewer people lost their lives to respiratory conditions, as air pollution was reduced due to the cessation of manufacturing activity.[87]

»———————→

The emergence of pandemics is not restricted to the twentieth century or to respiratory illnesses caused by coronavirus or influenza, of course. Dramatic outbreaks of infectious diseases have afflicted human beings for a long time. Pathogens are just as important to our species as the predators we faced in our distant evolutionary past. And infectious diseases, like other major forces—from the invention of agriculture and cities to the occurrence of economic crises and wars—have shaped our societies in our historical past.

The original plague referred to a particular condition—namely, bubonic plague. This condition has what historian Frank Snowden has called "four protagonists." First, there is the causative bacterium itself, *Yersinia pestis*.[88] Then there is the flea by which it moves. Then there is the rat that transports the fleas and that serves as a host. And, finally, there are the unfortunate humans who, like the rats, are wiped out by the pathogen. The bacteria move from one animal to another animal and from one species to another species (for example, from rats to us) through the exchange of fleas. The plague is therefore best understood as a disease that primarily affects wild rodents and only accidentally affects us. In this regard, it resembles the other zoonotic (meaning "originating in animals") pathogens we have been considering.

Because the plague also afflicted the rats, they also died horrible deaths, typically just before the humans. In medieval Europe, it was often noted that they would come above ground looking for water and then die in the streets in what are called "rat falls." Rats frequently appear in paintings of plague scenes for this reason— they are an evil omen.

And once a flea carrying *Yersinia* hops from a rat to a human, the pathogen can spread like wildfire. Close physical contact between humans allows fleas to spread directly from person to person. In a deep irony, fleas have evolved to find hosts in part by detecting the warmth of their bodies. Therefore, once their host died and the corpse turned cold, the fleas began leaping away to find a new host. Often, this meant they landed on people tending to the body of recently deceased loved ones.

In contrast to other infectious diseases that typically afflict the very young and the very old, plague killed indiscriminately. And another of the special qualities of bubonic plague was the extraordinarily high level of fatality and the speed with which it killed. Every time the plague came, the force of death was so great that contemporary observers spoke of how all of humanity might be annihilated.

The plague caused very gruesome deaths that were often painful and dehumanizing, with visible manifestations on the body. People dying of the plague were said to have a foul odor. Certain accounts even suggested, to my amazement, that it was the stench that drove caregivers away more than the fear of contracting the disease. The pathogen would travel to lymph nodes and cause painful swellings, called buboes, from which we get the adjective *bubonic*. Here is how Michael of Piazza described the Black Death during an outbreak in Messina in 1347:

The burn blisters appeared, and boils developed in different parts of the body: on the sexual organs, and others on the thighs, or on the arms, and in others on the neck. At first these were of the size of the hazelnut and the patient was seized by violent shivering fits, which soon rendered him so weak that he could no longer stand upright, but he was forced to lie on his bed, consumed by a violent fever and overcome by great tribulation. Soon the boils grew to the size of a walnut, then to that of a hen's egg or a goose's egg, and they were exceedingly painful, and irritated the body, causing it to vomit blood by vitiating the juices. The blood rose from the affected lungs to the throat producing a putrefying and ultimately decomposing effect on the whole body. The sickness lasted three days, and on the fourth, at the latest, the patient succumbed.[89]

Most infectious diseases do not "want" to kill us because it hinders their ability to spread, so why was plague so deadly? The explanation relies on the presence of a third party: the flea. The germ could not ordinarily spread from one person to another on its own. In order for it to get from one host to another, it first had to be consumed by a flea and wind up *in the flea's body*. However, a flea, even when it has a big meal, ingests only a tiny amount of blood. For the microbes to spread, they had to be pervasive enough in human blood so that even that tiny flea meal had enough of the pathogen to infect the next host. This means that the bacteria had to grow fast and overwhelm the body quickly so that there would be enough bacteria to actually make it into the flea. And so the bacteria evolved to create enormous levels of *bacteremia* (the presence of bacteria circulating in the bloodstream, also known as *sepsis*), with as many as 100,000,000 germs per cubic milliliter of human blood.

However, in some circumstances, it was possible for *Yersinia* to spread directly from person to person without the necessity of fleas. This was one of the most feared and deadly forms of plague, called *pneumonic plague*. Here, the bacteria found their way into the secretions of a person's lungs, so if he or she coughed or sneezed, droplets would go from that person's respiratory system into the respiratory system of another. As a result, outbreaks did not depend on either rats or fleas. This may help explain perplexing outbreaks of plague in environments where those creatures would not easily thrive, such as in Northern Europe.

The bubonic plague is estimated to have killed around two hundred million people over the fourteen hundred years it has ravaged worldwide populations, primarily in three major pandemics beginning in the sixth, fourteenth, and nineteenth centuries.[90]

Across a variety of sites, archaeologists have found little evidence for variation in plague mortality related to demographic or health attributes. The bubonic plague spared no one.[91] The first of these pandemics, known as the plague of Justinian, started in 541 CE, and it had eighteen successive waves before it ended in 755 CE. It is thought to have originated in Africa. The reasons it stopped are not well understood, and more recent historical, archaeological, and epidemiological evidence suggests that, though severe, the Justinian plague was not as disruptive of Mediterranean or European history as previously thought.[92]

The second plague, one of the most catastrophic pandemics ever, began in Central Asia and reached Europe in 1347.[93] It lasted on and off for nearly five hundred years, then disappeared in the 1830s. The last epidemic of plague in England occurred in 1665, but there was a substantial outbreak in Italy in 1743. The first wave of the second plague, which lasted from 1347 to 1353, is what we think of as the Black Death, although that term was not used at the time. During this first wave, driven by a favorable environment of densely packed towns and cities and substantial poverty, as much as half of the population of Europe was wiped out. This force was so powerful that it even acted as a selection pressure, changing the course of human evolution. As we will see in chapter 8, many people may have genetic features today that reflect the fact that their ancestors were the ones who survived.

Finally, in 1870, there was the third occurrence, largely in India, in what is known as the "modern plague." That epidemic caused between thirteen and fifteen million deaths, primarily in Mumbai, between 1898 and 1910. In the twentieth century in the United States, the CDC estimates that there were just over one thousand cases of plague between 1900 and 2016, concentrated mostly in the Southwest and California, largely among hunters and others exposed to wild animals.[94] We do not really even think of it as a plague anymore; we just call it a *Yersinia* infection. And it can be cured in a straightforward fashion with streptomycin or tetracycline, standard antibiotics in a doctor's tool kit.

Doctors in medieval Europe were at special risk during the plague—as health-care workers always are in epidemics—but so were a hodgepodge of people in other professions, such as priests, gravediggers, bakers (because the storage of grain attracted rats), and even street vendors. As I stumbled on this list of risky occupations for the plague, I could not help but notice the analogy to our predicament in 2020. Essential workers who sell us food always seem to be at risk.

Ships were also a locus of risk, and at sea, there was no escape. Sometimes the plague would take so many lives that a ship would drift on the water untended, all of its passengers and crew dead. The same had happened during the plague of Justinian. One ancient account noted "ships in the midst of the sea whose sailors were suddenly attacked by (God's) wrath and (the ships) became tombs for their captains and they continued adrift on the waves carrying the corpses of their owners."[95] This too, albeit with greater destruction back then, anticipated the coronavirus-afflicted cruise ships of 2020.

In his famous book *A Journal of the Plague Year*, written in 1722, Daniel Defoe noted that "the danger of immediate death to ourselves took away all bonds of love, all concern for one another."[96] People were often abandoned to die alone. The sudden death that plague brought posed religious problems to societies so dependent on clerical absolution. The inability to prepare for death, to atone for sins, and to have last rites were seen as particular indignities. Of course, people everywhere want to have a chance to prepare for death, and one study my lab did found that 84 percent of Americans thought that "feeling prepared to die" was "very important at the end of life."[97] Little has changed when it comes to outbreaks of serious infectious diseases. Plague in medieval Europe also repeatedly gave rise to episodes of scapegoating and witch burning. And, from the same impetus of fear, it prompted religious revivals as people sought to pacify an angry God.

Ironically, even though the emergence of modern nation-states, which are characterized by urban living and far-flung trade, contributed to the enormity of the outbreaks of infectious diseases, those same factors also equipped humans with tools to respond. For instance, in the fifteenth century, the Office of Health in Venice constructed facilities on outlying islands where all arriving ships had to sequester for forty days—which is the origin of the term *quarantine* (based on the Italian word for "forty," *quaranta*). This duration had its foundation in the Bible, which frequently refers to the number forty in the context of purification, such as the forty days and nights that the flood in Genesis lasted, the forty days that Moses spent on Mount Sinai before receiving the Ten Commandments, the forty days of Christ's temptation, and the forty days of Lent.[98]

For overland travelers, however, quarantine was much more difficult to enforce. Efforts to do so gave rise to the *cordon sanitaire*—a ribbon of forts and military outposts interspersed along a border and patrolled by troops in an attempt to prevent the movement of contaminated goods and people. Outsiders were denied entry or were subject to quarantine. There was clearly some recognition, if not full understanding, that cases could be imported from elsewhere:

> Some Genoese, whom the disease had forced to flee, crossed the Alps in search of a safe place to live and so came to Lombardy. Some had merchandise with them and sold it while they were staying in Bobbio, whereupon the purchaser, their host, and his whole household, together with several neighbors, were infected and died suddenly of the disease.[99]

In the late eighteenth and early nineteenth centuries, such a cordon, built by the Austro-Hungarian empire and reinforced when outbreaks were known to be happening, stretched over a thousand miles, from the Adriatic Sea to the Alps.[100] These actions suggest that epidemics have deeply concerned our predecessors and motivated enormous efforts at control.

The scale of mortality of the plague is truly difficult to comprehend. As noted above, it is estimated that 30 to 50 percent of the *entire* European population died in a five-year period during the first wave, from 1347 to 1351.[101] In some locations, whole villages and populations were wiped out completely. It is estimated that 60 percent of the population of both Florence and Siena died.[102] So many people died so fast that the workers digging the mass graves could not keep up. Here is how Baldassarre Bonaiuti, a Florentine historian, described it in 1348:

> At every church, or at most of them, they dug deep trenches, down to the waterline, wide and deep, depending on how large the parish was. And those who were responsible for the dead carried them on their backs in the night in which they died and threw them into the ditch, or else they paid a high price to those who would do it for them. The next morning, if there were many [bodies] in the trench, they covered them over with dirt. And then more bodies were put on top of them, with a little more dirt over those; they put layer on layer just like one puts layers of cheese in a lasagna.[103]

And even after such devastation, the plague just kept returning.

One of the theories as to why the plague eventually ended in Europe had little to do with the actions of the human protagonists. Instead, it had to do with competition among rats. In the early eighteenth century, a larger variety of rat, the brown rat, appears to have arrived in Europe and driven the native black rat from its former habitat. The brown rat was aggressive with its competing rat species, but it was fearful of humans and sought to avoid them. Another factor that is thought to have played a role was a particularly severe cold snap in 1650 that coincided with the end of the plague.[104] As we have seen, epidemics depend on a complex interaction of pathogen, host, and environment.

»———————→

Many other plagues we have not considered have loomed large in their own times, from outbreaks of smallpox that afflicted Americans in the seventeenth and eighteenth centuries, to cholera and yellow fever in the nineteenth, to polio, syphilis, and HIV in the twentieth.[105] Still, pandemics on the scale of COVID-19 are rare.

There is something about a threat that reoccurs at the dim reaches of living memory, every fifty or one hundred years, that makes our species seem particularly small. When such a threat reappears, human suffering is combined with the sad realization that we should have seen it coming. Epidemics generally take advantage of the deepest and most highly evolved aspects of our humanity. We evolved to live in groups, to have friends, to touch and hug each other, and to bury and mourn one another. If we lived like hermits, we would not be victims of contagious disease. But the germs that kill us during times of plague often spread precisely because of who we are. And so for centuries, our response in a time of plague has been to rediscover the necessity of surrendering these aspects of our nature for a while.

We forget the lessons of past pandemics for different reasons. In some cases, they are simply too far back in our collective memory or too obscured by other events. Those plagues have become objects of inquiry to small groups of academic historians or scientists or they are subjects of oral traditions or myths. During Passover in early April 2020, several of my Jewish friends observed that biblical plagues had always been abstract for them, but now they felt more real; the point of the story at the seder was more manifest. In other cases, the reasons for forgetting are more prosaic, more epidemiological, more related to numbers: the particular pandemic disease was not fatal enough (2009 H1N1 influenza), or it did not afflict enough people because it was not infectious enough (MERS), or it burned out too fast (SARS-1), or it afflicted a confined subgroup of the human population (Ebola), or it was brought low by a vaccine (measles and polio), or by treatment (HIV), or by eradication (smallpox), allowing most people to simply push the disease out of their minds.

While the way we have come to live in the time of the COVID-19 pandemic might feel alien and unnatural, it is actually neither of those things. Plagues are a feature of the human experience. What happened in 2020 was not new to our species. It was just new to *us*.

3.

Pulling Apart

They resolved to leave means neither of ingress or egress to the sudden impulses of despair or of frenzy from within. The abbey was amply provisioned. With such precautions the courtiers might bid defiance to contagion. The external world could take care of itself. In the meantime, it was folly to grieve, or to think. The prince had provided all the appliances of pleasure. There were buffoons, there were improvisatori, there were ballet dancers, there were musicians, there was Beauty, there was wine. All these and security were within. Without was the "Red Death."
—Edgar Allan Poe, "The Masque of the Red Death" (1842)

Though plagues have long been a part of the human experience, when the coronavirus gained its footing in early 2020, it met a scientific and medical environment wholly different from those of prior pandemics. Scientists can rapidly sequence the genome of the virus, which allows them to identify new variants of the virus and track its spread. We can invent and deploy genetic tests to quickly and accurately diagnose infection. We can track the flow of people around the globe using mobile-phone data. We have ICUs and computerized ventilators that were not even imaginable in the past. We have whole new classes of antiviral drugs and a deep understanding of virus biology and pharmacology. And we can use the internet to share information instantly and widely.

But how much did all of that really help us? We have not done much better at stopping a virus than our forebears did, and they had fewer resources. Despite all the advances in science and medicine in the past century, it's humbling and shocking to think how little things have changed. With our bans of public gatherings and our use of masks, it feels like a return to the primitive tools of an epidemiological stone age. And yet familiar threats call for familiar measures.

There are two broad ways to respond to an epidemic. The first is via *pharmaceutical interventions* such as medications and vaccines, which scientists around the world are now racing to develop. On that front, we are certainly on better footing than those before us. For centuries people have tried to treat disease with appalling concoctions that have no scientific basis, and almost without exception, they did not work. During the Black Death, the least dangerous treatments that unlucky souls were subjected to were rubbings with chopped onions (or dead snakes, if available). If that failed, self-flagellation, arsenic, and mercury were options. The last item might have even made some sense. In fact, in 1918, doctors noted a puzzling immunity to the flu in soldiers sick with syphilis who were being treated with mercury in a French venereal disease clinic. The notorious toxicity of the mercury, however, ensured that the cure was in some ways worse than the disease, an unfortunate refrain in the history of medicine.[1]

However, contrary to what many people think, medicine has actually played a surprisingly small role in the decline of most infectious diseases across time. One of the most powerful ways to illustrate this is to plot the rates of death for various infectious diseases starting in 1900. The rates drop at some point, and then there is a long, relatively flat line of low rates of death thereafter. It would seem reasonable to imagine that rates of measles deaths, for example, were high and then suddenly dropped after a measles vaccine was discovered and released in 1963. We might expect to see the same pattern in other infections, like scarlet fever, tuberculosis, typhoid, and diphtheria. But if you place an arrow above the year in which a specific vaccine or treatment for each of those conditions was invented, again and again, we see that the arrow actually winds up on the long flat part of the curve, well after the initial descent (as shown in figure 11). If you imagine a young child sledding down a steep hill, the vaccine or drug therapy for the disease would arrive during that long period of coasting on the level part of the sled run at the end.

Plots like this were first propounded by the British physician and historian Thomas McKeown in 1966 in order to illustrate this very point—that modern medicine was *not* the primary force that caused infectious diseases to go away.[2] The actual cause, McKeown argued, was socioeconomic improvements and the implementation of public health measures. For instance, the great rise in wealth worldwide during the past two centuries has increasingly afforded humans access to the clean water and safe food that technological and scientific advances have yielded. These advances and also improvements in family planning and education have played a bigger role in reducing the spread of infection than vaccines or treatments. This became known as the McKeown hypothesis, and it has been thoroughly validated. To be clear, this is not to say we should not use medicines! They still prevent much loss of life—just not to the extent imagined.

Similar complementary factors—socioeconomic and pharmaceutical—play a role in how we confront acute epidemics at a given

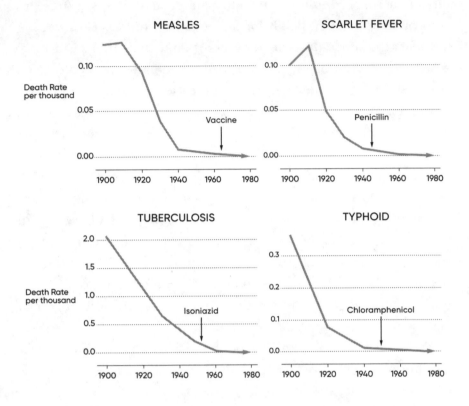

Figure 11: The McKeown hypothesis, illustrated here, shows the relatively small contributions of specific medical interventions, occurring at different points in time for each condition shown, in the conquest of infectious diseases.

point and not just infectious diseases across time. In addition to pharmaceutical interventions, we have a second way to respond: *nonpharmaceutical interventions*, or NPIs. NPIs come in two broad categories: individual and collective. At the individual level, this includes efforts like washing hands, wearing masks, self-isolating, and forswearing handshaking. By definition, these actions involve a certain level of personal choice, and although in extreme cases people have been punished for flouting rules (as was the case in 1918 San Francisco when members of an anti-masking group were jailed), individuals often have some control over how much mitigation they are willing to adopt to keep illness at bay.

Collective actions, however, are usually coordinated (and mandated) by governments. While they may not be to everyone's liking, they involve and affect everyone. These actions include closing borders, shutting schools, banning large gatherings, disinfecting public spaces, instituting testing and contact tracing and quarantines, providing public education, and issuing stay-at-home orders. Because those kinds of NPIs often impose burdens on citizens who remain (or at least appear) uninfected, these efforts can provoke resentment and even resistance.

An alternative way of thinking about the broad array of NPIs is based on the method by which they are intended to tamp down the epidemic. Some interventions, whether individual or collective, achieve their effect by reducing the transmissibility of the pathogen—for instance, wearing masks, handwashing, and sanitizing public places. These are transmission-reduction interventions. Other interventions work by modifying the pattern of human interactions to deprive the pathogen of opportunities to spread—for example, self-isolation, quarantine, and school closures. These are contact-reduction interventions, and they constitute the social-distancing measures everyone began talking about in 2020.[3] To be clear, we need to distance ourselves *physically*, not socially. The last thing the public should be advised to do is create more *social* distance between their friends and families at a time when physical proximity is restricted.

Individual and collective approaches are not mutually exclusive, and they work best if used in combination, like employing both chemotherapy and radiation to treat cancer. Let's consider various types of NPIs and how they combine to form an effective strategy to fight a viral pandemic and then examine how those strategies developed as the COVID-19 pandemic grew.

»———▶

On March 15, 2020, Dr. Anthony Fauci, the head of the National Institute of Allergy and Infectious Diseases (NIAID), publicly stated that a "national lockdown" might be necessary.[4] Soon afterward, governor after governor imposed stay-at-home orders, starting with California on March 19, then New York on March 22, and, last, Missouri on April 6.[5] Even before this, countless firms had moved to implement work-from-home rules, and Americans reduced their mobility on their own, as we saw in Seattle. Most people are not stupid; they would just as soon not leave the house when a deadly pathogen is lurking about. An analysis of cell phone mobility data that the *Washington Post* released on May 6 revealed that the peak of our collective efforts to stay at home came on April 7, when Americans spent 93 percent of their time in their homes, up from 72 percent on March 1.[6] So, during the first few months of the pandemic of 2020, people stayed at home as much as possible unless they had no alternative or worked in health care, sanitation, food production or distribution, or other essential industries.

Americans took this extreme step and other significant measures (such as closing schools, discussed below) for a reason, and it was not just to save our own skins. These interventions were intended to interrupt chains of transmission and thus for us to help one another. The point of the NPIs was to *flatten the curve*. This phrase, like the virus itself, leaped from obscurity to public consciousness around this time, migrating from academic journals and unvisited CDC websites to the front page of every newspaper and the mouth of every person on the street. At Dan and Whit's, the compendious country store in Vermont where I live (their longtime motto—"If we don't have it, you don't need it"—seemed increasingly apt), I heard a young clerk muttering that he was "so tired of flattening this curve" as he readjusted his mask.

But what exactly does flattening the curve mean? Suppose that ten million Americans are going to contract an infection in the first ten months after an epidemic starts. It makes all the difference in the world if they all get sick during the first month or if one million people fall ill each month for ten months. If all the patients fell ill during the first month, our health-care system would collapse. Furthermore, had we not made efforts to flatten the curve and blunt the force of the epidemic wave hitting us, it is possible that over a million Americans would have died in the first few months of the pandemic. Figure 12 illustrates how this works. The number of cases per day is shown (time is on the x-axis and the case count is on the y-axis). The case count is compact in an unflattened epidemic and spread out in the flattened one.

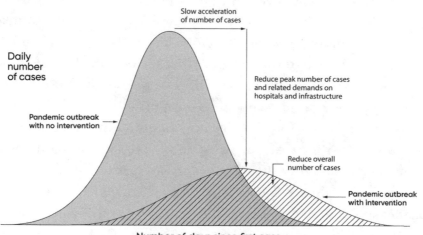

Figure 12: The process for flattening the curve of an epidemic outbreak results in a number of benefits, including a diminution of peak demand for health care, postponement of the peak, and reduction of the overall number of deaths.

Early studies from China, Italy, the United Kingdom, and the United States all showed that roughly 20 percent of people infected with SARS-2 needed hospitalization and roughly 5 percent needed ICU care.[7] If 5 percent of ten million people needed ICU care, that means we would need five hundred thousand ICU beds. The

United States has only one hundred thousand such beds. No nation has the ICU-bed capacity to cope with that many seriously ill people at once. The United States also has fewer hospital beds per capita than other industrialized countries; the U.S. has 2.9 beds per 1,000 people, whereas South Korea has 11.5, Japan has 13.4, Italy has 3.4, Australia has 3.8, and China has 4.2.[8] A further, absolutely essential requirement is the medical and nursing personnel to care for all the patients. But other shortages can hurt too, like a paucity of personal protective equipment (PPE), coffins, and even refrigerated trucks to transport bodies. In New York City, the curve was not sufficiently flattened to avoid running out of those things, and news reports showed nurses wearing garbage bags as improvised PPE and nursing homes and hospitals filling up with stacks of bodies at risk of decomposition. However, New York City was able to flatten the curve enough to avoid running out of ventilators, which would have necessitated some very hard triage decisions. And it was possible to expand ICU-bed capacity enough to meet the elevated demand. But a serious epidemic severely strains any health-care system. Flattening the curve allows the system to work, and avoids exhausting supplies of equipment and overwhelming health-care personnel.

Reports in January and February indicated that the COVID-19 pandemic devastated the hospitals in Wuhan and Milan. Indeed, as we noted in chapter 1, the Chinese built a whole new hospital in Wuhan in ten days to care for thousands of patients. Medical personnel were brought in by the busload from other regions to staff local hospitals.[9] In late February, the Italians were doing everything Americans would soon be doing at hospitals throughout the country: canceling elective procedures, repurposing operating rooms as ICUs, closing the doors to some overrun emergency rooms, and pressing health-care personnel from other fields into service.[10] The Italians called doctors out of retirement to help.[11] "We'll take anyone: old, young. We need personnel, especially qualified doctors," said Giulio Gallera, a health official in the afflicted area of Lombardy.

Flattening the curve buys time to save more lives. Ventilators and medications do not run out. Doctors and nurses are not exhausted, so they can do a better job of caring for the people who are sick, which means fewer deaths. With less cramped conditions, fewer health-care workers contract the disease themselves, keeping them on the front lines. Buying time also allows us to get the public health system ready; we can develop tests and procedures for contact tracing and learn more about the virus in our laboratories. When we flatten the curve, we also push some of the infections into the future to a point when, simply by virtue of the passage of time, scientists might have invented a vaccine or developed effective medicines or learned how to better care for people suffering from the disease— all of which reduces deaths too. Finally, it's always possible that the virus could mutate and become less dangerous, so people contracting it later on would get milder and less lethal versions of the disease. This is usually the case with pandemics, but it can take quite some time and it does not always happen (as we saw in the influenza pandemic of 1918). All these benefits are reflected in figure 12 by the fact that the total area under the curve, capturing the overall number of deaths, is smaller in the flattened version than in the unflattened version.

To be clear, flattening the curve means slowing the spread of the virus, not eradicating it. Even after the curve had been successfully flattened, the virus came back to China, to the United States, to every country. Many Americans did not fully appreciate this eventuality in the spring of 2020. There were victory laps. People thought that once the curve had been flattened, we were done. Unfortunately, in the United States, a flatter curve still meant we had as many as a thousand deaths per day in the spring and summer of 2020. That was hardly a victory, although it could have been worse.

»———————→

The shutdown of the United States in the spring of 2020 may have prevented sixty million cases and probably more than three hundred thousand deaths during the acute shock of the first wave of the pandemic.[12] Flattening the curve did not come without medical, social, and economic costs, however. Stay-at-home orders reduced SARS-2 transmission and COVID-19 deaths, but social isolation is mostly bad for mental health and might be reflected in increased rates of depression, suicide, and intimate-partner violence. It might also increase deaths for other serious conditions, such as asthma exacerbations and strokes, if symptoms are left untreated (although we will discuss some health benefits of forgoing medical treatment in chapter 7). In the early weeks of COVID-19, many clinicians noted an eerie and puzzling absence of non-COVID-related emergency department visits and acute surgical cases. Elective and nonemergency surgeries were rescheduled to prepare for COVID-19 cases, but few physicians anticipated that people would avoid care for dire health problems too. A colleague of mine at Yale, Dr. Harlan Krumholz, described a troubling disappearance of heart disease patients and surmised that people with chest pain might be avoiding going to the hospital.[13]

Physical-distancing measures also impose serious social and economic costs. If schools are closed, many children experience the loss of academic learning, and their parents lose childcare. If places of business are closed, adults lose their jobs. And the job losses were staggering. On May 7, 2020, the Department of Labor reported that the seasonally adjusted unemployment rate in the United States was 15.5 percent.[14] Cumulatively, as of that date, thirty-three million Americans had lost their jobs, erasing a decade of employment gains. Reading the report, I had the sense that even the functionary who was summarizing the data for public release was stunned: "The total number of people claiming benefits in all programs for the week ending April 18 was 18,919,371, an increase of 2,416,289 from the previous week. There were 1,673,740 persons claiming benefits in all programs in the comparable week in 2019." A survey released in late April by the Pew Charitable Trusts found that, overall, 43 percent of adults reported that they or someone in their household had either lost a job or taken a pay cut as a result of the pandemic. Of course, the impact fell especially hard on lower-income adults and ethnic and racial minorities.[15] Headlines from this period ranged from "US Unemployment Rate Soars to 14.7 percent, the Worst Since the Depression Era" to "US Job Losses Reached Great Depression Levels."[16]

Just going to any downtown district or reading a local paper would tell this story. In Hanover, New Hampshire, home to Dartmouth College, store after store had gone permanently out of business by May. "Retail Meltdown Will Reshape Main Street," read a local headline.[17] The owner of a beloved gelato shop (dubbed the best in America by *Forbes* magazine) put it starkly when she explained her permanent closure: there was no viable business model for success-fully selling ice cream without closely packed people in lines snaking around the block. Town officials inventively scrambled to repurpose parking spaces for outdoor dining (as is so common in Europe). But it was not enough to help local restaurants.

The Dow Jones Industrial Average plunged 37 percent, going from 29,551 on February 12 to 18,591 on March 23, though it returned to 25,734 by July 1, 2020. In early April, not long after widespread physical distancing had begun in the United States, it was estimated that in the second quarter, the quarter-over-quarter contraction would be roughly 10 percent, which ultimately proved to be the case.[18] The previous largest quarter-over-quarter change since data collection started, in 1947, was the first quarter of 1958, when the contraction was 2.6 percent. Perhaps not coincidentally, this was after the last major respiratory pandemic to strike the United States. This impact eclipsed even the financial crisis of the fourth quarter of 2008, which had a contraction in the economy of 2.2 percent.[19]

These trade-offs require coldhearted calculations and assessments, and they are difficult, especially where schools and workplaces are concerned. Many of the interventions officials implemented to flatten the curve therefore became highly politicized in 2020 in ways that compromised the public health response and made it even harder for us to respond. What gets lost in the din of argument over physical distancing is that the efficacy of these interventions is so well established that they should not be controversial. For instance, the U.S. Centers for Disease Control and Prevention released the aptly titled report "Community Mitigation Guidelines to Prevent Pandemic Influenza—United States, 2017" *three years prior* to the pandemic. It was full of sound advice that had been offered for decades. Nonetheless, a CDC report offering similar basic advice was suppressed by the White House on May 7, 2020, for reasons that were not given. CDC scientists were simply told that the recommendations they had prepared on how businesses could reopen after the shutdowns (with basic suggestions on hygienic practices and physical distancing) "would never see the light of day."[20] Of course, the report was leaked anyway. Possibly, the underlying political rationale for suppressing the report was to deemphasize the role of the federal government so the states themselves would be held responsible for reopening the country for business, which meant, cynically, that they would shoulder the blame for the inevitable return of outbreaks.

Implementing the NPIs in a timely manner and gathering public support for doing so when things do not yet look so bad are important challenges for public health officials and politicians. NPIs are inconvenient, unnatural, and often extremely costly, so many people understandably wish to avoid this necessity, especially if they have not yet seen deaths up close. In any epidemic, a basic educational task of leaders is to help people understand what is actually happening.[21] In fact, maintaining public trust can be seen as its own nonpharmaceutical intervention, not just a way to boost the efficacy of others. To gain the public trust necessary to implement disruptive but lifesaving interventions requires honest communication about the rationales for all suggested policies, including a discussion of the difficult trade-offs that are needed and the uncertainty involved. This helps improve a sense of civic purpose as well as public health.

In 1918, collective will was harnessed with greater ease than it was in 2020 because the Spanish flu had erupted in the midst of World War I. Members of the public supported ordinances that limited their freedoms because these rules were seen as a way to protect American troops abroad. One announcement from the Red Cross explicitly stated, "The man or woman or child who will not wear a mask now is a dangerous slacker." The governor of California described mask-wearing as the "patriotic duty" of every American.[22]

One of the most fundamental, but devilish, details about epidemic disease is the concept of exponential growth. While this makes flattening the curve all the more important, it also makes galvanizing public will all the harder. As many of us learned in high-school mathematics, exponential growth involves a curve that has a long flat portion and then a slightly rounded part and then a steep ascent, as shown in figure 13. With exponential growth, for a very long time, nothing seems to be happening. And then all of a sudden, a lot happens. It reminds me of Vladimir Lenin's obsevation that "there are decades where nothing happens; and there are weeks where decades happen." But it's very hard to get people to take action while they are on the flat part of the curve.

Yet, as we saw in chapter 2, early action (or inaction) makes a big difference—for instance, there was a rapid increase in flu cases when Philadelphia failed to cancel the Liberty Loans Parade. Therefore, helping the public grasp during the early part of the epidemic what is likely to happen later is essential. But this is also one of the reasons it's so difficult to sound the alarm. If we say that many people will be sick and that our world will be changed "soon," people will look around and conclude that everything seems normal enough, so no interventions are necessary, thank you very much. And it seems normal the next day too, so the Cassandras are seen as

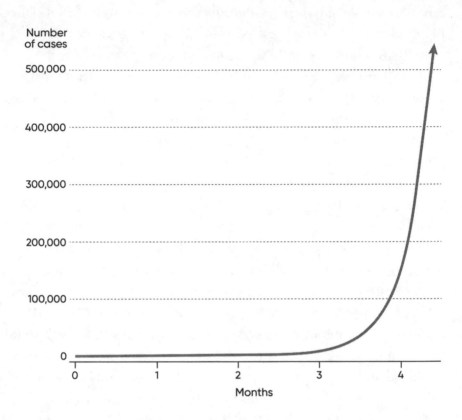

Figure 13: Exponential growth, with an approximate doubling time of one week, can lead to a sudden and very large increase in cases.

merely alarmists. By the time people start to die, it's too late to gain the full benefit of the more effective interventions, although they do still help.

»————→

Some of the objections to the implementation of the NPIs at the outset of the pandemic in the United States arose from a desire to adopt a more "natural" strategy and allow the epidemic to run its course in order to more quickly reach the point of herd immunity.

The proportion of people who need to be immune for a society to have herd immunity depends on how infectious the disease is; the less infectious the disease, the fewer individuals need to be immune. And the higher the R_0, the higher the fraction of people who must be immune to stop an epidemic. This is the reason that measles, which is among the most contagious diseases known, requires such high levels of vaccination in a population to avoid outbreaks. About 6 percent of unvaccinated people in an area is enough to give measles an opportunity to create an outbreak, as we have seen in recent years in regions where a large number of people have refused to get vaccinated.[23] That is, 94 percent or more of the population must be immune, either naturally or via vaccination, to stop measles epidemics.

However, for pathogens with a lower R_0, a smaller percentage of immune people will get the job done. The formula $(R_0-1)/R_0$ gives this percentage. Using an R_0 for SARS-2 of 3.0 means that the calculated percentage of the population that must be immune is 67 percent. But this percentage is an overestimate to some degree. The epidemiological basis for calculations of the reproduction number of pathogens makes an assumption that every person in a population has an equal chance of interacting with every other person, but we know this is not true in the real world. Some people have few social connections and few social interactions and others have many, as we saw in chapter 2. If popular people become immune, either naturally or via vaccination, this means that fewer people must be immune to reach herd immunity than the foregoing calculation suggests.

It's difficult to estimate the precise size of the effect since it depends on many factors. But the implication for reaching the herd-immunity threshold is that if the more popular people are overrepresented early on during the COVID-19 pandemic, a *lower* percentage of the whole population must become immune— perhaps only 40 to 50 percent.

Herd immunity is one reason that epidemics end without everyone becoming infected. In essence, when enough people who are conduits for infection become immune, the pathogen has no way to continue moving through the population. This is part of the explanation for why the 1957 influenza pandemic, for instance, ended with a smaller fraction of the population being affected than predicted by its R_0 alone.

Another important issue in herd immunity is vulnerability to infection. One modeling exercise by epidemiologist Marc Lipsitch and his colleagues suggested that we might want to deliberately engage in intermittent periods of easing up the NPIs when the cases fall to a certain level. This would allow some members of our population (particularly the less vulnerable members) to be exposed to the virus to develop immunity. As the cases rose, we would have to hunker down again. Depending on the seasonality of the germ—that is, the extent to which the spread and lethality of the virus is materially stronger in the winter than in the summer—we would have to engage in physical distancing between 25 percent and 75 percent of the time to build immunity while not overwhelming our health-care system and while continuing to sequester the vulnerable members of our society. Of course, as time passed and more and more of us became immune, the periods when we had to engage in physical distancing would become shorter and less frequent. In this way, we could, at least theoretically, regulate the number and vulnerability of people who acquired immunity in order to get herd immunity at the lowest overall cost in terms of infections and deaths.

Because of all this, questions arose after the start of the pandemic. Should we just let the epidemic hit us like a big wave? Why not get it over with? This line of reasoning feels sound in some ways; we cannot really stop the germ from infecting many people unless we have a vaccine, and efforts to stop it require us to devastate our economy and might possibly create many more deaths because of our response. After all, poverty is deadly too.

Some countries did consider this approach in 2020. The British toyed with, but ultimately rejected, the idea of "taking it on the chin."

Sobering estimates of a rapid increase in deaths prompted a belated national lockdown in the UK.[24] Prime Minister Boris Johnson himself contracted the disease after affecting an air of nonchalance for months, and he ended up in intensive care for several days in April 2020.[25] Sweden, alone among its Scandinavian neighbors, adopted a tactic like this, aiming to isolate the vulnerable and aged while allowing the young and healthy to go about their business as sensibly as possible in order to achieve sufficient levels of population immunity. By May 2020, Sweden had at least four times the rate of COVID-19 deaths as its Nordic neighbors with similar demographics and economies, and it began to gingerly correct its course.[26] Ultimately, the architect behind the strategy admitted that the policy had not worked out as planned.[27] Crucially, the Swedish economy was just as hard-hit as the economies of its neighbors who had pursued full lockdowns. Physical and economic health are inextricably linked.

But early on, some politicians and pundits in the United States made noise about how the country should follow the Swedish model or otherwise charge its way toward herd immunity. But it's ludicrous to compare Sweden to the United States. The former is a country of 10.2 million people with a cradle-to-grave social-welfare system, relatively few health problems, a culture of outdoor living, and a rule-abiding, collectively oriented citizenry. None of this resembles the United States, whose citizens have vastly higher rates of poverty, poor health, and other risk factors. Later in 2020, glib exhortations to "open up" the United States—as if it were as easy as opening a valve—grossly underestimated the complexity of the delicate dance required to manage the pandemic. And anyway, flattening the curve does not just postpone deaths; it prevents some of them as well, as we saw earlier.

Still, if we are unable to develop a vaccine, herd immunity is ultimately the path we will take as a society and worldwide, whether advertently or inadvertently. Some countries will undoubtedly manage it better than others, and my guess is that the nations that fare best will be the ones with high public trust and strong science-based leadership.

»———————►

When a pathogen has run through a population completely, many containment measures are no longer needed, and the epidemic is finished. Until then, in order to extinguish an epidemic, the R_e must be brought to below 1 (meaning that each extant case can no longer reproduce itself). But until then, containment and harm reduction are the goals. Let's consider various NPIs at our disposal to achieve these objectives.

To start, let's look at hygiene measures. Individual activities like handwashing and collective measures like sanitation are both crucial to fighting contagious disease. Americans born after 1940 have little appreciation for the importance of good hygiene because they have (mostly) had clean food and water and do not remember the havoc wrought by infectious disease. Life expectancy in 1900 was low largely because so many children died before their fifth birthday. Back then, people were not dying from heart disease and cancer as much as from pneumonia, diphtheria, and diarrhea.

At the outset of the 2020 pandemic, handwashing was emphasized, and many people stopped hugging and shaking hands when greeting. A large national survey conducted at the end of April 2020 reported that a suspiciously high percentage of Americans—96 percent—claimed to closely follow recommendations to frequently wash their hands, and 88 percent reported closely following recommendations to disinfect surfaces that were frequently touched.[28] Some people wore gloves. But the most visible and dramatic change in hygiene was mask-wearing, which made a comeback after a hundred years. The same national survey found that by the end of April 2020, 75 percent of Americans reported following recommendations to wear a mask when outside their homes.

Face masks were points of contention and confusion early in the epidemic, but they have been used for a long time to fight respiratory diseases, as even a glance at photos of street scenes during the 1918 pandemic reveals. Even then, people understood the utility of this simple intervention, and detailed scientific studies of mask effectiveness were published a century ago.[29] By May 2020, masks had proliferated in some parts of the country but faced a vicious backlash in others. Many businesses required them for entry, and municipalities and states passed ordinances requiring them, again only to be met with pushback (in some cases leading to reversal) due to public outcry.

Early on, many authorities, including the CDC and the surgeon general, recommended against universal mask adoption because the United States had such limited supplies of personal protective equipment. It was feared that recommending universal mask-wearing would deprive health-care workers of precious supplies. The World Health Organization also discouraged masks as late as April 2020.[30] If conserving masks for health-care workers was the true policy motivation, it made no sense to mislead the public, who were understandably confused by the mixed messages in the statement "This mask is so valuable, it has to be given to a health-care provider, but in any case it would not help you and you do not need it." Needed credibility was lost over this shifting story.

In addition to worries about PPE supplies, there was a fundamental misunderstanding of the role masks play in stemming the tide of the virus. Since COVID-19 can be transmitted by asymptomatic individuals, a key function of wearing a mask is not so much to protect oneself from becoming infected but rather to prevent oneself from transmitting infection to *others*.[31] Masks can be worn to protect a wearer from the *ingress* of viral particles, but this typically requires a more specialized mask like an N95 respirator (though, to be clear, cloth masks do help). Still, any mask can be worn to protect others from the *egress* of viral particles by dampening the propulsive force of droplets leaving a person's mouth. A mask near the exit to one's body stops the droplets, just like blocking the outlet of a hose is more effective at stopping the spray of water than trying to catch all the droplets in the spray in midair using a bucket.

Stopping droplets from spreading does not require any special medical equipment. Even homemade cotton masks lessen pathogen transmission. Cotton masks can dramatically reduce the number of viral particles expelled when sneezing, coughing, or even talking by perhaps as much as 99 percent. Moreover, encouraging the home production of masks—at least among those households that, unlike mine, have members capable of sewing—does not threaten medical PPE supplies or promote mask hoarding.

Masks also protect the wearer in other ways. For one, masks keep us from touching our faces. People touch their faces roughly once every four minutes.[32] Facts like this made me wonder whether the strange costumes worn by seventeenth-century plague doctors—waxed coats and long, beaked, bird-like masks—might have served a similar function. The explanation for the beak is that it held herbs to drive away the miasmatic fumes that, people thought, carried the infection. But perhaps it kept the doctor from touching his face too. Another beneficial effect of wearing a mask is that it prompts *other* people to keep their distance from the wearer, as shown by an ingenious experiment that involved surreptitiously tracking the distances people kept from experimenters randomly wearing masks in lines at supermarkets and post offices.[33]

Some colleagues of mine at Yale and I released a policy analysis regarding the utility of masks on April 6, 2020, in part to help address ongoing confusion and resistance to mask usage.[34] Our analysis of forty-six countries around the world showed that in the early months of the pandemic, those countries where mask-wearing had always been the norm (such as Taiwan) had many fewer deaths than those where it was not. Of course, it's difficult to make comparisons across countries because there are so many other things that vary, and mask-wearing might be a proxy for the uptake of other disease-fighting behaviors (like handwashing and having and heeding an efficient public health system). Nevertheless, we found the growth rate of death was 21 percent in countries without mask norms and 11 percent in countries with such norms. If every household in the United States used cloth face masks, it would conservatively generate at least three thousand dollars in value per household from the reduced mortality risk arising from the reduction in spread of the virus. Indeed, even if masks reduced the transmission rate of the virus by only 10 percent, our models indicate that hundreds of thousands of deaths would be prevented around the world, creating trillions of dollars in economic value. This is a big effect of a small thing.

Masks alone can therefore have a large effect on respiratory pandemics by bringing the R_e down. For instance, a mask with just 50 percent efficacy in reducing droplet transmission worn by just 50 percent of people can reduce the R_e from 2.4 to about 1.35— roughly the level of seasonal influenza. This means that, if there were one hundred cases of such an infection at the beginning of the month, in a no-mask scenario, there would be 31,280 cases at the end of the month; but in a mask scenario, there would be only 584. This reduction allows medical personnel to take better care of the smaller number of patients and to deploy contact tracing and quarantine measures more effectively. Of course, if the masks were even more efficient and the adherence even higher, the epidemic could be brought to heel, with an R_e lower than 1.0. If 70 percent or more of the population in a typical urban situation used masks of decent quality (that is, about 70 percent effective), it would prevent a large-scale outbreak of a respiratory pathogen of a moderately contagious disease such as COVID-19.[35]

For citizens of countries without preexisting mask norms, it can be jarring to see people out in public with their faces covered. But norms can change. The Czech Republic managed to go from masks being nonnormative to masks being nearly universal within ten days, largely due to the impetus of celebrity influencers and a viral video with the hashtag #Mask4All. A government decree followed in late March.[36] The slogan for the campaign—"My mask protects you; your mask protects me"—was a pithy way to appeal to people's altruism, and it changed the meaning of wearing a mask. It gained wide international circulation in the early days of the pandemic. Even nudists in the Czech Republic were admonished by the police to cover up their mouths and noses.[37] This converted a purely voluntary individual action into an accepted collective one.

Still, there has been a proliferation of refusals and protests in some parts of the United States. Some people framed it as an issue of personal liberty, holding signs that read MY BODY, MY CHOICE, a cynical reference to pro-choice arguments in the abortion debate. In the summer of 2020, confrontations in stores and violence rose as an already irritable public tried to balance the new normal.[38] In Georgia, Governor Brian Kemp, in defiance of good public health practice, explicitly banned cities and counties from enacting rules requiring masks.[39] Of course, as we have seen, the reason people can be required to wear masks is that it prevents them from giving sickness to *others*, analogous to laws protecting individuals from secondhand smoke. Wearing masks is a "public good," something that everyone contributes to and benefits from, like paying taxes to build a firehouse.

In order to set an example of making small sacrifices for the common good, political officials need to wear masks in public. I could not believe it when, in early May, I saw the vice president visit the Mayo Clinic and appear in photographs without wearing a mask while everyone else around him was wearing one. It was a stark contrast to the appeals to patriotism a century ago during the 1918 flu pandemic. Wearing a mask should be seen as a civic duty, like voting. Indeed, it is extremely unlikely that Americans will be able to return to more typical economic interactions *without* adopting the practice for a couple of years.

But there is only so much that individuals acting on their own can do, no matter how much handwashing and mask-wearing and physical distancing they engage in. Collective action is needed to stem the spread of the pathogen. Let's consider a sequence of possible collective interventions arrayed from the least to the most burdensome: border closings, testing and tracing, bans on gatherings, school closures, and the stay-at-home orders the United States ultimately deployed in the early weeks of the pandemic.

One of the most intuitive and ancient responses to an epidemic is to close borders.[40] If diseases are brought in from other countries, it seems to make a lot of sense to restrict travel, which is the reason that this idea has occurred to everyone and has been tried with every plague. One observer reflected on this during the Black Death:

> As a result, the inhabitants, frantic with terror, ordered that no foreigners should stay in the inns, and that the merchants by whom the pestilence was being spread should be compelled to leave the area immediately. The deadly plague reigned everywhere, and once populous cities, because of the death of their inhabitants, now kept their gates firmly shut so that no one could break in and steal the possessions of the dead.[41]

Indeed, in a large national survey of Americans conducted at the end of April 2020, 94 percent approved of restricting international travel. And 86 percent even favored restricting travel within the United States.[42]

But as we saw in chapter 1, in most circumstances, closing borders alone does not work. It may delay the arrival of a pathogen, but, with very rare exceptions, it does not stop it, even in island nations where it's easier to implement. In 2020, both New Zealand and Iceland tried.[43] The first SARS-2 infection in Iceland was confirmed on February 28, 2020, in a person who had returned from Italy, before that country had even been identified as a high-risk source. Less than three weeks later, Iceland designated all travel from outside the country as high risk.[44] But still the virus took root because of other unseen importations. New Zealand, however, which was very capably led by Prime Minister Jacinda Ardern, did extremely well with the first wave of the pandemic and eventually declared itself virus-free in June 2020, though small new outbreaks soon followed. To my knowledge, the only moderately successful attempt at total border closure was in four South Pacific islands in the 1918 pandemic; they were able to delay—but not prevent—the importation of the pathogen by between three and thirty months.[45] There is not really any such thing as totally sealing off borders—citizens abroad can still return home, and people can move illegally.

A pathogen with asymptomatic transmission, such as SARS-2, makes border closure even harder. The furtive nature of such germs helps explain why the seemingly sensible and widely used procedure of closing borders tends not to work. For instance, an analysis of the role of fever checks and mandatory isolation of people upon entry into China during the 2009 H1N1 pandemic, which started in Mexico that March, showed that, at best, border controls delayed the spread of the pandemic by less than four days.[46]

Furthermore, by the time policymakers think to close borders or are even aware of the epidemic, the virus has typically already crossed into their lands. And once community transmission of an epidemic begins in earnest, the prevention of further importations has little impact. Indeed, imported cases are thereafter usually just a fraction of the existing cases. One formal model that evaluated what would happen if all airplane flights were canceled on the thirtieth day after the onset of a pandemic (which would actually be incredibly speedy) concluded that, even if 99.9 percent of all flights were canceled, it would postpone the peak attack of a moderately transmissible disease (with an R_0 of 1.7) by just forty-two days.[47]

Restrictions on *internal* air travel in a country like the United States are relatively ineffective given how much road travel there is.

Attempts by state governors in late March 2020 to close their borders to other states, in addition to being constitutionally suspect, seemed mainly a kind of security theater. When the governor of Florida, Ron DeSantis, suggested that individuals with New York State plates be stopped at the border, it was viewed by many as simply an effort to shift responsibility for the pathogen onto outsiders. This may be a common political way of coping with pandemic disease, but it makes no public health sense.

I live in Norwich, Vermont, a rural town of three thousand people. When I became aware of the pandemic in January 2020, I foolishly thought to myself, *Okay, it's going to come here, but we'll get it six months after all the big cities in the USA.* And what happened? In late February, a medical resident returned from Italy to Hanover, New Hampshire, home of Dartmouth College and just one mile away from my town, across the Connecticut River that divides the two states. Although he was a doctor himself and had been experiencing respiratory symptoms for which he sought care, he failed to obey instructions to self-isolate and went to a large (aptly named) "mixer" of graduate students and faculty, where he infected other people in my neighborhood. On early maps of the outbreak in the United States, a little red dot appeared in my safe little corner of the world right from the beginning.[48]

Since it's generally unrealistic to stop importation, an alternative is to try to contain the epidemic through testing, tracing, and isolation, especially of early cases arriving or being detected in a given region. Though physicians in the early sixteenth century demonstrated a preliminary understanding of the principles underlying such contact tracing by attempting to track the movement of syphilis through their society, the earliest recorded instance of contact tracing as we would recognize it today occurred later in that century.[49] In

1576, Italian physician Andrea Gratiolo was treating bubonic plague patients near Lake Garda in the town of Desenzano when he heard rumors circulating that a woman had carried the disease to the area from her home in Trento. Noticing that the woman had traveled on a small boat in close contact with over a dozen others, Gratiolo decided to investigate the other passengers and found that none exhibited signs of infection. Gratiolo used his evidence to argue that the woman could not have carried the disease from Trento or else other passengers would have gotten sick.[50]

In the late eighteenth century, English physician John Haygarth traced the connections of individuals infected with smallpox to demonstrate that the disease spread only through close contact with infected individuals or materials rather than over long distances, as some physicians then thought.[51] Perhaps the most infamous instance of contact tracing in the history of public health was the case of Mary Mallon, or "Typhoid Mary," in the early twentieth century. By tracing Mary's previous employment contacts, a sanitary engineer named George Soper realized that seven other households where she had worked had also experienced typhoid cases.[52] Mary later became the first identified asymptomatic carrier of typhoid in the United States, and she was forced to spend much of the rest of her life in involuntary quarantine.

Mary Mallon's story stands out as an example of the enduring tension between civil liberties and public health. And it was in the early twentieth century that contact tracing first became implemented as a widespread, enforced policy. At this time, epidemiologists were making progress in elucidating details of infectious diseases' incubation periods, immunity, and biological causation that influenced policymaking.

The use of public health detectives who identify infected individuals and trace their contacts has been a fundamental disease-control measure employed by public health officials for many decades. English schools were some of the first institutions to adopt rigorous contact tracing in response to measles. Tuberculosis became the most common reason children were kept from school for long periods in England, prompting the reach of contact tracing to expand further into domestic life. By the time of the tuberculosis outbreak in England during World War I, contact tracing had become common practice in many urban schools.[53] In the United States, contact tracing was first proposed as a program in 1937 by surgeon general Thomas Parran to stymie the spread of syphilis among troops, and the program was solidly established by the late 1940s (though Parran was also involved in the infamous Tuskegee syphilis study that deliberately left black men untreated for this disease).[54] Over the course of the twentieth century, contact tracing has played an important role in the eradication of smallpox and in efforts to control infectious diseases as varied as HIV, tuberculosis, influenza, Ebola, and, of course, COVID-19.[55]

Nowadays, when people infected with a contagious disease are identified, either because they have symptoms or because tests have come back positive for active infection, specialized personnel interview them comprehensively—even intrusively—to help them recall everyone with whom they might have come into contact during the appropriate time frame of infectiousness. Those contacts are then tracked down and warned that they may have been exposed. This is done rapidly and anonymously (the contacts are not told who may have infected them). These individuals are then educated about the disease, including its symptoms, and they are told to self-isolate, monitor themselves for signs of the illness, and stay in touch with public health authorities. Or they may be placed in quarantine.

As epidemics expand, contact tracing creates a huge amount of work. When I spoke to members of the Singapore Health Department in April, for example, I was astonished to learn that they employed five thousand contact tracers in a population of about five million people. One person per thousand in their whole nation was employed for this purpose alone. At the time, Singapore had accumulated just 9,125 cases. In our country, this would translate into having 330,000 people engaged in this work.

The sheer overwhelming labor involved in this process makes it clear why so many tech firms and other entities rushed to offer technological solutions for contact tracing. Even rivals such as Apple and Google teamed up to develop technology to facilitate it. In Singapore, Israel, China, Taiwan, and South Korea, people's phone and bank-card records, and even facial-recognition cameras, were used for this purpose. China introduced color codes (green, yellow, or red) that could be checked on people's phones by scanners; the codes indicated whether people were uninfected and unexposed, exposed, or infected.[56] People even received messages based on thresholds set by computer models telling them that they had come into contact with someone who was sick and advising them to self-isolate. Quite apart from the issue of many false positives, the threat to civil liberties here is enormous.

Unlike China, where all *suspected* contacts were immediately taken from their homes and isolated in separate facilities while awaiting diagnostic confirmation, in America, contact tracing in the early days of the pandemic took a less intrusive, though still thorough, approach, at least in principle. The CDC defined a close contact of a person with coronavirus as someone who was "within 6 feet of an infected person for at least 15 minutes starting from 48 hours before illness onset until the time the patient is isolated." This is a somewhat arbitrary criterion, given that the virus can spread much farther than six feet.[57] Nonetheless, it captures most circumstances of transmission. Contacts were told that they should stay at home, maintain physical distancing, and monitor themselves for fourteen days. But with such a broad definition of *close contact*, you can see how labor-intensive it would be to find and counsel so many contacts.

If we miss our chance to detect patients and trace their contacts when there are only a few imported cases, the system can become too overwhelmed to make contact tracing feasible, which is what happened in the United States by May 2020. When the first case of coronavirus was detected in Vermont, in March, it took two public health officers in the state a day to track down the thirteen people who had come into contact with that patient. Those individuals were then put under quarantine and told to monitor their symptoms. None of them became sick. Daniel Daltry, one of the two officers who had done the tracing, wryly noted that "it was a tidy bow."[58] But within days of tracing the contacts of this patient, he and his colleague were overwhelmed with new cases that started "coming in like dominoes." There was no way for his tiny team to trace them all, even in a state with the lowest case count in the country.

When the COVID-19 pandemic struck, states introduced a hodge-podge of ways of doing contact tracing and sourcing the requisite labor supply. In Massachusetts, the governor decided to build a "contact tracing army," and he turned to a local nonprofit, Partners in Health, founded by two friends of mine from medical school, Paul Farmer and Ophelia Dahl. Using hard-won expertise from doing contact tracing for outbreaks in other parts of the world, the group planned to hire and train a thousand contact tracers who could work from their homes (where they were stuck anyway), each making twenty to thirty calls a day, which meant they could cover up to twenty thousand contacts. In Utah, government workers serving other functions were reassigned to contact tracing. San Francisco tried to build a contact tracing team of one hundred fifty people using city librarians and medical students. Some proposed transforming the Peace Corps, which had suspended global operations, by using their seven thousand repatriated volunteers to form a national contact tracing corps. The editor of the *Journal of the American Medical Association* even proposed suspending the first year of training for America's twenty thousand incoming medical students in order to deploy them for this purpose.[59]

All of these efforts presumed that the people contacted would answer their phones and be willing to talk or self-isolate or take a test, and there were discouraging signs in the summer of 2020 that this was not the case. The local efforts were also hampered by the lack of a robust federal approach to contact tracing, and states were unwisely left on their own to manage a threat that did not respect borders.

Tracing often goes hand in hand with *testing* to identify sick or susceptible individuals. But testing serves further purposes. It can be used in a focused manner to identify reservoirs of infection, such as in health-care workers, homeless people, or prisoners. It can be used to help manage outbreaks in particular places, such as factories or nursing homes. And by tracking the epidemic, testing is incredibly helpful in ensuring that the nation is not flying blind. But in all of the fevered discussion around testing—or lack thereof—the public often misunderstood the efficacy and limitations of testing. Testing is an important but imperfect weapon.

Two types of tests can assess coronavirus infection. The first is a test for the virus itself; it involves taking a swab of mucus (or soon, saliva) from inside the patient's throat or nose (often quite deep inside—one patient described it as a "stab in the brain") and processing this mucus to extract the RNA that comes from the virus. This RNA has its sequence assessed and compared to a reference standard of the RNA for the pathogen. In the very early days of the epidemic, all of the tests in the United States were conducted in Atlanta at the Centers for Disease Control.

A second type tests not for the virus but for antibodies to it—the proteins that our bodies make to fight the virus. This requires taking a sample of the patient's blood, via fingerstick or venipuncture, and testing it for the specialized antibodies to the virus (saliva tests have also been developed). The two key types of antibodies are IgM, which is made as quickly as three days after exposure to the virus and lasts only temporarily, and IgG, which is made somewhat later, around five days after exposure, and that circulates within the bloodstream for perhaps as long as a year.

If these antibodies are detected, it indicates that the person has previously been infected with the pathogen. People who test positive for the antibodies are usually not infectious, since the virus is typically gone from the body by that point. Many months into the pandemic, the duration of infectiousness was still not known precisely. Most people cleared the virus from their bodies within a couple of weeks, but there were a few people who continued to test positive for the virus four weeks later despite having mounted an antibody response.[60]

Unfortunately, the United States botched the rollout of the first test for the virus—if the country were a student in one of my courses, I would not hesitate to hand it an F. And we did only moderately well with the antibody test rollout, which began in earnest in late April—I would give our nation the grade of C. Scientists in China and Singapore developed both of these types of tests by late January 2020 and deployed them widely.

Many other countries around the world had started testing for the virus by early 2020, but the United States had not. We made three types of mistakes. First, and most important, the CDC released a test kit that was flawed, and when the error was detected, the response was unnecessarily slow. Second, the FDA refused to allow hospital labs to develop their own tests, even though most elite hospitals in the United States could do this and were eager to do so. Third, the Department of Health and Human Services took its time to work with outside labs to increase the availability of commercial tests, for which there was a huge market, and did not get it done until it was too late. These mistakes were identified in real time by many experts from across the political spectrum, and I read about them with rising alarm during the months of February and March. It's hard to overstate the Keystone Cops flavor to this stage of the country's pandemic response. People outside the United States watched with incredulity and dismay as the world's richest nation, with its illustrious CDC (which has provided the model for disease-control centers across the globe) and the most sophisticated medical care, failed to provide this most basic public health intervention. It's not as if the CDC had not succeeded in similar situations before; during the 2009 H1N1 pandemic, the CDC developed and shipped over one million tests throughout the United States just two weeks after the virus was discovered.[61]

The lack of coordination at the federal level, prompted in part by the undue politicization of the pandemic, severely hampered the United States. Dr. Tom Frieden, the former CDC director, characterized the federal response in February and March as an "epic failure," and I completely agree.[62] The lack of testing early on in the 2020 pandemic meant the United States was unable to detect cases of community transmission and therefore was not able to implement procedures to isolate infected or exposed individuals and contain the virus. Without testing, officials could not identify reservoirs of infection. As a result, its presence continued to grow exponentially to the point where it became impossible to do contact tracing at all. Throughout much of the spring of 2020, researchers simply did not know how many people were infected, where they were, and whether the nonpharmaceutical interventions designed to flatten the curve were making a difference. Without testing, public health officials were obliged to target the whole population for restrictions in social interactions rather than just those who were sick or exposed. Without testing, it was impossible to take a more surgical approach to quarantine. This failure imposed massive economic burdens on many.

In April, it became clear that there was no national plan to fight the pandemic, and governors began to work together, along with disease experts and former officials of multiple administrations, to develop a coordinated strategy. Unsurprisingly, that strategy involved ramping up testing and tracing efforts in order to narrow restrictions to those who were infected and allow others to return to work.

But even then, as the epidemic progressed, we had no policies in place to test random and representative samples of Americans either with direct testing for the virus or with antibody testing for immunity. This, too, was crucial. All the testing that was being done was on individuals who had symptoms or who had had contact with people who had tested positive. But if we test only people with symptoms, the fraction of positive test results will be high, and it will look like more people are infected than actually are. Conversely, if we test only thooo pcopl‹ wliu are anxious about the disease and who have no risk factors, we may underestimate the prevalence of the disease. The only way to be sure is to test random samples of the population.

Furthermore, since the test for the virus tells us only if someone is actively infected, whether or not they have symptoms, it does not shed light on the cumulative exposure—that is, how many people have ever been exposed. Only the antibody test can do that. This type of information is crucial in knowing whether our responses to the pandemic, stringent as they eventually were, were making a difference or were warranted. This information also helps to determine the infection fatality rate, because we have a more reliable indicator of who may have had asymptomatic infections. And it tells us whether we are getting closer to having enough immune people in a population to reach herd immunity.

Still, as necessary as testing is, it has limitations. One issue with testing is that it is dependent on the base rate of people actually afflicted with a condition. No test can be perfect, and all tests can yield both false-positive and false-negative results. But the situation can become even more complicated when the test is being used in a population with a low rate of the condition being measured. Suppose a pregnancy test has an error rate of 5 percent false positives (that is, 5 percent of the time, the test will incorrectly report that a person is pregnant when she is not). In a group of pregnant women, the error rate of the test does not matter because there are no non-pregnant people in this group, and it will correctly label all one hundred of them as pregnant. However, if we apply the same test to a sample of one hundred six-year-old boys, the fact that none of them could possibly be pregnant means that the test will incorrectly label five of them as pregnant.[63] So, if a test has some rate of yielding positive results for people who actually have a condition but also, inevitably, some rate of yielding positive results in people *without* the condition, the test accuracy will be affected by the underlying rate of the condition in the population. Even a small percentage of false positives can have a dramatic effect when few people in the population of interest actually have the condition. For conditions with low prevalence—such as the presence of antibodies against a virus early in the course of an epidemic—many of the positive tests for the condition would actually be false positives, and so testing will overestimate the prevalence of the virus (unless the test is exceptionally good).

Furthermore, testing and tracing rely heavily on each other to be truly effective. Testing on its own does little to help control a virus. It tells us only who already has the virus, not where it has spread. Similarly, tracing alone does not help much if we cannot identify asymptomatic individuals. Tracing is also compromised if it's not feasible to quarantine those individuals whom we find, either because they are unwilling or unable to self-isolate at home or because we lack places where they can stay—for example, if they are homeless or if they live in a crowded environment. It's only when you can test, trace, *and* isolate that you can break the chains of transmission and bring a virus under control.

When the number of cases gets too large, however, either because containment has failed or because of the nature of the pathogen itself—as happened in many parts of the country by late March—contact tracing is no longer possible at the scale that is needed. At that point, widespread implementation of physical distancing is necessary in order to slow the epidemic and reduce the number of cases so that contact tracing might once again be realistic. One of the reasons the South Koreans were able to avoid employing more stringent physical-distancing measures is that they rapidly deployed extraordinary testing and tracing procedures (along with universal mask adoption).

»——————→

If testing, tracing, and isolating fail to stop the spread of the pathogen, what can be done? For this objective, we must decrease social mixing using physical-distancing measures. The fewer people that the average person comes into contact with each day, the better. And school closures have historically been one of the most effective ways to interrupt transmission chains. By the end of March 2020, 94 percent of American schools had been shuttered, most of them for the duration of the school year.[64] This reduced social mixing among the nation's 56.6 million children and roughly three million teachers (numbers that do not include the many other personnel affiliated with schools, from custodians to bus drivers to food-service providers, or the many children in preschool and day-care programs and their two million caregivers).[65]

Early in the coronavirus pandemic, I was asked to provide advice regarding school closures to one large southern state and to the principals of several schools around the country.[66] All those school leaders were struggling mightily with the very difficult decision to close schools. The smooth operation of schools—with children playing in the playground, the rounds of the school buses, and the morning rituals of countless working families packing lunches and sending their children off for the day so that they themselves may go to work—is very hard to abandon. Many millions of children in the United States rely on schools not only for lunch but also breakfast and sometimes dinner. These children come from environments in which the safest place for them is actually the school (an argument that was also made explicitly in 1918 in New York City). Closing the schools can harm children who are neglected or unsafe at home and who benefit from the caring eyes of a teacher. Moreover, health-care workers and first responders are needed in a crisis; closing schools may mean they are not available to help in the epidemic because they are stuck at home watching their own children.[67]

Despite some objections, schools did close across the nation. The total closure of schools and the pivot to remote learning was a momentous event that most Americans had never experienced before, and families around the country reacted with varying degrees of incredulity, distress, relief, and resignation. Teachers and parents alike found the transition to remote learning difficult or, in some cases, impossible—the nation's patchy infrastructure left some teachers conducting videoconferences from their cars, and plenty of families had neither internet nor laptops. Videos of baffled and over-taxed parents quickly went viral, including one of an Israeli woman losing her cool (to put it mildly) at the prospect of homeschooling multiple children while holding down a job. Others coped, once again, with gallows humor; one photograph of a liquor-store bill-board that read HOMESCHOOLING? GET YOUR SUPPLIES HERE went viral. The school shutdown laid bare many life-work balance realities that were present before the shutdown, including the differential impact on men versus women and on the rich versus the poor. Once again, the pandemic highlighted and amplified long-standing societal challenges.

So was shutting schools down the right thing to do? Most of the research for other respiratory infections supports the claim that school closures soften the force of the epidemic, postponing the peak and reducing the case count.[68] This is likely true in the case of the COVID-19 pandemic too, though the evidence that is emerging is mixed, and it is very difficult to be sure.[69] An even more difficult issue relates to the usual utilitarian calculus of public health—whether the lives saved by closing the schools are, in fact, worth the short-term and long-term cost to children and to society. Complicating matters, there is an admittedly small percentage of families and children for whom the school closures in the spring of 2020 appear to have been beneficial, or at least not harmful, from an educational or socioeconomic point of view. This group includes children with supportive, stimulating home environments (some of whom have a nonworking parent or grandparent available during the daytime) and older students of all backgrounds who have experienced anxiety, fatigue, bullying, or burnout from traditional school schedules and programming and manage quite well on their own time.

The obstacles to safely *reopening* schools that were shuttered during the spring of 2020 also seemed insurmountable in many communities, and they included, to name but a few: the shockingly decrepit physical condition of school facilities, which put occupants at risk for respiratory illnesses; a lack of funding to implement new protocols necessary to meet recommended public health guidelines; and the need to adapt curriculum and teaching practices to account for the much wider range of learning outcomes and psychological needs children would likely present when they did return to school. School districts all over the country also wrestled with how to balance appropriate public health measures in a mixed-age population of adults and children with equally important objectives to promote health and well-being (such as a kindergartner's need to see her teacher's face, a tenth-grader's need for extracurricular activities, or older teachers' need to avoid contracting the infection). By July 2020, with cases rising sharply around the nation, polls showed that 71 percent of American parents thought it was risky to reopen schools.[70]

While there is much we still do not know about the long-term impact of the 2020 school closures on our society, I think it's fair to assume there will be many unforeseen and primarily (but not exclusively) negative consequences. Here, however, I want to focus not so much on the issue of whether the costs of school closure are "worth it" in economic or social terms but rather on how school closures work to stem an epidemic in the first place, the types of school closures societies adopt, and why they are considered the penultimate step just short of stay-at-home orders in the range of nonpharmaceutical interventions at a society's disposal.

There are two types of school closures. The first is *reactive* closure, which is closing a school (or all the schools in a district) when a case (or cases) has been diagnosed at the school. This is relatively uncontroversial, and almost everyone—teachers, parents, politicians—clamors for this when a case is discovered in a school. Detailed models show that if school closure is reactive, in the case of a moderately transmissible virus, the cumulative cases of the condition might decline 26 percent and the epidemic peak could be delayed by sixteen days.[71]

But reactive school closure, while rational and often helpful, is not enough. In my view, if decision-makers are prepared to close schools at all, they should be closed *before* the first case in a school, when disease cases begin to appear in the community or in nearby areas. This is called *proactive* closure and it's more controversial. Rigorous analyses show that proactive closure is one of the most beneficial interventions that can be employed to reduce the impact of epidemic disease.[72]

But if a school is prepared to close in a reactive fashion once there is an outbreak in the school, why not close a little earlier in a proactive fashion and get more of the benefit? If there is community transmission of the germ *near the school*, it's a certainty that it will be *in the school* before long. Hence, closing the school proactively, one or two weeks before one might have closed it in a reactive fashion, offers substantial advantages. Waiting to implement reactive closure places all of the same burdens on schools and parents but offers fewer of the benefits with respect to epidemic control.

It can be hard to estimate the benefits of school closures precisely. Closures are clearly more beneficial in outbreaks where the children themselves are substantially affected by the disease (as in the case of polio). Still, it's important to note that the primary purpose of school closure is to reduce social mixing, not necessarily to protect children from infection. By radically decreasing social interactions in a community, closing schools can have a powerful effect (even if, as in the case of SARS-2, children are relatively spared becoming sick). In part, it works by keeping the kids from acting as vectors (which they can indeed be with SARS-2), and in part it works by forcing *parents* to stay home. When epidemiologists develop models to assess the impact of school closures, they sometimes include a parameter that captures what fraction of the parents in a community must stay home as a result. In principle, adults could of course be encouraged or even ordered to stay home while their children still attended school; in reality, this does not happen because stopping work is economically devastating and, I believe, because we tend to put less emphasis on the needs of children compared to adults.

In American schools, the millions of children and adults are in far closer daily physical proximity, and for longer durations (thirty-five or more hours per week), than are adults in most workplaces. The impact of banning occasional large gatherings like sporting events or religious services does not even come close to that of school closure. Because of this, school closures are the most consequential NPI that can be employed, short of requiring everyone to stay home.

One inventive study examined the impact of school closures and other NPIs, and their precise timing, during the 1918 influenza pandemic. It looked at forty-three major U.S. cities and concluded that the *earlier* that schools were closed (ideally in advance of outbreaks), the lower the number of excess deaths.[73] Moreover, the *longer* that school closures (and other nonpharmaceutical interventions, such as bans on gatherings) were maintained, the lower the ultimate mortality rate was. As shown in figure 14, St. Louis closed its schools before local cases had already doubled and kept them closed for longer (one hundred forty-three days) than Pittsburgh (fifty-three days). By the end of the pandemic, St. Louis had less than half the excess deaths that Pittsburgh had (358 out of 100,000 people compared to 807 out of 100,000). Of course, there could have been other pertinent differences between these cities, and such observational studies are limited. But we cannot randomly assign whole cities to different sorts of NPIs in order to do a true experiment, so we must take what data we can.

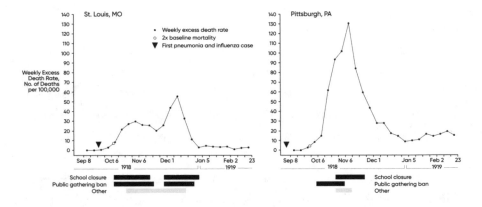

Figure 14: The different timing and nature of nonpharmaceutical interventions implemented in St. Louis and Pittsburgh during the 1918 Spanish flu pandemic was associated with a much better outcome, in terms of mortality, in St. Louis compared to Pittsburgh.

In 2020, we can see several different approaches to school closings across the world. Japan closed all the schools in the entire nation at the end of February 2020 and planned to keep them closed through April.[74] The Japanese had learned from prior outbreaks, including the H1N1 outbreak in 2009.[75] By contrast, Singapore did not use school closures and instead successfully implemented a rigorous and elaborate procedure of temperature checks and handwashing at their schools.[76] Italy closed schools nationwide in a reactive fashion in early March, far too late to blunt the force of the epidemic.[77] In the United States, the CDC guidance was still very gentle and avoided recommending closure as late as March 5, 2020.[78] But by early March, at least 46,000 schools had closed (or were scheduled to close or had closed briefly and reopened), affecting at least 21 million children out of the 56.6 million K-through-12 students in the

United States. By the end of the school year, 55.1 million students in over 124,000 schools had their schools closed.[79]

Americans would struggle with the decision about whether schools should reconvene for in-person classes in the fall of 2020. Yet if officials are unable or unwilling to keep schools closed, they have other choices to at least reduce physical contact. First, not every family will be comfortable sending their child to school during an outbreak—in some school districts in hot zones, like New York City in March 2020, many parents were already keeping their children home before school closures were announced.[80] Allowing these families to make this choice decreases transmission to others and can even be seen as a public service. Schools are ordinarily concerned about truancy (especially for at-risk students such as homeless children). But such rules might beneficially be suspended, or schools can offer some form of optional remote instruction to families who need it.[81] Teachers and administrators can also implement other practical measures to minimize risk. An important adaptation would be to give children of all ages much more outdoor time or hold classes outside, since disease transmission is so much less likely outdoors than indoors.[82] Many early-childhood teachers have discovered the myriad benefits of outdoor learning, so this could be helpful to children in other ways.[83]

Still other measures, such as increased custodial cleanings, temperature checks, and more frequent handwashing, will be time-consuming and expensive but should be implemented as much as possible. Anyone who has actually observed children will know that teachers need to provide a high level of surveillance around personal hygiene, even for older children. Increasing physical distance between students is a challenge in overcrowded schools where instruction sometimes takes place in hallways, but it's important. Canceling gym classes and band practices and prohibiting mixing in common areas are obvious, if regrettable, measures. Schools could also limit outside visitors, field trips, and nonessential social events. Changes in schedules and hybrid models of in-person and remote learning can help. But a school with such extensive constraints is not a school as we ordinarily understand the term. These are tough dilemmas.

»———————→

To understand the challenges of implementing NPI strategies in the midst of a pandemic and the dangers imposed by the failure to do so, let's consider New York City's 2020 response in depth. Amazingly for such a sophisticated city with a storied public health department and with a history of mostly successful actions in 1918, New York City stumbled in its response to the pandemic, lagging behind other cities in implementing physical-distancing measures and paying the price in sick and dead patients. Of course, it's difficult to be sure whether any sort of intervention, unless implausibly early and extreme, truly could have stopped the epidemic in an enormous, heterogeneous, and densely packed transportation hub like New York City. But moving more quickly would surely have saved lives during the first wave of the COVID-19 pandemic. Dr. Tom Frieden, the former head of the CDC, would later estimate that if New York had adopted widespread physical-distancing measures even a week or two earlier, the death toll might have been reduced by 50 to 80 percent.[84]

Based on an analysis of national and international travel patterns alone, it's very likely that the virus was already on the loose in New York City by mid-February 2020 at the latest.[85] Genetic analyses show this to be the case. Phylogenetic analysis of eighty-four distinct genomes of SARS-2 specimens collected in the city through March 18 confirmed that there were multiple, independent introductions of the virus to the city, primarily from Europe, Italy in particular, but also from other parts of the United States.[86] If I had to guess, I would say that the virus arrived in the city in January.

The first confirmed case of COVID-19 in New York City was not diagnosed until March 1. The patient was a thirty-nine-year-old woman, a health-care worker and resident of Manhattan, who had just returned from Iran.[87] By that point in time, just thirty-two people had been tested for COVID in the state of New York, but a model based on travel patterns estimated that there may have already been over ten thousand infected people in the state.[88] The day after this first known case was announced, the governor of New York, Andrew Cuomo, appeared with the mayor of New York City, Bill de Blasio, to say that contact tracers would look for every person on the woman's flight. It was later revealed that this did not happen, because it was the CDC's job to do that, and they did not prioritize it.[89] This was an inauspicious opening act for the known part of the outbreak in New York. Events would move very rapidly over the next four weeks, in keeping with the exponential growth of epidemics.

In his statement about this first case, Governor Cuomo sought to prevent panic, saying, "There is no reason for undue anxiety—the general risk remains low in New York. We are diligently managing this situation and will continue to provide information as it becomes available."[90] While the desire to avoid panic is commendable (though I had seen no evidence of New Yorkers panicking), such false reassurance was misguided given what we knew about the impact of the virus in China and Europe.

Two days later, on March 3, New York City confirmed its second case of COVID-19, in a fifty-year-old lawyer named Lawrence Garbuz, who had no recent travel history and who had been ill since the middle of February. Garbuz went to the emergency department with a fever on February 27, 2020, quickly lapsed into a coma, and woke up three weeks later.[91] This was the first case of community transmission to come to light in the city.[92] It was a sign that the virus was on the loose. Officials just happened to observe the tip of the iceberg.

Garbuz lived in the suburb of New Rochelle and worked in midtown Manhattan. In the wake of the discovery of his case, multiple institutions—ranging from a synagogue to Garbuz's daughter's school—were shut down, and many people, including health-care workers who had initially cared for him, would be quarantined.

To help prevent the further spread of infection, the National Guard was mobilized. On March 10, the governor announced a one-mile-radius "containment area" centered in New Rochelle that would be in place from March 12 to March 25.[93] The National Guardsmen cleaned schools and delivered food to quarantined individuals, but they also established a zone where all schools, places of worship, and large gathering facilities were closed and gatherings were banned. Still, this was hardly a Wuhan-level quarantine—people were free to move about and go into Manhattan for work, for instance.

By March 6, there were twenty-two more known cases in New York City. Of the forty-four total cases in the state at that time, twenty-nine were suspected to be linked to Garbuz. The city pleaded for more tests from the federal government as the case count rose. Fewer than one hundred people had been tested on that date, and city officials notified the federal government that a shortage of tests had "impeded our ability to beat back this epidemic."[94]

Testing in New York City had been severely limited since February due to the broader national limitations we discussed earlier. Infectious disease experts in the city had been informed, on a call on February 7, that the federal criteria would restrict testing to patients with a fever severe enough to require hospitalization and with a history of travel to China in the preceding fourteen days. "It was that moment that I think everybody in the room realized, we're dead," one doctor would later recall.[95] Contact tracing would simply be impossible due to the lack of testing and very limited manpower to follow up on results. New York City had fifty contact tracers at that time; by comparison, Wuhan had nine thousand.

New Yorkers, the *New York Times* reported, began to feel anxious: "That sentiment reverberates along the subways and sidewalks of New York City, where the usual throngs and random interactions with strangers—the very things built into the magic and texture of this city—are approached with an unsettling caution in the age of the new coronavirus."[96] Perhaps to promote feelings of normalcy, as of March 5, Mayor de Blasio was still encouraging subway travel, and a photo showed him grinning in a packed subway car.[97] "I'm here on the subway to say to people: nothing to fear, go about your lives, and we will tell you if you have to change your habits, but that's not now," he said. In my view, this was supremely irresponsible.

On March 9, deliberations about whether to close the schools proactively were still ongoing.[98] But some schools began to have to close reactively. The New York prep school Horace Mann closed after notifying parents that one of its students was being tested for coronavirus.[99] Experts were worried and continued to attempt to sound the alarm. Thirty-six New York City infectious disease doctors strongly endorsed closing New York City schools in a public letter to the mayor dated March 12, 2020.

But de Blasio planned to keep schools open, he announced on March 13. "We are doing our damnedest to keep schools open," he said.[100] De Blasio was concerned that poor families would have nowhere for their kids to go, just as officials were in 1918. "If you suddenly in one day close down the schools, how do you make sure that you are providing for these kids and their parents?" one of his aides later said. "We're not the suburbs. We can't tell people to stay at home and play around in your yard."[101] But another of de Blasio's advisers, Demetre Daskalakis, the head of disease control for the city, threatened to quit if the schools were not closed. Public-school teachers were also in rebellion.

School closure was inevitable, so it is unclear what the delay achieved. De Blasio relented two days later, on March 15, and finally closed the schools. Plans were made to provide childcare to health-care workers and first responders. At the time, the city had 329 confirmed COVID-19 cases; in comparison, San Francisco had closed the schools three days earlier, when it had only eighteen confirmed cases.

Decision-making was difficult during this period in March. Developments were fast-paced and leaders were in reactive mode. On March 11, after much outcry from public health experts, including myself, the St. Patrick's Day parade, which draws about a hundred and fifty thousand marchers and two million spectators, was prudently canceled—for the first time in two hundred fifty years—and New York thereby avoided making the mistake that Philadelphia made in 1918.[102] On March 12, both Governor Cuomo (for the state) and Mayor Bill de Blasio (for the city) declared a state of emergency and banned large gatherings.[103] On March 17, theaters, concert venues, nightclubs, and restaurants were closed.[104]

Finally, on March 20, the governor issued a shelter-in-place order, effective March 22, closing down all nonessential businesses and basically compelling New Yorkers to stay at home as much as possible. By that point, New York had 8,452 cases; while the state had 6 percent of the U.S. population, it had roughly half of all known cases.[105] By comparison, the state of California had gone to this level of physical distancing on March 19, just a few days before, when it had 1,009 cases.

The governor's order stated that:

- "All non-essential businesses statewide will be closed"
- "Non-essential gatherings of individuals of any size for any reason (e.g., parties, celebrations, or other social events) are canceled or postponed at this time"
- "Any concentration of individuals outside their home must be limited to workers providing essential services and social distancing should be practiced"
- "When in public, individuals must practice social distancing of at least six feet from others"
- "Businesses and entities that provide other essential services must implement rules that help facilitate social distancing of at least six feet"

- "Sick individuals should not leave their home unless to receive medical care and only after a telehealth visit to determine if leaving the home is in the best interest of their health"

The streets of New York City became eerily deserted. Later, Cuomo would extend his order through the end of April.[106] Some New Yorkers ignored the rules, of course, crowding public spaces (since sheltering at home in New York City's small living spaces was a hardship for many).

But it was too late. By March 22, New York City had become the epicenter of the pandemic in the United States, accounting for approximately 5 percent of the world's confirmed cases at the time.[107] At the outset of the epidemic, both the mayor and the governor had put a lot of stock in New York's large and sophisticated health-care system, reasoning that surely it could cope with the crisis. Hospital executives also expressed confidence. But on March 18, Mayor de Blasio reported that more than a thousand retired medical workers in the city had answered his appeal the prior day to volunteer to help, as had occurred in Italy.[108] Emergency departments and intensive care units "teemed with 'redeployed' dermatologists, ophthalmologists, and neurologists."[109]

Hospitals began to report "apocalyptic" conditions by March 25.[110] And on March 27, similar to what had happened in Wuhan, the Army Corps of Engineers deployed to New York City to convert the 1,800,000-square-foot Jacob Javits Convention Center into a 2,910-bed civilian hospital.[111] The city pleaded with federal authorities to nationalize the production of PPE and ventilators, which were rapidly running out; the mayor estimated that they had only ten days of supplies left. In a resigned statement, the governor noted that "40 percent, up to 80 percent of the population will wind up getting this virus. All we're trying to do is slow the spread, but it will spread."[112] As of that date, the city had 1,800 patients hospitalized, with 450 of them in ICUs, and just 99 deaths. This was quite a change in tone and substance from Governor Cuomo's comments just three weeks earlier—which is what happens in bad epidemics that move fast.

In the span of three weeks, since his subway ride, Mayor de Blasio had also changed his tune. He now warned, "The worst is yet to come. April is going to be a lot worse than March. And I fear May could be worse than April."[113] In early April, China and Oregon sent ventilators to New York City.[114] The mayor urged anybody in the city with a ventilator, including veterinarians, to step forward, saying, "We could use every single one of them."[115] Indeed, by April 21, given the overload of COVID-19 patients, New York State issued drastic new guidelines urging EMTs not to bother trying to revive people without a pulse when they got to a scene. "They're not giving people a second chance to live anymore," observed the head of the city union representing paramedics. "Our job is to bring patients back to life. This guideline takes that away from us."[116]

The health-care system strained under the load, and very worrisome reports reached me from my medical friends in the city. They described the situation as "gruesome" and "unreal" with the incessant arrival of very sick patients, and the ICUs bursting. It was like "a refugee camp in a war zone," one nurse observed.[117] Working conditions were extremely difficult, requiring providers to be in layers of protective equipment. They were uncomfortable, but they also feared the imminent unavailability of PPE. Doctors and nurses improvised equipment from office supplies or brought PPE in from home. Hospital morgues filled up. On March 25, some eighty-five refrigerated trailers were sent by FEMA to provide places for the bodies.[118] Rules were relaxed so that local crematories could "work around the clock."[119]

Some health-care personnel began to get sick, and many died, poignantly after being admitted to their own hospitals. Photographs emerged of exhausted health-care workers in what looked like white space suits resting outside hospitals—getting a bit of sun before returning to battle. In some photos, their faces were blistered and bruised from wearing tight-fitting masks all day. The photos reminded me of the famous photo of firefighters collapsed and asleep not far from a forest fire that still raged. And the situation also reminded me of the SARS-1 epidemic in the hospitals in Asian capitals years before. I have cared for patients in elite hospitals with plentiful supplies in every bedside cart, and I have been a citizen of this rich nation since I was born, and I could not believe these events. Our nation spends 17.7 percent of its GDP on health care, and this was our level of preparedness?[120]

Initially, the virus struck New York City in a seemingly even-handed way, claiming lives of both mass-transit workers and celebrities. But soon, of course, the burden was higher on the less privileged. Working-class immigrant communities, such as those in central Queens, were especially hard-hit. The adjoining neighborhoods of Corona, Elmhurst, East Elmhurst, and Jackson Heights emerged as an "epicenter within an epicenter." With a combined population of 600,000 people, these neighborhoods had 7,260 coronavirus cases on April 8, whereas Manhattan as a whole, with three times the population, had 10,860 cases[121]

As with past plagues, those able to do so fled the city. An analysis of cell phone records for a large sample of New Yorkers during the month of March showed that, while the city's population decreased by only about 4 to 5 percent overall, the population declined more than 50 percent in some of the richest neighborhoods.[122] People began to leave in early March, before the NPIs were officially implemented. While most of the richest residents fled to nearby places in upstate New York and Connecticut, many others were tracked to Arizona, Michigan, and Southern California.[123]

On March 31, Governor Cuomo observed, "I am tired of being behind this virus. We've been playing catch-up. You don't win by playing catch-up."[124] And he was right. As he spoke, there were 76,946 cases in the state.[125] By April 6, there were 72,181 confirmed cases and at least 2,475 deaths in New York City alone, and the city had 25 percent of the COVID-19 deaths in the United States. The statewide shelter-at-home order was extended through April 29.[126] Finally, on April 15, the pandemic peaked in New York's hospitals, a result of the NPIs that had been implemented three weeks earlier—on schedule given the epidemiology of how COVID-19 spreads and the timing of clinical progression of the disease (with up to two weeks for a patient to feel symptoms and a further week to fall critically ill).

By the end of April, the curve had been flattened. But the result of the outbreak was staggering: a statewide antibody testing survey found that 21.2 percent of the city's residents had contracted COVID-19 in the first wave.[127] And analyses revealed that travel from New York City had seeded a wave of outbreaks of coronavirus throughout the rest of the United States.[128]

»————→

While NPI strategies did eventually work in New York City, there are still many unknowns about how the city's responses might have yielded a more or less devastating outcome. One of the greatest mysteries of pandemics is why some regions are stricken and others are spared. In the 1957 flu pandemic, as we noted in chapter 2, there was an enormous range in the number of cases and deaths from place to place, and the same was observed in the United States and around the world in 2020 with COVID-19. No doubt when the dust settles, colored maps of countries in the world will show hot spots where the burden of death was particularly severe. Some of this variation will have to do with different policies implemented by various governments at the national and local level. Some of it will have to do with the vagaries of the number of imported cases that start the local epidemics, as we saw in both China and the United States. Some of it will have to do with environmental conditions or demographic profiles. Some might even relate to genetic differences in the strain of virus that strikes a particular area. A very tiny amount might possibly have to do with genetic variation in humans from place to place; some populations might happen to have genetic resistance. But most of the variation will just be due to chance—like that seen in photographs after a tornado where many houses are blown apart right next to a few that unaccountably are left standing.

In the first few months of the pandemic, many experts tried to discern why some countries did better than others. What did China do and how did it do it? How did South Korea and Taiwan manage to contain the pandemic? As rich democracies, could the latter two be models for other European countries or for the United States? Every day, people would follow the graphs showing a kind of leaderboard, country by country, often in different colors, indicating the upward trajectory of the pandemic. The Asian countries—China, South Korea, Japan, Singapore, and Taiwan—veered off in long and low plateaus.

But what struck me was not so much how to account for these differences that were possibly related to the timing and nature of the NPIs adopted or to other features of these societies, but rather how the trajectories were so similar in all the countries *before* the interventions were implemented. At the outset, the virus just killed us, and it was mostly unaffected by the political system, religion, health-care system, media environment, and myriad other attributes of our societies. After all, we are human, and the virus really does not care about the details. The *upward* trajectory as the pandemic initially took root in each nation was grimly similar.

Furthermore, people themselves knew what to do, notwithstanding the failures or successes of their country's responses. People began to physically distance before being told or ordered to do so. For example, analyses of foot traffic in stores and bookings at restaurants across the world revealed that those began to decline a couple of weeks before collective NPI policies were implemented. The synchrony of the decline in restaurant bookings in OECD nations, seen on the online app OpenTable, was noteworthy. Each country had different distancing policies, different laws and cultures, and different rates of COVID-19; but restaurant bookings all fell to zero over the course of fifteen days as the epidemic struck.[129] Parents also began to withdraw their children from school before official closings were announced. By the time the New York City public schools were shuttered, as we saw earlier, a substantial percentage of children were already staying home.[130]

Thus, it did not matter what people did or who people were until they collectively responded by separating from one another and physically distancing. This raises the possibility that what is required to deflect the trajectory is simply achieving some overall level of response. No matter the specific combination of nonpharmaceutical interventions, as long as a certain threshold is reached, the pandemic can be brought to heel.

This physical distancing and economic collapse, this *slowing down*, are all features of plagues. Priest and historian John of Ephesus noted as much during the plague of Justinian, over fifteen hundred years ago:

> And in all ways everything was brought to naught, was destroyed and turned into sorrow.... [And] buying and selling ceased and the shops with all their worldly riches beyond description and moneylenders' large shops closed. The entire city then came to a standstill as if it had perished.... Thus, everything ceased and stopped.[131]

Such accounts of the effects of serious epidemic disease are eerily familiar. An economy involves exchanges and these depend on social interactions. It's hard to have an economy—or a functioning society—if people are unable to interact. Plagues are a time of loss not just of lives, but of livelihoods—and of routines, connections, liberty, and much else.

4.

Grief, Fear, and Lies

> The plague was nothing; fear of the plague was
> much more formidable.
> —Henri Poincaré, *French Geodesy* (1900)

Wanda DeSelle, a seventy-six-year-old nurse from Madera, California, had no family members by her side when she died of coronavirus on March 30, 2020. Her daughter, Maureena Silva, had said goodbye to her via FaceTime a few days earlier, asking her to blink if she could hear. DeSelle had contracted the virus three weeks before when she attended a funeral of a young colleague in the medical practice where she had worked for forty years. At least fourteen others contracted the disease from that same funeral, in a super-spreading event. DeSelle subsequently transmitted the virus to her daughter, who fell ill while caring for her and who then passed it along to her own daughter. Both the daughter and the granddaughter recovered in time to attend DeSelle's funeral. But it was a funeral few could have imagined only a few weeks before: a drive-by burial during which the grieving family and a handful of friends sat in their separate cars at an eighty-foot distance from the casket, which was lowered into the grave by four funeral-home staff in masks and white gloves.[1]

The members of the close-knit ultra-Orthodox Jewish community in Crown Heights, Brooklyn, are deeply immersed in shared daily rituals and face-to-face interactions, so the high mortality rate from COVID-19 and the physical distancing it imposed was a particularly harsh blow. One rabbi organized a special hotline to play ancient Jewish prayers for the dying on a speakerphone placed next to isolated ICU patients' bedsides. "My fellow Jew," a voice said reassuringly, "your family very much wants to be with you in person. However, due to the current circumstances, it just isn't safe or allowed."[2] And then the rabbi recited Psalm 91:10: "No harm will befall you. Nor will a plague fall upon your tent. For He will command His angels on your behalf. To guard you in all your ways." The Talmud calls this passage the "song of plagues." One interpretation states that Moses himself composed it while ascending into a cloud hovering over Mount Sinai, reciting these words as protection from the angels of destruction.

It is hard for me to describe how hard my fellow hospice doctors and I would work, when I was still seeing patients, to keep people from dying alone. We took pains to arrange for families to have a warning about impending death so that they could be there when the patient died—for their sake and the patient's sake. We honed our prognostic skills to help with this objective, which we regarded as essential. In fact, I wrote two books about prognosis because it is so central to ensuring patients a "good death."[3] If patients had to die without their loved ones, we doctors often sat at the bedsides ourselves and held their hands. I did this many times, watching the particular kind of waxing and waning breathing pattern patients often showed near the end of life (known as Cheyne-Stokes respiration) and dreading the awful soft sound of the death rattle that sometimes happened as the patient finally stopped breathing. I had the sense, more than once, that the patient's hand had gotten strangely softer just before he or she died. I cannot explain it. Maybe it marked a surrender.

On other occasions, we hospice doctors would engineer intricate ways to get patients discharged from the hospital, circumventing bureaucratic obstacles and figuring out how to administer medications in non-intravenous ways so that the patients might die at home, among their loved ones. And so, given the norms I had internalized, the many accounts I read of families prohibited from visiting their loved ones during the early days of COVID-19 (in order to reduce the spread of infection or to conserve personal protective equipment) struck me as not just inhumane, but immoral.

But, as with every other serious epidemic, many people died alone in 2020. Families were deprived of the chance to say goodbye or properly grieve. In one study I had done with some colleagues, published in 2000, we documented that 81 percent of patients and 95 percent of family members felt that the presence of family when the patient died was very important (the lower percentage among patients may reflect their concerns for burdening their families).[4] But the U.S. health-care system was unable to honor these desires during the pandemic.

As a physician, I hated having to call a family to say that their loved one had died. I remember one call in particular that I had to make when I was a resident physician at the University of Pennsylvania in 1990. My patient's spouse simply could not understand what I was calling to say; it was as if my words were in a foreign language. During the first wave of the COVID-19 pandemic in New York City, another doctor explained how he had tried to comfort the husband of a dead patient over the phone. "He was having a really hard time. He was an older man, he was at home alone, and he didn't have any family with him." The doctor said that the encounter reminded him of prior experiences in Africa: "This happened with Ebola. The family is kept away. There are no funerals. This feels like that."[5]

Serious pandemics are a time of grief. When the death rate of a plague is particularly high, the situation is devastating. Here is an account by Petrarch of the Black Death:

What are we to do now, brother? Now that we have lost almost everything and found no rest. When can we expect it? Where shall we look for it? Time, as they say, has slipped through our fingers. Our former hopes are buried with our friends. The year 1348 left us lonely and bereft, for it took from us wealth which could not be restored by the Indian, Caspian or Carpathian Sea. Last losses are beyond recovery, and death's wound beyond cure.[6]

Thankfully, COVID-19 is not remotely as bad as the bubonic plague. But deadly epidemics always spawn parallel epidemics of a psychological or existential nature—less tangible but equally virulent. Grief, anger, fear, denial, despair, and even anomie are not unexpected emotional reactions to the personal and collective loss in a serious outbreak of infectious disease.

It can sometimes be hard to get a deep feeling for this suffering, either because people living through a pandemic are hunkered down trying to avoid their own calamities or because of the sheer scale of the losses. The old saying about the indifference that can arise in the face of mass mortality comes to mind: "One death is a tragedy, a million deaths a statistic" (a variant of which is usually attributed to Stalin). Author Laura Spinney, in her account of the 1918 flu pandemic, described the many millions of deaths that occurred from influenza (amplified by the massive casualties and dislocations of World War I) as the "dark matter of the universe, so intimate and familiar as not to be spoken about."[7]

A hundred years later, Americans have less experience with untimely death, and they are also less resigned to it. Many people live for decades without seeing death up close. While most Americans died at home a century ago, witnessed by all their loved ones, this is less common now.[8] But even so, there seemed to me to be an unsettling acquiescence to the COVID-19 deaths of 2020. When the death count reached 100,000 (coincidentally on Memorial Day), there was a brief spasm of collective mourning. On May 24, 2020, the front page of the *New York Times* consisted entirely of names of the dead. But the news of the milestone was fairly quickly eclipsed by political scandals and dramatic protests. Many people did not really have a picture of the disease or of the process of dying from it. Was it painful? Undignified? Lonely?

Of course, grief is not felt only for the lives lost, but also for the loss of our way of living. Adults lost their jobs or even their careers. People who were unable to pay their mortgages lost their homes. Scientists I know who had invested years in building scientific apparatuses had to abandon their research. Children were denied schooling, friendship, and outdoor play. Countless weddings and vacations were canceled, some permanently. Entrepreneurs lost their businesses. Online religious services, classes, counseling sessions, and Alcoholics Anonymous meetings were all pale replacements for the authentic connections humans evolved to crave but had to give up.

In the new reality imposed by the pandemic, nearly everyone experienced some deprivation. Sometimes the losses were indeed permanent or serious—lost lives, relationships, careers, or businesses. But even comparatively minor losses felt distressing, and they were widespread. We were not supposed to shake the hands of our neighbors or see them for dinner. We could not go out to restaurants, bars, coffeehouses, nightclubs, salons, or gyms in the same way, if at all. Even small, everyday occurrences could remind people of loss. I am not a clotheshorse, but after just three months of lockdown, of wearing a T-shirt and jeans every single day, I would look into my closet, see my suits, and miss wearing them. I often complain that I have to do too much professional traveling, but one day in May, I spotted my luggage in the closet and had a wistful feeling.

The damaging psychological impact of plague has been appreciated for a long time. In the fifth century BCE, regarding a plague in Athens of unknown etiology (possibilities range from typhus to Ebola), Greek general and historian Thucydides observed:

By far the most terrible feature in the malady was the dejection which ensued when anyone felt himself sickening, for the despair into which they instantly fell took away their power of resistance, and left them a much easier prey to the disorder, besides which there was the awful spectacle of men dying like sheep, through having caught the infection in nursing each other.[9]

The Roman emperor and philosopher Marcus Aurelius noted in the second century that what he called the "corruption of the mind" was much more dangerous during an epidemic than "any such miasma and vitiation of the air which we breathe around us."[10]

In April 2020, one national survey conducted to gauge emotional wellness revealed areas of significant distress for a sizable percentage of the population. Comparing the responses to 2019 polling data, people felt worse on several axes in 2020. The percentage of people reporting feelings of enjoyment was 64 percent, compared to 83 percent in 2019. Worry (52 percent versus 35 percent), sadness (32 percent versus 23 percent), and anger (24 percent versus 15 percent) were all higher.[11] Significant percentages of people in 2020 also reported boredom (44 percent) and loneliness (25 percent). People were worried about both the disease and its effects. For instance, another survey conducted at the end of April 2020 reported that 67 percent of Americans were "somewhat concerned" or "very concerned" about getting coronavirus themselves. But Americans were even more concerned that their family members would get the virus, with 79 percent expressing this fear.[12] Another more clinically focused study found that, whereas in 2018, 3.9 percent of Americans had severe psychological distress, in April 2020, 13.6 percent did, placing them at serious risk for longer-term psychiatric problems.[13]

Americans' feelings during the pandemic differed according to variables such as household income and gender. Adults with annual household income of less than $36,000 were less likely to feel happy (56 percent) and more likely to feel worried (58 percent), bored (49 percent), and lonely (38 percent) compared to those earning more than $90,000 (75 percent, 48 percent, 39 percent and 19 percent, respectively). Women reported feeling about as happy as men (71 percent versus 73 percent), but they were more worried (51 percent versus 44 percent) and more lonely (27 percent versus 20 percent).[14]

Over time, in another survey conducted after mitigation efforts to flatten the disease curve were well under way, Americans became slightly less fearful of contracting the coronavirus (57 percent were worried in April compared to 51 percent in May), but they became slightly more concerned about severe financial hardship (48 percent in April compared with 53 percent in May).[15] Despite these anxieties, however, Americans generally understood that the virus had to be controlled before things could return to the way they had been. Overall, at least two-thirds of Americans rated as "very important" the following conditions for resuming normal everyday activities at that time: (1) mandatory quarantine for anyone testing positive for coronavirus, (2) improved medical therapies to prevent or treat COVID-19, and (3) significant reduction in the number of new cases or deaths.[16]

»———→

In addition to grief, epidemics also bring fear. Fear can itself be contagious, forming a kind of parallel epidemic. Contagions of germs, emotions, and behaviors can act independently or they can intersect.[17] And fear has an advantage over even the most contagious pathogens—people can contract a disease only through contact with other infected individuals, but they can contract fear through contact with either infected individuals or fearful ones.[18]

We respond to the fear brought on by epidemics in various ways, many of which are directed at asserting control over the threat. For example, people have a tendency to blame others for the disease, which makes them feel like they have some influence over the force that is affecting them. It is more soothing to feel that there is a human agent responsible for the problem, because this means human effort might be effective in response. It's much more frightening to imagine that the plague originates from a vengeful, implacable god or from an uncaring and remorseless natural world.

This desire for a sense of control can be destructive, especially since the objects of people's blame are often minority groups or those seen as outsiders. To mitigate this, an important challenge for public health authorities and leaders during a pandemic is to acknowledge widespread negative emotions and feelings of powerlessness and to help people effectively respond to them in constructive ways by offering outlets for their emotions.[19]

This is one of the reasons it makes sense for public health authorities to encourage the use of masks, and it's also why people can find mask-wearing beneficial: it gives them something concrete to do in the face of the threat, regardless of the exact benefits of masks (which are considerable). This helps restore their sense of control. Another example is the act of collective public cheering for health-care workers, seen in New York City and London, which also gives people a sense of control and social solidarity.[20] This psychological benefit is important in itself, and such acts may in turn foster a willingness to engage in other more burdensome, difficult, and challenging practices. Getting fear and anxiety under control is critical.

But it is easier said than done. Sometimes, the lengths people go to displace fear, restore a sense of control, and assign blame can be quite baroque. During the 1916 outbreak of polio in the United States, the public faced a conundrum: the virus disproportionately struck young children in suburbs and more sparsely populated rural areas. How could this be squared with deeply entrenched stereotypes about urban, immigrant tenement-dwellers as vectors of disease? A New Jersey newspaper made a conceptual leap when it published a drawing of a giant horsefly seen menacing a helpless baby with this caption:

> *I am the baby-killer!*
> *I come from garbage-cans uncovered,*
> *From gutter pools and filth of streets,*
> *From stables and backyards neglected,*
> *Slovenly homes—all manner of unclean places,*
> *I love to crawl on babies' bottles and baby lips;*
> *I love to wipe my poison feet on open food*
> *In stores and markets patronized by fools.*

Creating an *imaginary* vector connected the sickness of innocent children to groups deemed worthy of blame while also suggesting a way to avoid risk and restore a sense of control. A germ cannot be seen, but a fly can, and the actions required to deal with flies were positive and helped people feel more in control (use screens and swatters). The fly theory also helped to explain how the sickness could have appeared in clean suburban homes: the danger was transported from a risk source composed of outsiders. In reality, of course, polio primarily spreads by the ingestion of fecal material from an infected person (for instance, through contaminated water or food). As Anne Finger explains in her moving autobiographical and cultural history of the polio virus, "the fly became a carrier, not just of contagion, but of displaced emotions."[21]

Fear brought on by epidemics can even cause people to shun those trying to help control the disease. On March 30, 2020, emergency physician Richard Levitan reported for duty at New York's Bellevue Hospital. The New Hampshire resident, like other volunteers from around the country who had answered Governor Andrew

Cuomo's call, had rushed to help relieve the overwhelmed ICU staff in the city's hospitals. Dr. Levitan had trained at Bellevue himself and was an expert in intubation, the thorny task of placing and managing the endotracheal tube in critically ill patients undergoing mechanical ventilation. "This [pandemic] is the airway challenge of the century," he said. "I'm an airway guy. I'm not going to sit this one out."[22]

Dr. Levitan was planning to stay in his brother's empty apartment, but he soon faced a vexing obstacle: one of New York's notoriously unwelcoming co-op boards, which refused him temporary residence on the basis that he would be working with infected patients. He noted that the building was a ghost town and some portion of those remaining were likely already infected, whereas he had come to New York from a low-prevalence state. But it did not make a difference.

Other stories emerged of travel nurses (who do temporary stints in hospitals of a few weeks or months at a time) being evicted on short notice by landlords, sometimes with hostility and threats that their belongings would be thrown away. As a nurse in Hawaii cogently pointed out, however, "[We] are the same people who are going to take care of you if you wind up in the hospital…and if I'm sleeping in my car, I'm not functioning my best."[23]

Doctors and nurses are trained to maintain safety precautions for themselves and others, and they take extraordinary measures to keep others safe, both inside hospital walls and beyond. In fact, as Dr. Atul Gawande observed, "In the face of enormous risk, American hospitals have learned how to avoid becoming sites of spread."[24] Out of a hospital staff of 75,000 in his Boston health system, there were so few workplace transmissions in the spring of 2020 that he argued hospitals should be providing a model of how to successfully reopen other sectors of the economy. But this did not stop a Florida judge from ordering temporary full-time custody of a four-year-old to the ex-husband of an emergency medicine doctor, despite the doctor's argument that she was doing her utmost to keep her daughter safe and that, if she were still married, no one would have removed her child from her home.[25]

We clearly do not want health-care workers sleeping in cars or losing their children, but people can act counterproductively from both an individual and collective point of view when in the grip of strong emotions. It is hard not to see these examples as anything but a product of an atmosphere of fear generated by an epidemic disease.

Yet fearful people can also play a positive role if their fear paralyzes them to the point that they self-isolate indoors or put on a mask or take other actions helpful to themselves and the broader society. Such a response will, in general, reduce the impact of the epidemic by reducing the number of people in circulation. Frightened people are not exactly eager to mingle at the neighborhood bar. However, this benefit can be fickle; a premature "recovery" from fear may result in reintroducing previously fearful individuals into the population at an inopportune time—such as when an outbreak has not been sufficiently controlled—thus contributing to subsequent waves of the epidemic. To further complicate matters, if the fear results in individuals fleeing—which, as we have seen, is a common response to outbreaks—then this can add to the overall intensity of the epidemic, providing new seeds for outbreaks in previously unaffected areas. And fear that results in stigmatization or scapegoating or that increases anxiety to the point where people cannot take in useful information can also worsen the impact of the epidemic.

An extraordinary example of the harm caused by an epidemic of fear occurred in India in 1994 with an outbreak of bubonic plague. Bubonic plague has been mostly conquered in India in the modern era, although it formed part of a deadly "trinity" of infections, along with cholera and smallpox, for much of India's history. Starting in early August 1994, after a typical sequence of increasing fleas and the observation of rat falls, a few cases of bubonic plague were diagnosed in the village of Mamla in the Beed district of the

Maharashtra State.[26] On September 14, the government of India publicly announced that Beed was affected by plague.

Within hours of this announcement, there was widespread panic. And on September 23, 1994, reports of pneumonic plague, the even deadlier variant, suddenly appeared in the industrial town of Surat in neighboring Gujarat State. Between September 21 and October 20, 1,027 patients with serious pneumonia were admitted to hospitals in that city, and 146 of the cases were for some reason presumed to be pneumonic plague.[27] Investigators were skeptical, however, because no cases of the bubonic plague or of outbreaks in the rat population had been noted in Gujarat, both of which typically precede pneumonic plague. There were quite a few other epidemiological features that also made the diagnosis suspect. For example, almost no community transmission was documented. Individuals would become sick, but their family members would not. The cases were not clustered geographically within the city; they were spread out. A further curious observation in the Surat outbreak was that most of the initial cases were in young men in the diamond-cutting industry.[28] And primary pneumonic plague is extremely rare (it is typically seen in people handling infected animals).

Despite reasons for skepticism, the panicked response was extraordinary. Massive spraying of DDT and other insecticides was implemented. All restaurants and food vendors as well as all public gathering places were shut down. Almost all industrial production came to a halt for more than a month. The tourist traffic in India declined by nearly half. Extremely large numbers of people across India—some far away from Surat—self-medicated with tetracycline, often in haphazard and dangerous ways. In faraway cities like Bombay and New Delhi, people started wearing masks. As many as a million people—a quarter of the Surat population—fled in crowded railway cars.

It turned out that there had not actually been a material outbreak of disease. By some accounts, no one was infected with the plague at all.[29] Years later, there would be confirmation, based on laboratory and genetic analysis, of just eighteen cases of plague in the city, although they did not appear connected to each other.[30] The response, however, showed that fear of disease has its own epidemiology, its own spreading dynamics.

In the most fascinating epidemics of fear, referred to as *mass psychogenic illness* or *mass sociogenic illness*, otherwise healthy people fall ill in a psychological epidemic. These terms are nowadays preferred to the erstwhile *epidemic hysteria*.[31] In such outbreaks, people can develop physical symptoms that have no physiological basis, driven by anxiety and fear. In the "pure anxiety" type, people report a variety of symptoms, including abdominal pain, headache, fainting, dizziness, shortness of breath, nausea, and so on. In the "motor" type, people may engage in hysterical dancing or manifest pseudoseizures.

The seventeenth-century Salem witch trials were triggered by a bout of mass psychogenic illness when a group of Puritan girls fell ill with "fits" and laid the blame for their apparent possession on a number of local women. Historical records of such phenomena date back to at least 1374, when, in close succession to the Black Death, "dancing manias" broke out, initially in Aachen, Germany. These consisted of people who,

> united by one common delusion, exhibited to the public both in the streets and in the churches the following strange spectacle. They formed circles hand in hand, and appeared to have lost all control over their senses, continued dancing, regardless of the bystanders, for hours together, in wild delirium, until at length they fell to the ground in a state of exhaustion. They then complained of extreme oppression, and groaned as if in the agonies of death.[32]

In those days, symptoms were often blamed on demons and witchcraft, but in modern times, the trigger is usually identified as some kind of environmental contamination. Usually, such outbreaks occur in schools or occupational settings (perhaps this explains the Surat diamond cutters) where people are in close contact.

For instance, an outbreak occurred in 1998 at the Warren County High School in McMinnville, Tennessee. The school had 1,825 students and 140 staff members. One day, a teacher believed she smelled gasoline, and she complained of headache, shortness of breath, dizziness, and nausea. Some of her students soon developed similar symptoms. As the classroom was being evacuated, other students, witnessing the evacuation, began to report the same complaints. A school-wide fire alarm was activated, and the school was emptied. The original teacher and several students were transported by ambulance to a nearby hospital, in full view of other students and teachers who were outside because of the alarm. In the end, one hundred people went to the hospital that day, and thirty-eight were admitted. The school was closed.[33]

Extensive examination of possible environmental causes for the illness was conducted by the CDC. No physical sources were identified despite much testing and evaluation. The investigators concluded that psychogenic factors were responsible. They found that the illness was associated with directly observing another ill person during the outbreak in a contagious process.[34] Similar examples occur roughly every two years or so somewhere in the United States.

Other sorts of emotional reactions may foster other normative contagions. For instance, when people flouted rules of mask-wearing or physical distancing in parts of the United States in May 2020, they reinforced one another's perceptions of safety and control in a way that was counterproductive to their own interests and to those of our society.[35] The large parties I saw in photographs taken in Missouri, Michigan, and Florida in June 2020 reminded me of a miniature version of the dancing manias seen long ago. As noted above, in the case of the possible parallel epidemics of a virus and of fear, there can be both biological and social contagions. People generally copy the visible behaviors of those around them, so there can be tipping points in both directions. As more and more people start to wear masks and obey rules of physical distancing, more people follow suit. Conversely, as more and more people ignore these practices, fewer people take them seriously. As psychologist Matthew Lieberman put it, "Our brains are built to ensure that we will come to hold the beliefs and values of those around us."[36]

»———→

Truth is another casualty of plague. Some of the most damaging and self-injurious responses to an epidemic are denial and lies. Deadly epidemics always drag along these sidekicks. Unfortunately, modern media technologies provide a bonanza of disinformation, the sort of thing that charlatans of earlier epochs could only dream of.

In the case of coronavirus, the pandemic began with one particularly big lie that originated in China right at the start and lasted through the middle of January: the suppression of information about what was happening in Wuhan. This is part of the reason that, for many millions of Chinese, Dr. Li became a symbol of the desire for free and honest expression. Before he died of COVID-19, Li himself told a Chinese magazine from his hospital bed, "I think there should be more than one voice in a healthy society, and I don't approve of using public power for excessive interference."[37]

In March, something similar began to happen in hospitals around the United States. In Bellingham, Washington, not far from Seattle, during the peak of the outbreak, Dr. Ming Lin was terminated after seventeen years of working at PeaceHealth St. Joseph Medical Center because he posted pleas for personal protective equipment on Facebook and decried the fact that the hospital was not serious enough about protecting patients and health-care workers. Dr. Lin (who had previously worked in New York City during the 9/11 attacks) expressed his wholly appropriate concern about the hospital refusing to screen all patients for the virus outside of the facility before bringing them inside the crowded emergency department, which might contribute to spread of the infection.[38] When I first heard about this case, I marveled at the chutzpah and absurdity of firing a doctor for speaking out in the middle of an epidemic when he was actually needed most.

One orthopedic surgeon working in an urban COVID-19 hot spot in the Northeast reported, "We get a daily warning about being very prudent about posts on personal accounts. They've talked about this with respect to various issues: case numbers, case severity, testing availability, and PPEs."[39] An attending physician at a hospital in Indiana posted a plea on social media for N95 masks. Administrators warned him not to do it again, because it would make the hospital seem incompetent.[40] In Chicago, Lauri Mazurkiewicz, a nurse at Northwestern Memorial Hospital, was fired after sending an e-mail to her colleagues expressing the desire to wear more PPE on duty. One of her concerns was that she might bring the virus back to her seventy-five-year-old father, who suffered from respiratory disease.[41] There were far too many such incidents in the United States in the spring of 2020.

Hospitals around the country issued ham-fisted edicts trying to stop personnel from speaking out. At the New York University Langone Health System in New York City, the executive vice president for communications and marketing informed the faculty and staff that all media inquiries had to be directed to her office. She continued, "Anyone who does not adhere to this policy, or who speaks or disseminates information to the media without explicit permission of the Office of Communications and Marketing, will be subject to disciplinary action, including termination."[42] To put it mildly, the professors of medicine who worked in this hospital did not appreciate this note, as some of them informed me.

Over the past thirty years, there has been a steady shift from seeing doctors as independent professionals to seeing them as mere employees of large organizations that are often led by people who have little appreciation for clinical care. The pandemic brought this tension into sharp relief. A family doctor in Massachusetts observed, "There's been a loss of autonomy and a denigration going on for a couple of decades now," and the coronavirus epidemic is "causing it to erupt."[43] Around the country, administrators who should have been finding ways to improve the efficiency of their hospitals or obtain the equipment that physicians and nurses needed to care for patients were instead trying to manage the epidemic by censoring or suppressing bad news.

The muzzling of doctors also took place at the highest levels of our government. President Trump sidelined and nearly fired Dr. Nancy Messonnier, a CDC official who was the director of the National Center for Immunization and Respiratory Diseases, because during a press conference on February 25, 2020, she honestly noted that the CDC was preparing for a pandemic. She said, "It's not a question of if this will happen, but when this will happen, and how many people in this country will have severe illnesses." The president did not like the fact that her statement resulted in a slight decline in the stock market—as if keeping quiet about the impending pandemic would prevent either the disease or the economic losses it would inevitably cause. In fact, one of the great travesties of the COVID-19 pandemic in the United States has been the undermining and muzzling of the widely respected CDC.[44] On that same day, Health and Human Services Secretary Alex Azar absurdly claimed the virus was "contained" in the United States.[45] But we cannot beat the virus with silence or with lies. Only truth and a megaphone help in such fights.

In another example, on May 22, 2020, the CDC posted updated recommendations urging religious communities to "consider suspending or at least decreasing the use of choir/musical ensembles and congregant singing, chanting, or reciting during services or other programming, if appropriate within the faith tradition." Since

SARS-2 was known to spread by these means, such guidance was factual and potentially lifesaving.

The White House demanded this information be removed. As in China earlier, the flow of epidemiological information needed political approval.[46] In a specious misdirection, the White House insisted on adding language that the guidance "is not intended to infringe on rights protected by the first amendment." Yet, like any building where large groups assembled, churches could be venues for the spread of disease, and several outbreaks at churches had already proven deadly.[47] In any case, as the Supreme Court noted at the same time (in a different case involving California churches), the key issue was whether secular and religious organizations were being treated differently by the state during a public health emergency.[48] Providing information about the risk posed by singing or large gatherings with regard to contagious respiratory disease, even if the specific example is churches, could not be unconstitutional. These types of restrictions on the flow of information are often justified on the grounds that they reduce panic or foster consistency, but they so frequently result in harmful effects that it's hard to give those explanations any credence.

None of this was new, of course. Federal scientists have been silenced for politically inconvenient expression of basic scientific facts during epidemics before. During the HIV epidemic in 1987, surgeon general Everett Koop, who had been chosen by President Reagan for his opposition to abortion, surprised many in conservative and liberal circles alike by urging explicit language to describe HIV risks, including discussions of anal sex and condom use. As Koop argued:

> People criticize me because they see homosexuals or drug abusers or promiscuous people as being sort of beyond the pale and AIDS is no more than they deserve. My answer to that is that I am the surgeon general of the heterosexuals and the

homosexuals, of the young and the old, of the moral or the immoral, the married and the unmarried. I don't have the luxury of deciding which side I want to be on. So I can tell you how to keep yourself alive no matter what you are. That's my job.[49]

But Koop was overruled by the secretary of education, William J. Bennett, who proceeded to inject ideological considerations into this public health topic. Bennett insisted that AIDS education efforts be "values based" and not involve "morally ambiguous materials."[50]

As a consequence of Bennett's intervention, the government materials became epidemiologically vague and even misleading. Yet, since there were no treatments or vaccinations to prevent the disease back then, public education and public health interventions were indeed the only tools we had available. Forms of transmission reduction (for example, condom use) or contact reduction (for example, reducing the number of sexual partners) were required. President Reagan was not able to bring himself to mention the epidemic until six years into his administration—again, as if not mentioning an epidemic can somehow make it disappear.

Similar dynamics, psychologically and politically, have played out with SARS-2. It was not just average Americans who engaged in wishful thinking that the virus was "no worse than the flu" or that it would simply disappear on its own. Even the president of the United States, with the best epidemiologists and intelligence apparatus in the world available to him, engaged in deep and public denial. "Nobody would have ever thought a thing like this could have happened," the president insisted as he falsely accused the former administration of not making pandemic preparation plans while he himself had disbanded the epidemic-response team in 2018.[51]

We know from subsequent leaks that the president was indeed presented with information about the seriousness of the virus and its pandemic potential beginning at least in early January 2020.[52]

And yet, as documented by the *Washington Post*, he repeatedly stated that "it would go away." On February 10, when there were 12 known cases, he said that he thought the virus would "go away" by April, "with the heat." On February 25, when there were 53 known cases, he said, "I think that's a problem that's going to go away." On February 27, when there were 60 cases, he said, famously, "We have done an incredible job. We're going to continue. It's going to disappear. One day—it's like a miracle—it will disappear." On March 6, when there were 278 cases and 14 deaths, again he said, "It'll go away." On March 10, when there were 959 cases and 28 deaths, he said, "We're prepared, and we're doing a great job with it. And it will go away. Just stay calm. It will go away." On March 12, with 1,663 cases and 40 deaths recorded, he said, "It's going to go away." On March 30, with 161,807 cases and 2,978 deaths, he was still saying, "It will go away. You know it—you know it is going away, and it will go away. And we're going to have a great victory." On April 3, with 275,586 cases and 7,087 deaths, he again said, "It is going to go away." He continued, repeating himself: "It is going away....I said it's going away, and it is going away."[53] In remarks on June 23, when the United States had 126,060 deaths and roughly 2.5 million cases, he said, "We did so well before the plague, and we're doing so well after the plague. It's going away."[54] Such statements continued as both the cases and the deaths kept rising. Neither the virus nor Trump's statements went away.

In addition to denying the threat of the virus, Trump promoted misinformation about other essential responses to the crisis. On March 2, 2020, Trump claimed that a vaccine would be ready "over the next few months," although, in reality, the timeline would be considerably longer.[55] On March 6, Trump falsely insisted that "anyone who wants a test can get a test" despite widespread frustration among doctors and patients regarding the manifestly insufficient supply of such tests.[56] He repeated such statements often, noting the "beautiful" and abundant testing and boasting about American superiority while failing to compare testing rates in other countries on a per capita basis.[57]

What accounts for these departures from reality, and why do they fail to elicit outrage or even argument in so many people? In part, the president was playing to the desires of large numbers of Americans to simply wish the calamity away. But in addition, the denial of the reality of the COVID-19 pandemic is the latest manifestation of a fissure between science and politics that has been widening for decades in the United States, especially when scientific consensus has implications that are inconvenient to policymakers, as we will discuss in chapter 7.[58] But the timeline of a pandemic is much more rapid than that of other public controversies about science (like climate change), meaning that the reality of the situation and the consequences of political acts that are divorced from scientific reality present themselves much more immediately for all to see.[59]

In location after location around the country and around the world, people seemed shocked that the virus could devastate their population and flood their hospitals with patients, even though the virus had done this exact thing in other places just a few weeks before. Houston, whose hospitals were inundated in June, seemed to have expected a different fate than New York City, whose hospitals were deluged in March, just as New York City had thought it was different from Wuhan, which was overrun in February. This denial was itself a very dangerous aspect of the COVID-19 pandemic.

But denial is an old ally of pathogens. Here is an observation by a physician during the great plague of Marseille in 1720, which was the last major outbreak of bubonic plague in Western Europe:

Already the public, prone to delude themselves, and easy to believe what they wish to be true, attributed the malady of these persons to anything rather than to the plague, and began even to joke upon their own alarms. But the subtle destroyer, mocking alike the precautions of the wise and the jokes of the incredulous was secretly insinuating itself far and wide.[60]

All this is very human. And just as we saw that fear might offer some upside to pandemic response, I suppose denial might also offer some advantages—for instance, by allowing people to go about their lives despite the threat.[61]

A similar sort of wishful thinking emerged during the profound protests after the brutal murder of George Floyd in Minneapolis on May 25, 2020. The forty-six-year-old black man was already hand-cuffed and on the ground when a police officer knelt on his neck until he died. This, coming on the heels of many similar cases, led to massive protests erupting around the nation.[62] In large part, these were due to pent-up anger about racial inequality, but the high levels of unemployment, the long periods people had been cooped up at home, the emotionally demanding experiences (including deaths of loved ones) that many people had suffered, and the disillusionment with the president all surely facilitated the protests. The protesters seemed to represent a broad swath of Americans across ethnic and racial lines, judging from the video footage.

But many experts who had previously opined that schools had to close and that even small funerals were dangerous now seemed willing to overlook the risks of mass gatherings for a just cause they supported politically.[63] To be fair, most of the protesters wore masks and the protests were outdoors, which is much less risky. But the public health messaging was inconsistent. At roughly the same time, despite rapidly rising case counts, various governors reopened their states. In Tennessee in early June, the governor loosened restrictions, allowing fairs and parades. "Thanks to the continued hard work of Tennesseans and business owners operating responsibly, we're able to further reopen our state's economy," Governor Bill Lee intoned optimistically.[64] But the virus does not care whether our reason for assembling is a protest, a funeral, or a parade.

There are many people, and not just those in positions of power, who have persuaded themselves that reality is "socially constructed"—that there is no objective reality, there is only that which we define using human faculties. This is a deeply interesting philosophical idea. But this has also led to the belief that we can change reality by manipulating words or images: if we call something by a different name, it actually is different. This is only partly true, in a narrow sense. The virus is real and it does not care how we see it or what we say about it. Throughout the pandemic—from the beginning, when political figures across the world and across the United States thought it could be denied or wished away by positive statements, to right- and left-wing protests in the spring and summer of 2020 that seemed to reflect the belief that the virus was gone or would not affect people too much as long as their cause was just—this sort of thinking prevailed.

But reality matters. One analysis estimates that, if control measures such as physical distancing had been implemented just one week earlier in the United States, the nation would have seen 61.6 percent fewer reported infections and 55.0 percent fewer reported deaths by May 3, 2020.[65] Although responsibility for the pandemic cannot be placed solely on the shoulders of any single person, group, or institution—and the United States was not the only country to downplay early-warning signs of the virus—one of the great tragedies of the COVID-19 pandemic is that some of the worst outcomes could have been avoided had our predicament been acknowledged and acted upon at the appropriate time.

»———→

Misinformation has been everywhere during the COVID-19 pandemic. A survey of 8,914 people, released on March 18, 2020, revealed that 29 percent of Americans believed that SARS-2 was developed in a Chinese laboratory in Wuhan.[66] The Wuhan Institute of Virology, a high-security biosafety level 4 (BSL-4) laboratory for conducting research on the deadliest pathogens, is, in fact, located there. This lab was initially established in the 1950s and it was relaunched with fanfare as a BSL-4 lab in 2017, with the many precautions typical of such facilities. And it is true that, at the time of the relaunching, "some scientists outside China worried about pathogens escaping, and the addition of a biological dimension to geopolitical tensions between China and other nations."[67]

But the conspiracy theory that SARS-2 was *deliberately* genetically engineered appeared almost immediately after the virus did, in January 2020.[68] By late February, some commentators were supporting the theory based in part on public statements made by the Chinese government about improving safety precautions at microbiology labs.[69] The argument was this: Why would the Chinese authorities want to increase safety if SARS-2 had not escaped from a lab? But why, I wondered, would they announce such changes if SARS-2 had actually escaped? In February, Senator Tom Cotton of Arkansas publicly spread this theory about the origins of the virus.[70] President Trump did it well into May 2020, despite the fact that the American intelligence agencies themselves, along with knowledgeable geneticists, had concluded that the virus was not genetically engineered.[71]

There is indeed much that argues against this conspiracy theory. Since SARS-2 is especially fatal for the elderly and the chronically ill, it would not make a particularly effective bioweapon; to cause the most havoc, something that targeted young and healthy people would be better. But most persuasively, detailed genetic analyses of the pathogen show a pattern of descent from prior bat coronaviruses and of randomly occurring genetic mutations that are not compatible with deliberate genetic engineering.[72]

However, it is very difficult to totally exclude the possibility of an *accidental* release of a *naturally* occurring pathogen that was collected from bats and then taken to the lab for study. But since we know of many examples of zoonotic diseases leaping to humans in the normal course of events, including SARS-1, the balance of probabilities, at least to me and most experts, still leans heavily toward a chance move of a naturally occurring pathogen.

Another early dubious theory was that the virus was somehow spread by 5G cell phone towers. This led to the burning and destruction of many such towers in the United Kingdom. After initial discussion of this topic among fringe online networks in the United States, certain American celebrities amplified it, including actor Woody Harrelson; he directed his two million followers on Instagram to a video about this theory (though he subsequently deleted the post). British musician M.I.A. also shared her observations about the conspiracy. Later in April, Ineitha Lynnette Hardaway and Herneitha Rochelle Hardaway Richardson, the social media personalities and former Fox Nation hosts known as Diamond and Silk, also warned their millions of followers of a link between 5G and coronavirus.[73]

Not to be outdone, the Chinese government propaganda machine swung into action. A U.S. Army reservist and mother of two, Maatje Benassi, had gone to Wuhan to compete in a cycling competition in October 2019 as part of the Military World Games (a kind of Olympics for the armed forces). For unknown reasons, an American conspiracy theorist decided to connect the idea that somehow Benassi had brought the coronavirus to Wuhan to different rumors that the virus was an American plot and that the virus had come from the United States to China rather than the other way around. He proclaimed this to his one hundred thousand followers on YouTube. Media outlets in China affiliated with the Chinese Communist Party were more than happy to promote this newly embellished theory in their own country, generating enormous attention on many Chinese social media sites. The real-life consequences of this lie have been awful for Benassi and her family, who became the target of conspiracy theorists. U.S. Defense Secretary Mike Esper said it was "completely ridiculous and irresponsible" for the Chinese government to promote this claim, which was ironic, considering President Trump's own bogus statements about various aspects of the pandemic.[74]

All these types of misinformation multiplied during the pandemic. Indeed, conspiracy theories can resemble pathogens in their capacity to mutate, evolving to fit the landscape in which they survive and spread. And, like the virus, this misinformation can harm us.

Misinformation also abounded with respect to supposed cures for the virus. As if on cue, right from the beginning of the COVID-19 pandemic, hucksters of all varieties appeared to offer nostrums that did not, and could not, work.[75] Many took advantage of powerful modern media tools that allowed them to reach millions of people. An organization in Florida calling itself the Genesis II Church of Health and Healing was ordered by a judge to stop selling its "sacramental dosing for coronavirus," which consisted of a powerful bleaching agent ordinarily used as an industrial chemical.[76] The organization described its purported elixir as a "miracle mineral solution."

Media figure Alex Jones, the Sandy Hook massacre denier who appears to endlessly search for ways to profit from other people's sorrow, also got in on the act. His InfoWars operation, which consists of a blog, audio and video feeds, and an online store, begin to market products that contained colloidal silver to treat coronavirus. The substance has no known antiviral effect, though taking too much of it can turn your skin blue. "This stuff kills the whole SARS-corona family at point-blank range," he said in a livestream on March 10, 2020.[77] An Oklahoma company called N-Ergetics opined that "colloidal silver is still the only known anti-viral supplement to kill all seven of these human coronaviruses. . . . This Chinese Wuhan Flu Pneumonia has a non-traditional remedy that has successfully killed coronaviruses from the flu virus to pandemic diseases, in vitro, for over 100 years."[78] I have to say that accurately identifying the correct number of coronavirus species known to infect humans was a nice touch in this pitch. Even Jim Bakker, another fraudster (and televangelist-cum-convict from the 1980s) got in on the act, selling a similar silver-based product.

Many herbal-remedy companies, with names like Herbal Amy and Quinessence Aromatherapy, also sprang into action to exploit the opportunity. They, like the purveyors of bleach and silver, got warning letters from the FDA or FTC. California-based GuruNanda used online social media and its website to make its pitch: "Just what is this new Coronavirus, and how can you prevent and/or treat it?" it asked, before claiming that its frankincense product was a way to "decrease your chances of becoming infected."[79] In Los Angeles, an animal rights activist was enjoined by the FTC from illegally selling an herbal supplement that he said could treat coronavirus. That product, sold under the brand name Whole Leaf Organics, allegedly contained "16 hand-selected herbal extract strains" that both prevented and treated COVID-19.[80] On Instagram, "beauty influencers" like Michelle Phan promoted "essential oils" as a treatment for the disease. On the same medium, "wellness guru" Amanda Chantal Bacon suggested the use of "plant-based alchemy."[81] A pastor in California suggested oil of oregano. There were many others.

Sensing the deluge of phony products, the FDA tried to stem the tide by releasing an open letter on March 6, 2020, warning the public, "There currently are no vaccines, pills, potions, lotions, lozenges, or other prescription or over-the-counter products available to treat or cure coronavirus disease."[82] But a lie is halfway around the world while the truth is still putting on its shoes. And anyway, Donald Trump once again undermined the government's message. During an astonishing news conference on April 23, 2020, he speculated that bleach administered topically or even internally or via injection could help and that blasting people with ultraviolet light might cure the disease.[83] The manufacturers of Clorox and Lysol had to issue statements pleading with Americans not to inject or ingest their products, since that could kill them. There was a certain irony in this turn of events, given Lysol's old history of promoting itself as a "feminine" disinfectant (it was used, ineffectively, as a contraceptive).[84]

Even worse, Trump used his bully pulpit in a more sustained fashion to repeat claims that hydroxychloroquine, an antimalarial agent, could cure or prevent COVID-19. While a series of un-controlled studies had raised the possibility that this drug might help, many physicians, myself included, were deeply alarmed by this recommendation, not only because the drug can have cardiac toxicity, but also because there was no good evidence for its utility. Trump fervently supported the drug, despite repeated warnings by the scientific community that there was no conclusive evidence regarding its effectiveness. Like the wearing of masks, opinions on the efficacy of hydroxychloroquine became a political litmus test.

Chloroquine and its chemical cousin hydroxychloroquine have long been used against malaria and also to treat arthritis and lupus. Some papers by Chinese scientists appeared online in February 2020 suggesting the possible utility of these drugs in the care of COVID-19 patients.[85] But the connection between hydroxychloroquine and COVID-19 really got going on March 11, 2020, when a small group of investors and a philosopher discussed it as a potential cure for the virus in a Twitter thread. They released a paper on Google Docs (since deleted) suggesting that the drug could both treat and prevent COVID-19, falsely affiliating multiple universities and the National Academy of Sciences with the paper. Soon, they were invited on two different Fox News shows.[86]

Hours after their Fox News appearance on March 21, 2020, Trump tweeted that the combination of hydroxychloroquine and azithromycin (an antibiotic) might be "one of the biggest game changers in the history of medicine."[87] He announced from the

White House podium the following day that the "promising drug [hydroxychloroquine]" had been approved for immediate use, which was not actually true.[88] At the same late-March briefing, Trump's own coronavirus task force members, including Dr. Fauci, were hesitant to promote the drug. The FDA did approve it for emergency use on March 29, stipulating that it should be given only to hospitalized COVID-19 patients when participation in a clinical trial was not an option.

Unfortunately, the president's rhetoric had more impact, initially, than scientific publications. One large and representative national survey conducted at the end of April 2020 found that 40 percent of respondents indicated that they obtained information from his briefings in the preceding twenty-four hours—which compared favorably to outlets like CNN (37 percent), Fox News (37 percent), and MSNBC (19 percent).[89] Between March 23 and March 25 alone, Fox News promoted hydroxychloroquine 146 times.[90] On March 22, a couple in Arizona drank fish-tank cleaner after seeing that it contained chloroquine phosphate (which was not the pharmaceutical form of chloroquine), leading to the man's death. His wife recalled thinking, "Hey, isn't that the stuff they're talking about on TV?"[91]

From April to May, the scientific community and the FDA released numerous reports warning the public about hydroxychloroquine's harmful, even fatal, effects. On April 24, the FDA issued a safety warning, since the drug can result in dangerous heart rhythms. Medical journals also published papers warning about the same negative effects. These cardiac side effects can be particularly deadly for severely afflicted COVID-19 patients, who often are older or have heart disease.[92] Small initial studies using the drug in humans began to emerge with inconclusive outcomes or with an indication that the drugs could be harmful due to cardiac toxicity.[93] Several larger observational studies in May and June found no beneficial association between using the drug and a lowered risk of intubation or death, whether the drugs were used early or late in the course of the disease.[94] Subsequently, in early June 2020, a high-quality randomized trial involving 821 patients showed that the drug did not prevent COVID-19; in July, two other randomized trials, one with 4,716 patients and one with 667 patients, showed it did not help patients who were already sick with the disease either.[95] Nevertheless, on May 18, 2020, hydroxychloroquine made national headlines again when the president championed it in the strongest way possible, announcing that he was taking the drug preventatively. It's unclear why Trump did this, but several members of his White House team had tested positive for the virus in the preceding week.

The actions of snake-oil salesmen preying on our emotions at every level of society are not without consequences. They waste personal and collective resources. They lead to a broad under-mining of science and rationality when we need them most. And they lead to a false sense of security, encouraging risky behavior and therefore the spread of the virus.

Superstitions and desperate desires for preventatives and cures have always arisen during times of plague. For example, during the plague of Justinian, John of Ephesus noted:

> A rumor from somebody spread among those who had survived, that if they threw pitchers from the windows of their upper stories on to the streets and they burst below, death would flee from the city. When foolish women, [out of their] minds, succumbed to this folly in one neighborhood and threw pitch-ers out.... The rumor spread from this quarter to another, and over the whole city, and everybody succumbed to this foolish-ness, so that for three days people could not show themselves on the streets since those who had escaped death (in the plague) were assiduously (occupied), alone or in groups, in their houses with chasing away death by breaking pitchers.[96]

We can see many of the ideas circulating regarding COVID-19 in this light. One survey of a representative sample of Americans conducted in late March 2020 reported that, in the preceding week alone, 34 percent of Americans had seen someone on social media being told that they had shared misinformation regarding COVID-19, and 23 percent of respondents reported that they themselves had felt compelled to correct someone online. An even higher percentage, 68 percent, endorsed the idea that people should respond when they see others sharing falsehoods.[97]

An extensive analysis of Twitter data from January 16, 2020, to March 15, 2020, revealed that the sharing of sites full of misinformation was almost as common as the sharing of links to credible sites, like the CDC.[98] Online discussions of five specific myths increased noticeably in early March, including the efficacy of home remedies like "eating garlic, drinking ginger tea, drinking silver, or sipping water," and conspiracy theories about the virus being an engineered bioweapon designed by "the Chinese government, the US government, the liberal media, [or] Bill Gates."

Another analysis of two hundred million tweets about the pandemic collected from January through May 2020 found that 62 percent of the top one thousand retweeters were bots. There were more than a hundred different types of inaccurate information, including about quack cures, but the bots really weighed in on online conversations regarding ending stay-at-home orders and "reopening America." The researchers concluded that the bot activity seemed orchestrated, possibly by agents of the Russian or Chinese governments, and many of the tweets promoted by the bots referred to conspiracy theories such as the ones about coronavirus being linked to 5G cell phone towers.[99] There is increasing evidence that the Chinese and Russian governments have a long-standing and ongoing program in place to undermine Americans' confidence in science and scientists.[100] And formal analyses of a network of interactions among one hundred million people on Facebook show that users spreading falsehoods—for instance, about the risks of vaccination—are often better able to occupy positions of structural power in the network, dominating the conversation and increasingly winning out against true information.[101]

»———————→

The circulation of truth and lies during the coronavirus pandemic was also heightened inadvertently by the increasing use by scientists of a new tool for communication, known as preprint servers, that in some ways contributed to the confusion. For over sixty years, the method employed to get a journal article published has been the same: researchers submit their papers to journals, the editors of those journals circulate the papers anonymously to the researchers' peers, and those scientists attempt to find flaws in the papers or make suggestions for improvement.[102] The paper's authors then respond to these criticisms—a process that usually involves multiple rounds of review and substantial delays—and the paper is published (or, more often, rejected). Peer-review is not a guarantee of accuracy, but, like the jury system for criminal justice, it is—as some of my colleagues joke—the "worst system we have except for any other."

But beginning in the early 1990s, some scientists started releasing their papers prior to the peer-review process on preprint servers; this allows their peers a broad opportunity to comment on papers before they are sent out for formal review. There are many such servers and systems, among them arXiv, BioRxiv, medRxiv, socARxiv, psyARxiv, SSRN, and NBER. Since the systems are generally open to everyone, not just scientists, journalists and ordinary citizens have increasingly been accessing this information. On the one hand, prepublication access increases the speed of dissemination of useful information and facilitates the widest possible effort to weed out inaccuracies. On the other hand, the information may be false or incomplete or, at a minimum, not carefully vetted (even robust studies are improved significantly by undergoing peer review), and nonexperts usually lack the skills to evaluate its scientific validity. During the coronavirus pandemic, therefore, preprint servers contributed to dissemination of true information but also to what the WHO called an "infodemic" of false information.

Online platforms like Twitter and Medium also fed this phenomenon. Many scientists weeded out bad ideas and had productive conversations online. But we also saw the wide dissemination of falsehoods and crackpot ideas that could adversely affect both health policy and clinical care. For example, early on, many dramatic estimates of the R_0 for SARS-2 caught the public's attention. Although most likely the true value of this parameter lies in the neighborhood of 3.0, significantly higher estimates, as high as 7.0, appeared in late January. They were widely circulated and caused tremendous—and, as it turned out, false—alarm.[103] The truth of this pathogen was bad enough without exaggeration of preliminary information making things worse. Another glaring example was an erroneous preprint that claimed that the coronavirus—impossibly—had insertions of genetic material from HIV. By the time scientists had debunked this study, prompting its wholesale withdrawal, it had entered into wide circulation on Twitter and even in the mainstream media.

»——————➤

The epidemics of emotions and of misinformation intersect in worrisome ways with the underlying epidemic of the pathogen itself. And this in turn highlights once again the crucial role of public education during the time of a pandemic. Even though the science might be tenuous and findings can change given new observations, officials can and should be both sensitive and honest. It's of course acceptable for officials to change their minds and update or even reverse prior advice. But we can reduce cynicism and strengthen collective will if the reasons for such changes are offered, and the strength of the evidence—and the degree of uncertainty—is communicated whenever information is shared.

Given what the virus did to us and what we did to ourselves in response, we had many reasons for despair. And in addition to these twin biological and social shocks, we also faced the problem of uncertainty about the nature of the challenges before us. Our adversity elicited psychological responses that are common to our species: we felt sadness and grief; responded with anxiety, fear, and anger; and tried to hide the truth from one another and even from ourselves.

These emotional responses and behaviors can themselves rightly be seen as foundational parts of epidemics. We might even say that the definition of an epidemic disease should include the fact that it can have outsize psychological effects. And the public health response to epidemics must be driven not only by the medical, social, and economic aspects of the threat, but also by the psychological dimensions. To cope with a plague in the twenty-first century, we responded not only with familiar interventions, but also with familiar feelings.

5.

Us and Them

Whatever had to do with mutual assistance or pity had vanished from their minds; each one had thoughts only for himself. He who was sick was looked upon as a common foe, and if it happened that any one was unfortunate enough to fall down on the street, exhausted by the first fever-paroxysm of the plague, there was no door that opened to him, but with lance-pricks and the casting of stones they forced him to drag himself out of the way of those who were still healthy.

—Jens Peter Jacobsen, *The Plague in Bergamo*
(1882)

On Valentine's Day in 1349, the municipal authorities of Strasbourg decided that the city's roughly two thousand Jewish residents were responsible for the plague—they had poisoned the water or deliberately bred spiders in pots or...the reason did not matter, just that it was the Jews' fault.[1] Faced with the choice of converting to Christianity or being put to death, about half the city's Jews chose the former. The rest were rounded up, taken to the Jewish cemetery, and buried alive. Strasbourg officials also passed a law banning Jews from entering the city.

The impulse to blame others for causing infections or for being infected is powerful, and the historical record brims with devastating examples. Some Christians were also put to death in similar blood-baths in other cities, as described in the testimony of one witness:

> You should know that all the Jews living in Villeneuve have been burnt by due legal process, and at August three Christians were flayed for their involvement in the poisoning. I was myself present on that occasion. Many Christians have been similarly arrested for this crime in many other places, notably in Évian, Geneva, La Croisette and Hauteville, who at the very last, on the point of death, confirm that they distributed poison given them by the Jews. Some of these Christians have been quartered, others flayed and hanged. Certain commissioners have been appointed by the Count to punish the Jews, and I believe that none remains alive.[2]

Three centuries later, in Milan in 1630, another outbreak of plague needed a scapegoat. The target of the city's wrath was a group of four Spaniards who were accused of deliberately spreading the plague by smearing ointment on the doors of households. They were tortured, and they confessed. Their punishment consisted of having their hands cut off, being broken on the wheel, and finally, for good measure, being burned at the stake. A so-called column of infamy was erected at the site of their execution to deter others from spreading the plague in the city. Looking back, the real infamy was the temptation to blame others, typically outsiders or minorities, for calamity in a time of plague.

These examples are dramatic and seem relics of a faraway and bestial time. But such primitive thinking is ever present, a ready tool when a deadly infectious disease strikes. For example, even though it was common knowledge that Franklin Delano Roosevelt had emerged from a bout of polio with sufficient vigor to serve over three terms as president of the United States (during the Great Depression and the Second World War, no less), attitudes about polio survivors and their families were often ignorant and cruel. Announcements identifying child victims were routinely published in newspapers, ostensibly to protect the public. While some of these notices did inspire neighborly charity, the parents of Mike Pierce, who was five when he fell ill in 1949, were shunned by their Southington, Connecticut, community and "lost all their friends" when a group of petitioners ran the family out of town. In adulthood, Pierce himself had an opportunity to confront the petition's architect, a man he had dreamed of killing. When he finally met the old man, however, he took his hand instead. "I could have crushed it, and I just shook it. I said, 'Pleased to meet you.' When he looked at me, his eyes averted; I knew I won. He just shuffled away."[3]

Some severely disabled polio survivors were even used as props to instill terror in the healthy children who might have otherwise ignored precautions designed for their safety. Judith Willemy of Lowell, Massachusetts, recalls her elementary school being taken to see a child survivor:

> We all lined up, and outside the school was a large truck. We went up the stairway in the front of this truck, and in the middle of the truck was a child laying on a table in a capsule. The only thing showing was her head, and above her head was a mirror. You knew this child had polio, which meant that she couldn't walk, but she was in this large capsule, so how much of her body was left? We are wondering had she done that terrible thing, gone swimming? We just walked past her, gazing upon this scene, almost like a sideshow in a circus.... It just left a real impression of how we weren't taught that this was a

person, how this person must have felt as children were going by. It was a terrible experience, and I'm really sorry that child had to suffer that way.[4]

As we saw in chapter 4, this kind of fearful and dehumanizing thinking can also find expression in the seemingly reasonable desire to close borders, identifying "outsiders" as the source of a problem. In 2020, many people in the United States, including the president, fanned the flames of anti-Asian discrimination. It's one thing to say, accurately, that the virus originated in China, but it's quite another to falsely frame this event as an attack. In early May 2020, Trump said: "This is really the worst attack we've ever had. This is worse than Pearl Harbor. This is worse than the World Trade Center. There's never been an attack like this."[5]

Partly to reduce the risk of such discrimination, the World Health Organization for a number of years has employed the convention of no longer naming pathogens after where they come from.[6] This goes against tradition—pathogens ranging from Rocky Mountain Spotted fever to Lyme disease to West Nile virus to St. Louis encephalitis to Ebola to Middle East respiratory syndrome and many others have all been named for their place of origin or first discovery. Even the Spanish flu bore a place-name, although, as we saw earlier, evidence suggests that it emerged in Kansas, not in Spain. And even among the Chinese themselves, as analyses I did with my Chinese collaborators revealed, it was typical to call the virus by a place-name; in the early days of the pandemic, countless online searches done by Chinese citizens included the term *Wuhan pneumonia* as people searched for information.[7]

Of course, this matter assumed geopolitical significance due to the extraordinary medical and economic toll around the world. Given the devastation in terms of both lives and money lost in China, I hardly think the Chinese would wish the epidemic upon themselves. It's true that many pandemics have originated in China, as we have seen, and this does have some relationship with the agricultural and culinary practices of the region. But China is also large and populous (and densely settled in places), so pandemics may be more likely to emerge from there for these reasons alone. And pandemics have originated on every continent, including our own.

But in the early days of the pandemic in the United States, many people sought to blame the virus on those who were ethnically Chinese or even just looked Asian. There were news reports of discrimination by some people toward Asian-Americans.[8] This resembled the way that many Arab-Americans and Muslim-Americans were ostracized or attacked after 9/11.[9]

And the real threat is another living thing, a virus, itself a part of the natural world with no agenda other than its own continued existence. Unfortunately, that does not stop some politicians and religious leaders from framing epidemics as retribution for both the individual victims and our society at large. During the emerging HIV epidemic in the United States in the 1980s, I remember many such figures pursuing a crusade against people who were homosexual, the population in whom the disease was first identified. Ranging from Senator Jesse Helms, who shut off funding to the pathbreaking Gay Men's Health Collective on the grounds that it would promote "perversion," to Secretary of Education William Bennett, to the former beauty queen and orange-juice shill Anita Bryant, there was no end of public figures who used the disease as a convenient excuse to express hatred toward those they perceived as outsiders.[10] The HIV epidemic was also stratified by socioeconomic factors, not just sexual identity. It disproportionately afflicted African-Americans and the poor. They, too, were blamed for their predicament, as if they had done something to deserve it.

Ideas like this—blaming immigrant groups or feeling indifference to the poor or the elderly—made a reappearance in 2020. Even some ministers took up this soiled banner again. But why should we blame anyone for becoming infected with a mindless germ that does not distinguish among us? If the pathogen does not discriminate, why should we?

I need to be clear that I am not saying that people engaging in unprotected sex or avoiding wearing masks are not, via their behavior, playing a role in their own fate and in the fate of those around them. But our focus should be on the *behaviors*, not on the human beings who get sick or on what group they happen to belong to. This has always been a central tenet of successful public health campaigns.

The early days of the COVID-19 pandemic provided many disheartening examples of the levels to which some will go to draw spurious distinctions and mark arbitrary borders. On April 1, 2020, the cruise ship *Zaandam* was held offshore in Florida because 193 guests and crew had flu-like symptoms and eight had tested positive for COVID-19. In fact, four guests had already died since the ship had left Buenos Aires three weeks earlier, on March 7. The governor of Florida, Ron DeSantis, said he would not allow the ship to make port. According to maritime law, such a ship would have to return to the country where it was flagged (in this case, the Bahamas). However, many of the people on the ship were American citizens who wanted to be allowed into their home country.

Late on April 1, Governor DeSantis said that he would allow the forty-nine residents of Florida to disembark, but not anyone else.[11] There were roughly two hundred fifty other Americans on the ship. Did it really matter that they were not residents of Florida? Of course, there were also hundreds of other citizens of other countries, mostly Canada and Europe, who were desperate to disembark as well. What do artificial lines on a map have to do with the actual clinical and moral issues at hand?

Well before the pandemic had begun, in September 2019, three workers from New Jersey had taken up residence on a small island in Maine to work on a construction site. None of them had any symptoms consistent with coronavirus infection in late March 2020, but nonetheless, rumors began to circulate that these outsiders posed a risk of contagion. One day, the three men noticed that their internet had stopped working. They went out to investigate and found that a tree had been deliberately chopped down to block their road and prevent them from leaving, and it had brought down some wires. While they were outside, a number of armed individuals showed up. The men returned to their home and used a radio to contact the Coast Guard, seeking assistance. They also used a drone to monitor the activity of the armed group. State representative Genevieve McDonald, who represented the island, would later point out, "Now is not the time to develop or encourage an us-versus-them mentality. Targeting people because of their license plates will not serve any of us well."

Occasionally, an "us-versus-them" effort to close a border can make a bit more sense. On March 27, 2020, in Old Crow, a remote village of about 280 people in the far north of the Yukon, a young couple got off a plane. They had come from Quebec to avoid the emerging pandemic. Locals immediately recognized that they were out of place and insisted that they isolate themselves in an apartment above the only grocery store in this tiny hamlet. The community had only one nursing station and a doctor who flew in every two months or so, and they could not risk a coronavirus outbreak. Within forty-eight hours, the local police officer had escorted the couple back onto a plane so they could leave. A local community leader would later report that the couple had told him that they "made contact with the community through a dream."[12] Unfortunately, the leader noted, "dreams are not passports."

$$\gg\longrightarrow$$

In principle, at least, the indiscriminate spread of a ravaging pathogen would seem to decrease divisions, since a common threat might help people better appreciate their shared fate. This optimistic idea, that a plague might heighten the natural equality among human beings, has echoed throughout history, as noted by the ancient historian Procopius writing about the plague of Justinian:

> But for this calamity it is quite impossible either to express in words or to conceive in thought any explanation, except indeed to refer it to God. For it did not come in a part of the world nor upon certain men, nor did it confine itself to any season of the year, so that from such circumstances it might be possible to find subtle explanations of a cause, but it embraced the entire world, and blighted the lives of all men, though differing from one another in the most marked degree, respecting neither sex nor age. For much as men differ...in the case of this disease alone the difference availed naught.[13]

In 2020, some observers argued that the microbe could, by various means, be an equalizer. First, it might prompt people to see that they all have an interest in what happens to those around them, if only for selfish reasons. Those previously unconcerned about the homeless might want to improve their conditions just to reduce the risk that they might provide a reservoir for a pathogen. Second, some observers, such as economist Robert Shiller, speculated that a pandemic might decrease economic inequality over the intermediate term by reducing the wealth of those who had invested in the stock market or by leading to more progressive taxation.[14] Those at the top have more to lose than those at the bottom, so reducing wealth increases equality. Third, and more darkly and ironically, if a pathogen kills the very old and the very young, it leaves behind people who are middle-aged, thus reducing differences in age among survivors. Or if a pathogen preferentially kills those who are already chronically ill, it leaves behind people who are more uniformly in better health.

But in reality, pandemics mostly heighten and highlight inequality.

Affluent people are able to protect their health and livelihood more effectively than others. Remember that wealthy people have been fleeing to their country homes to avoid plague for thousands of years. And epidemic disease is not an equal-opportunity killer. Pathogens almost always affect the weaker members of any group, whether this is defined by a burden of chronic illness, advanced age, or substantial poverty

Only when the disease kills very large fractions of a population in a location—as in the case of the bubonic plague or Ebola or smallpox introduced to American Indian populations—do social distinctions cease to matter. This observation was used in the first wave of the Black Death by Pope Clement VI to push against anti-Semitism. He used impeccable logic:

> Recently, however, it has been brought to our attention by public fame—or, more accurately, infamy—that numerous Christians are blaming the plague with which God, provoked by their sins, has afflicted the Christian people, on poisonings carried out by the Jews at the instigation of the devil, and that out of their own hot-headedness they have impiously slain many Jews, making no exception for age or sex; and that Jews have been falsely accused of such outrageous behavior so that they can be legitimately put on trial before appropriate judges—which has done nothing to cool the rage of the Christians but has rather inflamed them even more. While such behavior goes unopposed it looks as though their error is approved.
>
> Were the Jews, by any chance, to be guilty or cognizant of such enormities a sufficient punishment could scarcely be conceived; yet we should be prepared to accept the force of the argument that it cannot be true that the Jews, by such a heinous crime, are the cause or occasion of the plague, because throughout many parts of the world the same plague, by the hidden judgement of God, has afflicted and afflicts the

Jews themselves and many other races who have never lived alongside them.[15]

This pope stayed in Avignon during the plague, supervising the care of the ill and dying. He never got sick himself even though so many people died that when the cemeteries filled up, he had to consecrate the entire Rhône River so that it could be considered holy ground into which to throw the bodies. Even this was not enough, so he proclaimed that all who had died of plague—no matter who they were—were forgiven their sins.[16]

Of course, the virus does not have desires and does not deliberately discriminate among people. But due to a variety of sociological and biological factors, who you are does matter. Plagues can amplify existing social divisions and often create new ones— between the sick and the healthy or between those considered clean or contaminated. And in times of plague, we witness a chasm between those deemed blameless and those considered blameworthy. Simpleminded Manichaean thinking surges—good versus evil, us versus them. Let's look at some existing divisions in our society that coronavirus highlights—such as age, sex, race, and socioeconomic status—and then at some new ones it will likely foster.

»——→

One of the unusual features of SARS-2 is the way its impact varies based on age. Most respiratory infections, such as the 1957 flu pandemic, manifest what is called a U-shaped curve on a graph that charts both age and probability of death among those who get the disease, as illustrated in figure 15.[17] Infants and small children and the elderly are at higher risk of death, but older children and working-age adults are at lower risk. The 1918 influenza pandemic famously had a W-shaped curve. The very young and the very old were at increased risk, but there was also elevated risk in the middle of the age distribution, spiking in patients around twenty-five years old. Scientists have been studying this for decades but are still unsure why it happened. It might have to do with prior pandemics that exposed and conferred some immunity to certain cohorts of people at specific times in their lives or with the particular way people were exposed in the run-up to World War I. Finally, there are epidemics with L-shaped curves. Some have a forward-L shape, in which there is elevated risk among the very young and a relatively flat risk for all other ages. This was the case with polio (not a respiratory disease), which ravaged children but largely spared adults. Or we can see a backward-L shape, which has a low risk for the young and an elevated risk for the elderly. This is the unusual pattern that we see in the 2020 coronavirus pandemic.

Ordinarily, children suffer tremendously from infectious diseases. It's their leading killer worldwide; over 58 percent of deaths in children under five are due to infections, with a list of horrors that includes tetanus, malaria, measles, HIV, pertussis, and countless pathogens causing fatal pneumonia and diarrhea.[18] I therefore found it poignant that they appear to escape the predations of COVID-19. As parents of four children ranging in age from ten to twenty-eight, my wife, Erika, and I—like so many others around the world—took great solace in the fact that that they were all at relatively low risk of death from SARS-2.

Age affects the infection process in several ways. First, there is the attack rate, which is the likelihood of someone in the population getting the condition.[19] The attack rate for COVID-19 varies by age, and younger people are less likely to get infected. This was clear early on. In Wuhan, no children at all tested positive for the virus between November 2019 and mid-January 2020.[20] Early studies in China tracked families living and traveling together and showed that, even in such close-knit groups, children were less likely to fall ill. For instance, children under age nine living with an infected family member had an attack rate of 7.4 percent, and adults ages sixty to

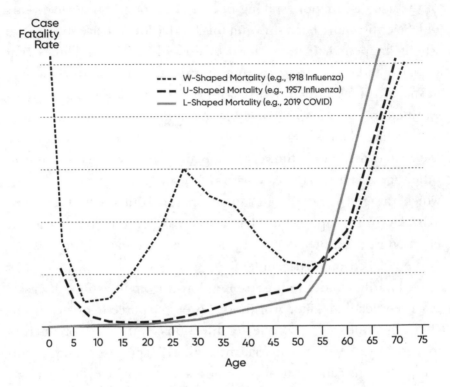

Figure 15: Mortality from respiratory pathogens can depend on age in a number of patterned ways, as shown in these hypothetical curves.

sixty-nine in a similar situation had an attack rate of 15.4 percent.[21] Many subsequent studies confirmed this observation.[22] Transmission from pregnant women to their children in utero (known as "vertical transmission") seems relatively rare for COVID-19 as well.[23]

Not only is the attack rate low in children, but the case fatality rate is also very low for those young people who do get infected (though there have been rare cases of serious complications).[24] In Wuhan, a very small proportion of those under nineteen developed severe (2.5 percent) or critical (0.2 percent) disease, one early study showed. Many subsequent analyses confirmed this.[25] A study of 2,143 pediatric patients in China, for example, found that just one child (a fourteen-year-old) died.[26] In the United States, a similar age pattern for COVID-19 was observed.[27] Overall, mortality among those under twenty is very low, on the order of between one and three people dying out of ten thousand who get sick. For patients in their late fifties, this rises to about one in one hundred. For patients eighty or older, it rises to about one in five.[28] This is the backward-L curve. To be clear, young people may have nonlethal complications that might leave them with long-term pulmonary, neurological, cardiac, or renal morbidity. And of course, with millions of people infected in the United States, there have been and will be cases of young people dying.

This mortality pattern was also seen in the 2003 SARS-1 pandemic; in Hong Kong, no patients under age twenty-four died, but more than half of the patients older than sixty-five died.[29] Incidentally, some of the geographic variation in overall mortality in the United States and globally may be related to different age distributions in different populations; the pandemic could be much less lethal in a country with a young population, like Nigeria, than a country with an older one, like Italy.

What explains children's relative lack of sensitivity to SARS-2? There are behavioral and environmental differences (such as less long-term exposure to smoke and pollution), but some prominent biological hypotheses suggest it also has to do with differences in the ACE2 receptor the virus targets for cell entry, differences in the immune system that fends off the virus, and differences in exposure to vaccines or other viruses that may collaterally influence COVID-19's effects.

Some studies have observed decreases in ACE2 expression associated with increased age, suggesting that the high abundance of ACE2 receptors or differences in the receptors' activity may somehow paradoxically help fortify kids' immunity against the virus.[30] Several risk factors for COVID-19—including older age, hypertension, diabetes, and cardiovascular disease—also share, to a varying extent, a degree of ACE2-receptor deficiency.[31] The precise distribution of ACE2 receptors in the pulmonary system may also vary with age and this may play a role too. The role of ACE2 receptors in COVID-19 pathophysiology will require much more research to clarify.[32]

Other hypotheses point to innate aspects of the immune system that differ with age. For example, children have more adaptive immunity (optimized for pathogens they have not previously seen), whereas adults have more memory-driven immunity (geared to pathogens they have encountered before).[33] The immune cells dedicated to fighting foreign agents in children may be more efficient than those in adults, enabling them to make more effective antibodies against pathogens, including SARS-2.[34] Younger immune systems may also be too immature to experience cytokine storms, the damaging immune-system overreaction that plays a major role in COVID-19 morbidity and mortality.[35]

Another proposition is that routine childhood vaccinations might provide kids with cross-immunity against SARS-2. In particular, the tuberculosis vaccine BCG (not currently used in the United States) has received substantial attention due to its nonspecific protective role that could have an effect on novel viruses.[36] Other experts speculate that children's exposure to other viruses may provide them with protective cross-immunity to SARS-2 (an idea that might also apply to help explain regional differences in the seriousness of the pandemic), while an opposite idea is that adults' past immunity to other coronaviruses might make COVID-19 worse for their age group by exacerbating immunological overreactions to a novel coronavirus.[37] Still another idea is that competitive virus-virus interactions due to the presence of other viruses in children's lungs limit SARS-2 growth and thus mitigate COVID-19 severity.[38] In short, there is a plethora of hypotheses, and more research is needed.

While the attack rate and mortality are clearly lower in children, there is still the question of whether young people can spread the disease as easily as older people can—that is, whether they are more or less *infectious*. To be clear, kids can indeed transmit the disease, which was one of the reasons school closures were adopted as an effective tool for tamping down the COVID-19 epidemic. But the question is how much less infectious children might be than adults. In the extreme, if children had been, by chance, entirely unable to transmit the infection, the impetus to close schools would have declined, though millions of teachers and parents could still be disease vectors there.

Children's infectiousness has been a subject of intense study, with most work concluding that children have likely not been the prime drivers of the COVID-19 pandemic.[39] However, there is still uncertainty about this, and the fact that most available evidence has been collected under lockdown conditions may mean that it cannot be extrapolated to normal circumstances. Few studies have directly addressed the transmission of SARS-2 by children through means such as detailed contact tracing, but those that have been done (for example, in Switzerland and France) have reported that children are either similarly infectious or somewhat less infectious as adults.[40] One potential explanation for lower infectiousness is that younger individuals' milder symptoms, including weaker and less persistent cough, could reduce their spreading of infectious particles. Or the fact that children are shorter than adults may reduce such spread, since respiratory droplets fall downward. Still another possibility is that school closures made children less likely to become infected and thus less likely to serve as index cases. However, children tend to have more social interactions than adults, which could offset the smaller chance they have of spreading the virus.[41]

Regardless of the magnitude of children's role in transmitting SARS-2, it might still not be fair to children to close schools, given the many hazards of missing school, as we saw in chapter 3. Moreover, even though death from COVID-19 is rare in children, they are still adversely affected by our *responses* to the COVID-19 pandemic, which involve unemployment, dislocation, and fear. Children suffer in disasters.

Despite my great relief that the young are spared in this pandemic, I found the narrative that "It's just old folks who die" to be very disturbing, especially when such "sacrifice" is actually avoidable. While older people do, of course, have fewer years left to live, comments like these give the impression that some lives are worth less than others or that older adults contribute little to society or their families. Given the ever-present temptation to see death from a contagious disease as affecting some group *other* than ourselves, it is unsurprising that people say such things. And yet we are speaking of everyone's parents, grandparents, friends, and neighbors.

Being elderly is a risk factor in another way as well. In early April 2020, many hospitals around the country were preparing policies related to triage that were based in part on age. Reports had emerged of the necessity of such triage in Italy in the previous month.[42] Should circumstances require, the very old were to be denied ventilators or taken off them so that the ventilators might be reassigned to younger people more likely to survive. The brutal prospect of battlefield triage had come to our major cities, to our glittering hospitals. To me, this was almost inconceivable. I was familiar with an infamous episode involving involuntary withdrawal of life support and euthanasia in a hospital in New Orleans during the exigent circumstances of Hurricane Katrina.[43] But the idea that we might have to engage in triage on a mass scale in the United States had not previously occurred to me.

Yet it would have been imprudent *not* to consider this possibility. The Yale New Haven Hospital, with which I am affiliated, circulated a well-thought-out policy on April 10, 2020:

[The protocol] was created with the goal of maximizing the survival of as many patients as possible. The protocol is aimed at mitigating the moral distress and isolation experienced by clinicians as we confront the inherent conflicts between individual patient advocacy and the imperative to allocate scarce resources as fairly as possible to maximize public health. These guidelines were heavily influenced by those developed elsewhere, and our guidelines are in turn informing those being put together at other hospitals across Connecticut....

All recalculation of triage scores and related decisions will be made based upon individual assessments of patients using the best available objective clinical data. In all cases, clinically irrelevant factors, including without limitation perceived social worth, race, ethnicity, gender, gender identity, sexual orientation, religion, immigration status, incarceration status, ability to pay, homelessness, putative "VIP" status, and disabilities unrelated to the likelihood of survival will not be considered.

In addition, the hospital noted that the protocol was circulated "for informational purposes—it is NOT active and will be activated only as a last resort, after all other avenues to expand capacity have been exhausted." While the first wave of the pandemic did not necessitate such steps at any hospital in the United States, at least through July 2020, the pressure on our health-care system was still immense, as accounts from New York City demonstrated.

Further controversy erupted when the federal government tried to suggest that age should *not* be a factor in such decisions because it would amount to age discrimination and a violation of the Americans with Disabilities Act. But the key issue in triage is deciding if the allocation of resources is likely to have enough of a benefit to be justified or if those resources could be better allocated elsewhere. Triage is by definition a hard and imperfect and utilitarian calculus. And we engage in triage in medicine all the time—with transplantation candidates, for instance, where some patients are prioritized to receive organs.[44]

We saw an example of this balancing act in the rules initially circulated for EMTs in New York City in late April 2020 (discussed in chapter 3), instructing them not to attempt to resuscitate certain patients so as to limit their risk and to preserve precious resources "to save the greatest number of lives."[45] The order was later rescinded, but it was a clear reminder of the trade-offs that, in more ordinary times, would not have been needed.

These circumstances of rationing health care all come down to survivability. Is the person likely to survive or not? And for how long? And in turn, survivability depends heavily on a person's age and prior medical conditions. It's not possible to triage patients and ignore such factors. In a time of plague, social distinctions like age and health status are unavoidably highlighted.

>———→

From the very first paper describing the initial patients with COVID-19 in Wuhan, published on January 24, 2020, it was clear that men outnumbered women as victims of the disease; they constituted 73 percent of the first forty-one cases reported.[46] And men were both more likely to show symptoms and more likely to die once they fell ill. A later study of 4,103 patients in the New York University Hospital system during the month of March found that, while equal numbers of men and women tested positive for the virus, men were more likely to be sicker and get hospitalized and also more likely to become critically ill and die.[47] Overall, men suffer 50 percent more mortality than women.[48]

When this gender asymmetry was first observed in China, it was ascribed to the dramatically higher rates of smoking in Chinese men compared to women, but subsequent analyses showed that this was not the primary reason. A more convincing explanation is that men at older ages are generally in worse health than women; they tend to have other risk factors for developing a serious case of COVID-19, such as hypertension, diabetes, cardiovascular disease, and cancer. In the New York City sample, adjusting for those factors eliminated the disparity by sex in both hospital admission and mortality.

But other distinct attributes in the immune systems of men and women may explain part of the difference. The sex hormones estrogen, progesterone, and testosterone modulate innate immune-cell responses to various infectious diseases, including viral infections, and they might affect COVID-19 susceptibility and severity.[49] Estrogen and progesterone (both found in higher levels in women) may help protect women from COVID-19 by promoting the expression or activity of ACE2 receptors that may help ward off the virus and by mitigating harmful immune overreactions. Testosterone (found in higher levels in men) may have the opposite effect.[50] Women may also experience superior COVID-19 immunity due to the high density of immune-related genes on the X chromosome. Whereas cells in men have just a single, maternal X chromosome, cells in women have both a maternal and a paternal X chromosome. Female cells inactivate one of the two X chromosomes, resulting in a mosaic of cells with distinct combinations of gene variants. This variety could result in an immunity advantage compared to the more fixed expression in males.[51]

»———→

Like other infectious diseases, coronavirus strikes differentially along socioeconomic lines. While this pandemic did not cause the structural inequalities in our society, it nevertheless brought them into stark relief.

We have already seen how, in New York City, the rates of infection were much higher in lower-income and immigrant-rich parts of town, such as central Queens. We have also seen how the wealthier residents of the city could simply flee the virus, continuing to work remotely away from the center of the outbreak. Most low-income jobs cannot be done remotely, however. Cooks, nurse's aides, grocery-store cashiers, construction workers, janitors, childcare providers, and truck drivers cannot work from home. Individuals in these occupations who did not lose their jobs due to the slowdown of the economy found themselves in environments that placed them at higher risk of contracting the infection. And since many of these positions lack adequate health insurance, such workers could not easily access health care (or even sick days) when they became ill. Therefore, many working-class people continued going to their jobs when sick and did not seek treatment. And people working while they were sick served to worsen the epidemic.

The intrinsic nature of a highly contagious infectious disease reveals the ways in which stark inequality and the lack of universal health care hurts society at large. Leaving infection undiagnosed, let alone untreated, creates what social scientists call *externalities*—a side effect that has consequences for parties who are not directly involved. If we are going to compel people to act in certain ways that decrease the spread of infection (like staying away from work) or that increase the risk of infection (like going to work sick), then surely providing access to health care should be part of the deal. Furthermore, the United States is unusual among wealthy democracies in not providing universal guaranteed sick leave. These observations highlight the fact that we should all care about what happens with respect to the health of our neighbors (and not just during a pandemic, when the issues are more obvious).

Sometimes the hardships faced by the poor during the pandemic were stupefyingly harsh. One particularly heartbreaking account involved Akiva Durr, a mother of two young girls living in one of the most economically deprived neighborhoods in Detroit. Because she was unable to pay her utility bill, her household had been without water for six months. Before the pandemic, she got water from neighbors and friends to bathe her children. "I'd give them a bath every other day, or do a sponge bath to save water," she reported. "It's depressing."[52] Her family had previously organized their lives to use the toilet and get drinking water while at school or work, but with the stay-at-home orders, this was no longer possible. This is not a rare situation—an estimated fifteen million Americans faced water shutoffs annually before the pandemic—but the lack of water for good hygiene is especially troubling for the whole community during an outbreak.

With stark clarity, the impact of the economic calamity unfolding in the face of the COVID-19 shutdown became apparent in photos of food banks from April 2020. Food banks were stretched to their limits, since many previous sources of support, such as monthly church food drives and restaurant donations, had ceased. A San Antonio, Texas, food bank served approximately ten thousand households on Thursday, April 9, 2020. The day began with a million pounds of food in twenty-five tractor trailers at the enormous parking lot where the staff did the distribution. But so many people arrived that the food began to run out, and Eric Cooper, the San Antonio Food Bank CEO, called his warehouse to send more trucks. Aerial photos of the parking lot showed thousands of cars lined up, snaking down the highway for miles. Each car received about two hundred pounds of groceries. It was the largest single-day distribution of food in the nonprofit's forty-year history. "It was a rough one today," said Cooper. "We have never executed on as large of a demand as we are now."[53] His organization could not keep up with the need, and he was hoping for help from "the National Guard or somebody."

It was not just in Texas. Food banks around the nation reported enormous spikes in demand as many millions of laid-off workers turned to them for help. As I looked at photographs from places scattered throughout the United States—in Orem, Utah; Carson, California; Pittsburgh, Pennsylvania; Hialeah, Florida; Baltimore, Maryland—I could not help thinking of similar images of food lines in 1918.[54]

»——————➤

Another factor that increased danger for those with low incomes is that they are more likely to live or work in crowded conditions. Many poor Americans could not easily engage in physical distancing within their own homes. Even people falling in the middle range of the country's income distribution had trouble quarantining a member of the family in a separate bedroom with a private bathroom. And for the over half a million homeless people in the United States, the directive to "stay home, stay safe" (rural Vermont's coronavirus message) must have seemed ludicrous, if not infuriating.[55] Unsurprisingly, a study of 402 homeless adults sleeping in a shelter in Boston on April 2 and April 3, 2020, found that 36 percent of them tested positive for SARS-2, whereas in the city as a whole at the time, the percentage was likely less than 2 percent.[56]

Crowded conditions also led to an explosion of infections in prisons and jails across the country. The United States has more people *per capita* in prisons than any nation on earth—by some measures, it rivals even the Soviet Union under Stalin—and the incarcerated are overwhelmingly from low-income backgrounds. In our prisons, it's impossible for inmates to maintain physical distance. The routines require all the prisoners to touch the same surfaces and jostle through the same narrow hallways. This is by definition a confined community, and once an infection takes root, it spreads like wildfire. Unsurprisingly, in a list of outbreaks maintained by the *New York Times*, thirty of the top fifty outbreaks recorded as of May 17, 2020, were in prisons, including the four biggest outbreaks in any type of location in the United States.

Meatpacking plants were also hit hard. Other than nursing homes and prisons (and the aircraft carrier USS *Theodore Roosevelt*, anchored in Guam), meatpacking plants were the most frequent locus of outbreaks (fifteen out of the top fifty outbreaks) as of May 17, 2020. Indeed, the largest outbreak outside of a prison was in a meatpacking plant.[57] The Smithfield meatpacking facility in Sioux Falls, South Dakota, alone had 1,095 cases. Sometimes, in the face of an outbreak, companies implemented truly foolish policies (from a public health perspective), such as paying employees more to show up to work. For instance, Smithfield offered a five-hundred-dollar "responsibility bonus" for workers who did not miss a shift for the whole month of April, which surely made the outbreak worse.[58] A better approach would have been to pay the workers more on a daily basis (so that they liked their jobs more and so that the plant could recruit more workers) while also increasing and liberalizing sick leave so sick workers could stay home (of course, in the immediate aftermath of an outbreak, production would have to be cut). Such changes in staffing would increase meat prices for consumers, but that is part of the cost of responding to a pandemic as a society. It's expensive for everyone when we have a virus in our midst.

The frequency of COVID-19 outbreaks at meatpacking plants grew so high that there was concern in some quarters that the nation's food supply would be affected.[59] On the White House website on April 28, the president said he intended to invoke the Defense Production Act (ordinarily intended for wartime emergencies) to force meat companies to stay open.[60] The president declared that the meatpacking plants were part of the nation's "critical infra-structure," given that "closure of any of these plants could disrupt our food supply and detrimentally impact our hard-working farmers and ranchers." Unfortunately, no mention was made of the detri-mental impact on the hardworking meatpackers who were facing COVID-19 infections.

Meatpacking plants have been notoriously bad places to work for a very long time, and they figured centrally in a famous novel by Upton Sinclair, *The Jungle*, which described the lives of immigrants toiling in Chicago meatpacking facilities in the early part of the twentieth century. The CDC released an analysis of the outbreaks in May 2020, and it noted that, among 115 facilities in 19 states involving over 130,000 workers, there were 4,913 cases (which was 3 percent of the workforce at that time), and 20 deaths (for a CFR of 0.4 percent).[61] This report highlighted worker proximity and crowded and shared living arrangements and transportation as important factors. In a May 7, 2020, press conference, Health and Human Services Secretary Alex Azar was at pains to blame social aspects of the workers' lives rather than conditions inside the facilities. But this explanation made little sense because meatpacking factories all over the world were also affected—outbreaks were reported in Brazil, Australia, Spain, Ireland, Portugal, Canada, Germany, and Israel— and those employees had different living arrangements. In addition, there were other industries with similar workforces who had similar living arrangements but who remained unaffected.

Why are meatpacking factories more susceptible to COVID-19 outbreaks than other industrial settings? As the CDC reported, crowded working conditions surely play a role. But there is more. Meatpacking is a dangerous profession, involving cuts and abrasions. The plants are deliberately kept cold, and very turbulent air conditions often prevail. Loud equipment requires workers (who are often standing close together and face to face) to yell in order to be heard, which forces virus out of their mouths (similar to what happened in other SARS-2 outbreaks in church choirs).[62] Meatpacking also involves the creation of aerosols (for example, through the use of saws), and this has possibly caused occupational outbreaks before.[63] And we saw in the case of the 2003 SARS-1 outbreak that several super-spreading events were associated with the aerosolization of contaminated body fluids. We have also seen cases in 2020 involving the possible aerosolization of viruses from patient secretions in ICU settings, via faulty ventilators or nebulizers, with consequent spread to health-care workers.[64] Alas, neglecting the outbreaks in the meat-packing plants might have led, within a couple of weeks, to spillovers in the surrounding communities. The pathogen, having brewed at the factories, spread outward.[65]

Ethnic and racial disparities in the attack rate and CFR of coronavirus also became clear not long after the pandemic struck the United States. The findings typified the usual patterns of differential burden of infectious diseases seen in the past.

Both cases and deaths in minority groups generally surpassed their proportions of local populations. According to data from the CDC through May 28, 2020, Hispanics and African-Americans in the United States were roughly three times as likely to become infected with SARS-2 and two times as likely to die from it as whites. These trends are evident across rural areas, suburban counties, and cities. For example, 40 percent of infected people living in Kansas City, Missouri, are black or Hispanic, though they make up only 16 percent of the state's population. Blacks and Hispanics make up 20 percent of the population in Kent County, Michigan, but represent 63 percent of COVID-19 cases.[66]

A national analysis of ethnicity and race in COVID-19 cases through July 8, 2020, also found substantial black-white differences in mortality. Blacks made up 12.4 percent of the population in the forty-five states and the District of Columbia that were studied, and yet they suffered roughly 22.6 percent of the deaths. Accounting for the age difference between blacks and whites made the comparisons even worse. That is, blacks are younger than whites, on average, and should have a lower death rate for this reason alone. In one national study, correcting for this revealed that, overall, they had 3.8 times the risk of dying as whites.[67] Interestingly, the differential impact of the virus according to race varied across states, even after adjusting for age. In Kansas, blacks were 8.1 times more likely to die than whites; in New York, 4.5 times; in Mississippi, 3.4 times; and in Massachusetts, 2.1 times.

Another way to appreciate this is to look at absolute mortality. In the hard-hit states of Connecticut, Michigan, New Jersey, and

New York, more than one out of every thousand black residents died of coronavirus in the first four months of the pandemic. Looking just at New York, two out of every one thousand black residents in that state died. This was driven largely by the mortality in New York City itself, where the rate approached three out of one thousand blacks dying within a few months! Just as a reference point, consider that an average forty-year-old American has a risk of dying from *all causes* in the *entire upcoming year* of about two per one thousand.

In bygone days, blacks were sometimes thought to be relatively immune to contagious diseases while somehow simultaneously warranting their predicament if they did get sick. For instance, during the yellow fever epidemic (caused by a virus transmitted by mosquitos) that struck Philadelphia in 1793 and that killed over five thousand people (out of a population of fifty thousand), African-American clergyman and abolitionist Absalom Jones had this to say:

> It is even to this day a generally received opinion in this city, that our colour was not so liable to the sickness as the whites. We hope that our friends will pardon us for setting this matter in its true state.
>
> The public were informed that in the West-Indies and other places where this terrible malady had been, it was observed the blacks were not affected with it. Happy would it have been for you, and much more for us, if this observation had been verified by our experience.
>
> When the people of colour had the sickness and died, we were imposed upon and told it was not with the prevailing sickness, until it became too notorious to be denied, then we were told some few died but not many. Thus were our services extorted *at the peril of our lives*, yet you accuse us of extorting *a little money from you.*[68]

In these details, Jones captures the reality that working-class people were seen as expendable; their services were *required,* even at risk to their lives, in a pattern not unlike that seen in 2020, when essential workers had jobs that placed them at greater risk.

Hispanic communities have also been hit hard by the virus. This is less evident in crude comparisons across the entire nation—their share of the population in the sample of forty-five states and the District of Columbia noted above is 18.3 percent, and yet they suffered 16.8 percent of the deaths. But when these figures were adjusted for age, Hispanics actually had a 2.5 times increased risk of death compared with whites.[69] In New York State during the first few months, one out of every one thousand Hispanic residents died from the virus, and once again, this was driven by mortality in New York City, where the rate exceeded two per one thousand Hispanic residents. In New York City as of April 11, 2020, Hispanics accounted for 34 percent of the deaths but were 29 percent of the population.[70]

Reports from the Indian Health Service suggested something similar was happening among American Indian tribes. As of July 17, 2020, there were 26,470 confirmed cases of coronavirus infection among American Indians, with over 9,000 of them occurring on the huge Navajo reservation that stretches across parts of New Mexico, Utah, and Arizona.[71] The Navajo Nation is home to two hundred fifty thousand people, and at the end of April, it had the third-highest per capita rate of coronavirus infection in the United States, after New Jersey and New York. A nationwide analysis revealed that American Indians represented 1 percent of the population but 2 percent of the deaths as of July 2020.[72]

The burden on many tribes was huge because, while the tribes are required to provide services on the reservations, they are not allowed to tax their citizens and must rely on casinos and other businesses that were entirely shut down as part of the physical-distancing measures. And American Indian families often do not have an economic safety net; the median income for American

Indian households is about $39,700, which is substantially lower than the typical American household income of $57,600.

Much of the difference in the destructiveness of COVID-19 along ethno-racial lines has to do with other risk factors of dying from or contracting the disease. For example, it's known that hypertension, diabetes, obesity, and cardiovascular disease all increase the risk of mortality among those infected with SARS-2, and these conditions are much more prevalent in most minority populations. There is no doubt that some, and perhaps even most, of the disparity in mortality outcomes might be explained by these factors. Furthermore, household arrangements in minority and nonminority groups might also help explain the difference; in the United States, 26 percent of African-Americans and 27 percent of Hispanics live in multigenerational families with co-resident grandparents; only 16 percent of whites do.[73]

Residential segregation plays a role in the burden of disease too. Residential segregation is rooted in a host of different systemic factors, but the practical effect contributes to a principle in human social organization known as *homophily*, which is a technical term meaning, essentially, "birds of a feather flock together." People do not all have the same chance of interacting with every other member of society. If people tend to preferentially spend time with members of their "own" group, however that is defined, then an infectious disease that takes root in a community will spread in that community, and rates will be higher in this group simply for this reason alone. Hence, if one group is, by chance, seeded with a pathogen, the pathogen will bounce around in that group for a while before it spreads to other groups.

Something similar happened when HIV was introduced into our society in the 1980s. Although sexual practices of gay men did play a role in the disease's spread—especially with respect to the much greater numbers of partners, the greater concurrency of partners, and the more common anal sex—the simple fact that the virus happened to take root in the gay community meant that rates were initially higher in that group compared to other populations, and the epidemic was more intense in this group for much longer. If HIV had initially started in heterosexual men and women, gay men would have seemed to be relatively immune, at least at the outset. And if a group is especially insular, for whatever reason, an epidemic will spread within it until its members have been heavily exposed and take a while to leak to other groups. But, inevitably, germs will spread. It's what they do.

Beyond these structural social factors, there could be other unknown biological reasons that some ethnic and racial groups have higher levels of the virus, perhaps related to prior exposures to other pathogens or variation in innate immunity to the germ, as we will see in chapter 8. But after we account for all the foregoing medical and social factors using statistical methods—by which I mean that epidemiologists can compare fatality rates between black and white people, say, who do not differ with respect to other variables, such as income, prior health status, education, occupation, residence, and so on—we often find that the ethnic and racial disparities go away.

This issue of controlling for other attributes that vary across groups in turn raises the issue of what statisticians call a causal model. If one adjusts for where people live, their wealth, their occupations, and their chronic health conditions, it's true that the "effect" of race may disappear or be minimized. But what does this really mean? On the one hand, we could take this parsing of COVID-19 data to be good news because it means that there is no net effect of race—that racial difference is simply a red herring, and the real difference relates to the differential prevalence of health conditions or variation in the nature of occupations and so on. On the other hand, if in fact the *way* our society is structured ensures that ethnicity and race are associated with worse outcomes in these social factors, then adjusting for those factors and concluding that race has no effect is precisely the wrong thing to do. When I teach about these issues in my undergraduate classes, I observe that dismissing racial differences in health outcomes on the grounds of their being mediated by other factors is a bit like saying that, after adjusting for the quality of ingredients, the ambience, the sophistication of the menu, and the existence of a good wine list, there is no difference between a meal at McDonald's and a meal at the fanciest restaurant in New York City.

»———→

And so the pandemic in 2020 highlighted long-standing differences and inequality. But it also fostered some new dichotomies.

In China, a huge divide opened up between those who were from Wuhan and those who were not. Discrimination against Wuhan residents was widespread. Even if a person was not infected and had not been to Wuhan in years, simply having come from there (as Chinese identity papers invariably indicated) meant that a person could not find an apartment to rent in Beijing, a thousand miles away.[74] In the United States, it would be like landlords in Iowa refusing to rent apartments to people with New York State driver's licenses. And yet, similar regional differences did appear in other ways in our country, such as when some governors tried to stop out-of-state drivers from entering their states.

Still other distinctions were also highlighted by the virus, such as between those who could work from home and those who could not or between those who were following NPI recommendations and those who were not (with respect to wearing masks and keeping a safe distance).

And the virus created another new division: those who are immune and those who are not. Over the initial years of this pandemic, as the virus circulates in our species, progressively more people will have the illness and recover. Based on what we know about other coronaviruses, this will confer immunity of some duration. And it seems likely that even if people are re-infected with the same virus, the second episode will be mild (though it is too early to say for certain).[75]

At the start of the epidemic, along with other experts, I advocated for widespread antibody testing to allow us to target our efforts to control the epidemic and also to identify people who were immune and who could more safely return to work (especially health-care workers). But it occurred to me that we could have a kind of dystopian scenario where employers bid up wages for those who could prove they were immune. Indeed, first in Europe and later in the United States, people considered the idea of testing for immunity and issuing some sort of certification to this effect. Holders of such immunity passports could then return to nonessential (not just essential) jobs and could participate in large gatherings with their similarly immune brethren.[76]

Certification of immunity (after vaccination or recovery from disease) is not without precedent. Hospitals and schools require workers to get vaccinations and prove that they do not have tuberculosis. When my wife, Erika, and I got married in Philadelphia in 1991, we were required to take a blood test to prove that we did not have syphilis. Veterinary workers may be required to prove they have been vaccinated for rabies. Many countries, including the United States, require immigrants to prove that they have been vaccinated for various contagious pathogens (in 2020, the list in the United States included fourteen different diseases).[77] And in antebellum New Orleans, people who were immune to yellow fever had a special status (they were said to be "acclimated").[78]

But an immunity-passport program would differ from such prior examples and raises a number of practical and ethical questions. First of all, during the early period of the pandemic, immunity can be acquired only naturally, not by using a safe vaccine. And, unlike the prior examples, such a program would not be restricted to particular professions or activities; the privileges afforded by such a certificate would be much broader, like the freedom to travel, return to school, go to places of worship, resume employment, or make use of online-dating services.

Differentiation based on immune status sounds creepy but is not *necessarily* ethically problematic. Unlike racial or gender discrimination, people are not permanently tied to any one group—as new people recover from COVID-19, they can join the immune group. Ironically, precisely because certain minority groups were differentially exposed to the pathogen initially, they would be at the front of the line for immunity certification. If some people were at higher risk of being infected, why should they be denied the benefits of being able to document immunity afterward? Immunity certification, if widely and freely available, might even partially redress preexisting socioeconomic disparities that placed people at special risk of getting COVID-19 in the first place.

But requiring an immunity certificate to access a broad range of coveted human interactions would necessitate implementing a transparent, fair, and affordable process for testing. This could be useful for other reasons. Even if a formalized immunity-certification program is not the right approach, it is a collective good to foster knowledge about who is immune. We all benefit when people know their immune status. As it turns out, the Coronavirus Aid, Relief, and Economic Security Act passed in 2020 required private and public insurers to cover coronavirus testing and specified that it would reimburse hospitals for certain kinds of testing in uninsured patients. This meant, in principle, that testing was available for free to all Americans.

The distinction between the immune and nonimmune also should, in theory, reverse the usual pattern: people who had been exposed to the illness would be privileged rather than stigmatized. But in early May, that was still not the case. For example, it was reported that the military would "permanently disqualify" recruits who had had COVID-19.[79] This seems ill-advised on several levels, not least that such people might well be immune, and having soldiers not at risk would accelerate the rate that our armed forces reached herd immunity (even if there were other costs associated with ignoring prior SARS-2 exposure).

Already by May 2020, some COVID-19 survivors were being shunned by neighbors and family members. Samantha Hoffenberg, who was a resident of Manhattan, had to watch her father die of COVID 19 in April; he had gotten it in a hospital where he had been admitted for dementia treatment. Later, she too fell ill from the disease. She was committed to keeping her distance from her loved ones until she was fully recovered. On April 23, there was a fire in her building and she was hospitalized for smoke inhalation. Even though she was no longer infectious, her family refused to visit her. "I have never been in such a sad dark place after that happened," she said. "And my own family is that scared of me that they are not even able to see through the fact that I am alone through this."[80] This is a timeless experience during plagues.

Finally, with all the advantages to having immunity, we have to consider the possibility that some people might deliberately attempt to get sick, which would be extremely counterproductive from a public health point of view. No doubt some individuals would do this, but the frequency would likely be small, since the disease can be deadly and any advantages to immune certification will disappear once an effective vaccine emerges.

Eventually, probably sometime in 2022, either our whole society will have herd immunity or vaccines will be widely available, so immunity certifications would be only a temporary state of affairs. This one division in our society, forged by the pathogen, will become less salient when the initial pandemic period ends.

»———➤

As we saw with the targeting of minorities during the Black Death, the divisions engendered by plague can transform into violence and social unrest. As if the germ itself were not dangerous enough, we have to worry about the fact that we might turn against each other. With the enormous economic toll and increasing burden of disease and death, it's reasonable to ask if American society might face serious strife. People were apparently concerned about this possibility as the pandemic bore down on us, since gun sales in the United States spiked. March 2020 saw one of the highest rates of firearms sales ever, with over 1.9 million guns sold.[81] Online searches for gun information were also at the highest level ever.[82] The purchase of guns was deemed as essential as the purchase of food and gasoline, and governors allowed gun stores to remain open. Fortunately, the world has so far been spared serious outbreaks of violence directly due to the virus.

One of the features of COVID-19 that made it hard for people to take the disease seriously was the lack of visible symptoms (in most cases). Cholera kills by copious diarrhea and dehydration, to the point that patients are gaunt. Smallpox is brutally scarring. The bubonic plague was disfiguring and odiferous. The 1918 Spanish flu made people black and blue, and they often died gasping. The *visibility* of the symptoms of these diseases, quite apart from their much higher lethality, galvanized public action. Furthermore, with COVID-19, what little the media could capture visually about the deaths—such as shrouded bodies piled on a nursing-home floor or in the back of a truck—had a surreal, disembodied feel. Thus, because so many sick people were sequestered in health-care facilities or were alone at home with no one to document their suffering when they died, and because reports focused mostly on visible signs of the economic collapse (with pictures of shuttered stores or lines at food banks), Americans did not see how the virus did its awful work. The deaths and even the mourning for COVID-19 victims occurred strangely offstage, making them harder to appreciate.

This in turn affected our ability to join together to fight the pathogen. To the extent that the risk of death seemed distant and abstract—a problem for *them* rather than *us*—the economic sacrifice and disruption to our lives seemed unwarranted. We could delude and reassure ourselves that "It's just a dozen elderly people in a nursing home in Seattle," or "It's just meatpackers," or "It's just New Yorkers."

And so, of all the societal divisions that emerge in the time of plague, perhaps the most meaningful is the divide between those who know someone who has died and those who do not. But as more people die and as more of us come to know someone who has died or see a death up close, the epidemic will seem more real and more worthy of a coordinated response. For every hundred thousand people who die, there are a million people who were close to them and ten million people who knew them personally.[83] Slowly but surely, as the deaths mount, we will see that this is a problem that affects us all.

The differences in attack rates of the pathogen along preexisting socioeconomic lines made the differences among us more prominent. And while these differences are real and often important, overemphasizing them can be harmful, both practically and morally. On a practical level, to the extent that we emphasize differential attack rates across groups rather than our shared vulnerability, it becomes easier to frame the problem as belonging to *someone else* and even to blame people for their predicament. What is most helpful, in my view, is to emphasize our common humanity. What is needed in order to confront a pandemic is solidarity and a collective will for disease control.

6.

Banding Together

What's true of all the evils in the world is true
of the plague as well. It helps men to rise above
themselves. All the same, when you see the
misery it brings, you'd need to be a madman,
or a coward, or stone blind, to give in tamely to
the plague.

—Albert Camus, *The Plague* (1947)

In the middle of March 2020, Yale's in-person classes were canceled,
and Liam Elkind, an undergraduate, was home in Manhattan. Looking for a way to help others as the pandemic began to rage, he and
a friend founded Invisible Hands, an organization whose goal was
to deliver groceries and other necessities to the elderly and other
at-risk people in their neighborhood and beyond. As soon as the
website went live, volunteer and media interest exploded. Within
four days, more than twelve hundred volunteers had signed up. To
reach clients who were unlikely to find it online, Invisible Hands
spread the word with old-fashioned flyers that volunteers translated
into many languages and tacked up in buildings around the city.
Within a month, Invisible Hands had signed up twelve thousand

volunteers, fielded over four thousand requests for help, and started a subsidy program to provide some free food too.[1]

Many such organizations sprang into action across the country.[2] Mutual aid societies, which have been around for more than a hundred years, have seen a resurgence. The Mutual Aid Disaster Relief network, for instance, is rooted in the idea of collective, reciprocal care and egalitarian social relations. "Mutual aid entails what's often called 'solidarity not charity.' It isn't a handout from some top-down entity, nor someone's paid employment. It embodies a spirit of empathy, generosity, and dignity," its website notes.[3] Other support networks, like COVID-19 Mutual Aid USA, compiled resources for those who wanted to help and those in need of help by curating detailed lists of all mutual aid networks in every state and town.[4] One radio station in San Francisco maintained a website of dozens of links for a dizzying array of local mutual aid societies; people could click on a link and fill in a form to request or offer assistance, from help with food or health care to assistance with tenants' rights or shelter.[5] Similarly, Mutual Aid NYC compiled organizations in multiple categories, including elder care, delivery and transport, internet and technology, mental health, safety, and pet care.[6]

Americans volunteered to help their neighbors and their communities in other ways by supporting or staffing food banks and helping with shopping. People donated their time to produce cloth masks or talk by phone to others who were struggling with loneliness. Librarians repurposed their facilities for an astonishing array of services, providing meals, free Wi-Fi, or 3-D printers to make PPE, or serving as temporary homeless shelters.[7] Many Americans continued to pay their employees even though they were not at work out of a desire to both sustain their businesses and help their workers. Countless people who could afford to, and many who could not, increased their charitable giving.

College students and medical students who were stuck at home organized to provide childcare to health-care workers and other employees crucial to the operation of hospitals, such as custodians and cafeteria workers. One group in Minnesota, called CovidSitters, matched hundreds of volunteers to people who needed childcare.[8] This model was subsequently copied around the country. "The people who are out there truly need help, and I just wanted to be part of the solution," said public-school teacher Danielle Chalfie, who volunteered to care for the children of frontline health-care workers. "I'm taking care of things for them at home so that they can go out and take care of people out there."[9]

According to a survey conducted in late May 2020, 37 percent of Americans reported having donated their money, supplies, or time to help others in their community, and 75 percent said they had supported local businesses since the start of the pandemic.[10] Many Americans also reported checking in on their elderly or sick neighbors (43 percent) and potentially exposing themselves to the virus to help others (17 percent). Conversely, 14 percent of people asked for help from others in their community and 16 percent reported having received help.

Americans are famously charitable. Alexis de Tocqueville admired the country's mutual aid societies when he traveled around the young country in 1831 compiling observations for his prescient book *Democracy in America*. In this respect, at least, not much has changed in two centuries; our country ranks first in the world for charitable donations, and on a per capita basis, Americans donate more than seven times as much as Europeans, perhaps unsurprisingly, given our wealth, religiosity, and favorable tax structure. Not even Canada, our geographic and cultural neighbor, comes close to the level of philanthropy seen in the United States. Even more striking, individual donations make up 81 percent of American charitable giving—far more than is donated by foundations (14 percent) or corporations (just 5 percent).[11]

It should not really surprise anyone, then, that exceptional acts of kindness are found alongside the baser and more self-serving actions we have seen during the pandemic. A survey of over eleven thousand respondents in March 2020 found that 46 percent of people agreed that COVID-19 was bringing out the best in Americans, which was about as many as who thought it was bringing out the worst.[12] And most Americans, 61 percent, reported having a lot of faith in the goodness and altruism of their fellow citizens, a percentage no different from 2018.[13]

But there is something unusual about the kind of generosity we see during an epidemic, when the act of generosity itself can pose material health risks. It's one thing to donate money, time, or blood after a hurricane; it's quite another to deliver groceries to a housebound neighbor when there is a possibility that you might get infected in the process. Altruism in the time of contagion asks us to put the needs of others—often strangers—ahead of our own. We see this kind of selflessness especially among doctors, nurses, firefighters, teachers, and the like, people who are trained to prioritize the needs of the more vulnerable.

Writer Ernest Hemingway saw this up close during World War I. At age nineteen, recovering from his own injuries incurred as an ambulance driver, he watched the nurse he had fallen in love with, Agnes von Kurowsky, attend to a young soldier dying of the Spanish flu. Hemingway was apparently deeply rattled to see the young man appear to drown in his own phlegm. In a letter about the scene (in which she referred to Hemingway as "the kid") von Kurowsky expressed intense sorrow at the loss of this "sweet" and "lovely and smiling" man who had died in her arms. She later explained that "trained nurses are usually noted for being reckless with their own health, while trying hard to keep up that of others."

Hemingway was so struck by her selfless courage that he based an untitled short story on the incident. In it, the narrator admits:

It was the first man I had seen die of the influenza and it frightened me. The two nurses cleaned him up and I went back to my room and washed my hands and face and gargled and got back into bed. I had offered to help them clean up but they did not want that. When I was alone in the room I found I was very frightened by the way Connor had died and I did not go back to sleep. I was frightened into a panic. After a while the nurse I was in love with opened the door and came into the room and over to the bed.[14]

But the narrator is afraid to be in the presence of his lover for fear of contagion, and he admits as much after she remarks, "You're too afraid to kiss me." After a pause, the selfless nurse declares that she would have "sucked [the mucus in the patient] all out with a tube if it would have done any good."

But "regular" people, too, are willing to take risks and contribute to society. For instance, many of those who recovered from COVID-19 returned to hospitals in order to donate their blood for a possible antibody treatment (which we will cover in more detail below). The Hasidic Jewish population of Brooklyn, who were especially hard hit by COVID-19, have made the most disproportionate contribution to this cause, according to a Mayo Clinic researcher. "By far the largest group is our Orthodox friends in New York City," he reported. When blood banks had filled up in the New York regions, members of this deeply insular community traveled through the night by car to Pennsylvania and Delaware to donate their blood, receiving permission from their rabbis to travel on the Sabbath if necessary. "We look at it as a gift that we recovered," a donor explained. "Everybody here has the gift of these antibodies, and they want to use them to save people."[15]

Even wearing a mask is partly an act of kindness, and there is no guarantee that the generosity will be reciprocated. Moreover, in some parts of the country, wearing a mask is grounds for exclusion from stores and restaurants because of the misguided belief that they reflect paranoia or some kind of un-American opposition to liberty. "I want [customers] that aren't sheep," one tavern owner said.[16] But the wearing of masks reflects altruism, since the mask-wearers derive more limited protection from their own masks (in most cases), as we saw in chapter 3.

These acts of kindness, solidarity, and cooperation are not unique to the COVID-19 pandemic; they have been seen in virtually every epidemic in human history. Polio survivor Anne Finger described the extraordinary generosity of local families when she was hospitalized in 1954. She received so many presents she felt "buried in sand up to her neck," she said. When her father went to buy gas, he was told that it was "already taken care of."[17]

Unlike war or famine or natural disasters such as hurricanes or earthquakes during which people can gather together, an epidemic is a collective catastrophe that must be experienced separately. An invisible force pushes us physically apart. For centuries in times of plague, people have sheltered at home, avoiding their neighbors and friends, and even died alone. And we have seen how epidemic disease can inflame darker tendencies, fostering fear, anger, and blame. But epidemics also offer us the opportunity—indeed, the necessity—to band together. They highlight our shared vulnerability and our common humanity. Like other collective threats, epidemics call for solidarity. Luckily, human beings have evolved traits that help: love, cooperation, and teaching.

>———→

Love and connection can make suffering more bearable. Experiments show that if a person is obliged to undergo something painful (like having pressure applied to an index finger) or stressful (like immersing a foot in three inches of cold water), the pain is tolerated better when his or her spouse is present. Sometimes, the analgesic and stress-relieving effects of love can be activated just by thinking about a partner or spouse.[18] And maintaining a romantic relationship in a time of crisis can be invaluable. Historian Miriam Slater vividly remembers the story of a married couple she knew as a child who had met and fallen in love in a concentration camp during World War II: "They were separated in the concentration camp. He told me that he would sneak out and go to the fence…and she'd come to the fence and they would see each other through the fence. I said, 'How would you manage that? That's very dangerous.' He said, 'The Nazis would beat you up if they found you there, but she was worth the beating.'"[19]

Research in disaster psychology suggests that the effect of natural and man-made disasters on partnerships depends heavily on context. In some instances, disasters can directly foster marital instability via economic or mental-health problems like job loss or depression. However, some people rely on spousal support during times of danger. When thoughts of death become more salient, it's quite natural to grow closer to loved ones in order to form cognitive or physical buffers against the fear of death. For example, one experiment found that after undergraduate students were asked questions that made thoughts of their own mortality more prominent ("Please briefly describe the emotions that the thought of your own death arouses in you" and "What do you think happens to you as you physically die and once you are physically dead?"), they reported higher commitment to their romantic partner than did those who were assigned to an emotionally neutral group or one focusing on physical pain.[20]

The impulse to marry during a crisis is sometimes fueled not only by a heightened sense of intimacy but by pragmatic concerns. On the eve of the global travel shutdown, in March 2020, Columbia University astro-engineering students Nathalie Hager and Mikhail Karasev were deeply in love but facing the prospect of an intolerable separation. Hager, a German citizen, was headed home to Berlin, but Karasev, a Russian citizen, was barred from entering the European Union. They struggled at first to find a solution but then arranged a quick marriage at city hall. Karasev was subsequently (after myriad hurdles) given a German residency permit so they could fly to Berlin together. The twenty-one-year-old explained that he did not see the accelerated marriage as a risk: "I really want to spend the rest of my life with this person. I felt this was really fast, but it was the only way we could stay together."[21]

Marriage rates have spiked following previous disasters too. To some extent, disasters can amplify romantic feelings because of a phenomenon known as the *misattribution of arousal,* in which people confuse emotional stimulation or even physical danger for a state of romantic arousal.[22] For example, in 1989, Hurricane Hugo hit South Carolina; the category 4 storm caused over six billion dollars in physical damage and affected 40 percent of all residences. That year, statewide marriage rates increased significantly (by an average of 0.70 per 1,000 people, which may sound trivial but is not), temporarily reversing the steady decline observed over two decades prior to the disaster. Birth rates also experienced a net increase of 41 births per 100,000 people in 1990, and the twenty-four counties declared disaster areas experienced a particularly notable uptick in birth rates.[23] Marriages increased during the early stages of World War I and II as well, and they spiked during the postwar periods. The year 1942, just after the United States' entry into World War II, experienced the highest recorded marriage rate to that point (13.1 marriages per 1,000 people) and peaked after the war in 1946 (16.4 marriages per 1,000 people). Marriages at younger-than-usual and older-than-usual ages also increased during that war.[24]

However, statewide divorce rates in South Carolina significantly increased following Hurricane Hugo as well, with the twenty-four counties that were declared disaster areas exhibiting the greatest increase in divorce rates.[25] And wars have also been shown to influence divorce. Divorce rates peaked (4.3 per 1,000 population for a total of 610,000 divorces and annulments) in 1946, following the end of World War II, and decreased to 2.8 by 1948. The divorce rate rose during and just after the Vietnam War (1965–1973) too.[26] These observations lend credence to the assertion by psychotherapist Esther Perel that crises are a "relationship accelerator." They tend to hasten happy couples to increase their commitment and nudge unhappy ones to part ways.[27]

Given the complexity of the link between catastrophe and romantic behavior, the impact of COVID-19 on how people fall in love or how their intimate relationships fare remains to be seen. For instance, it does seem that domestic violence spiked during the stay-at-home orders in the spring of 2020. While some estimates found a 25 percent decrease in total crime, domestic-violence rates were up by at least 5 percent in five of the nation's big cities.[28] In Chicago, domestic-violence calls to the police rose 7 percent during the city's shelter-in-place order when compared to the same period in 2019.[29] But in New York City, where police reports of domestic violence were down by 15 percent compared to the previous year, advocates cautioned that many victims might lack the privacy or freedom to report abuse during the city's dramatic restrictions. Police commissioner Dermot Shea said that he was concerned that "it's happening and it's not getting reported."[30]

The pandemic also influenced romantic behavior in other ways. Early anecdotal evidence suggests that the COVID-19 pandemic is stoking old flames and increasing intimacy. Americans are feeling lonelier than usual during the pandemic, which might be why self-isolated people are increasingly reaching out to their exes via text or through direct messaging on social media.[31] The increased production and consumption of social media content, the unsated desire for romantic connection imposed by physical distancing, and the fact that people are reevaluating their lives during the stressful times might all be playing a role in exes "coming out of the woodwork."[32]

A survey of 6,004 members of the dating site Match.com conducted in April 2020 found that only 6 percent of singles had dated through video chat prior to the pandemic, but after the pandemic struck, 69 percent said that they were open to video chatting to meet a possible partner. Because the pandemic is lengthening the dating process, people have more time to develop a more reliable type of "slow" love, which might contribute to more lasting marriages after the pandemic ends.[33] According to biological anthropologist Helen Fisher, "During this pandemic, singles are likely to share far more meaningful thoughts of fear and hope—and get to know vital things about a potential partner fast." Such revelations, and the vulnerability that they communicate, can foster intimacy, love, and commitment.

But for those simply looking for sexual connections, the New York City health department had some creative advice. In a startlingly frank and optimistic fact sheet about safe sex during the coronavirus pandemic, the department encouraged New Yorkers to "pick larger, more open, and well-ventilated spaces" for their group sex encounters and told them to be "creative with sexual positions and physical barriers like walls" when engaged in activities involving two people.[34]

»———→

When the pandemic began, our species' innate capacities for connection and cooperation were tested by people having to affirmatively work together to implement physical-distancing measures. While avoiding close contact can go against human nature, the very fact that we were able to band together to form a coordinated response to the threat we faced reflects other profound evolved capacities. Numerous examples appeared of physically distant individuals who nevertheless joined together to cope with the pandemic in symbolically powerful ways. One of my favorites emerged in late March and early April 2020. Orchestra musicians, all of them isolated in their own homes, recorded themselves individually performing their parts in a symphony; the videos were then edited together to show each musician playing beautifully. The New York Philharmonic's rendition of Ravel's *Bolero* reduced many people to tears. It was a perfect illustration of how our social species can cooperate when separated. Many ordinary people around the world did the same thing, jointly making music, at a distance, from balconies and on the streets. The spontaneity of some of these gestures was an even more powerful illustration of our innate capacity for cooperation.

A key point about physical distancing and staying at home is that people are not doing these things primarily to help *themselves* but rather to help *one another*. That took a while to sink in. Early in the pandemic, many people seemed to think that the brave and altruistic thing to do was to go about their business and show that they were not afraid of the virus. Some politicians also took this approach. As the pandemic was beginning to reach Texas in late March 2020, and before it crushed the state in July, Lieutenant Governor Dan Patrick, with an almost stereotypical Texas swagger, suggested that elderly people would be willing to risk death to help their community avoid economic hardship: "No one reached out to me and said, 'As a senior citizen, are you willing to take a chance on your survival in exchange for keeping the America that all America loves for your children and grandchildren?' And if that's the exchange, I'm all in," he said. "And that doesn't make me noble or brave or anything like that. I just think there are lots of grandparents out there in this country like me...that what we all care about and what we all love more than anything are those children," he added. "And I want to, you know, live smart and see through this, but I don't want to see the whole country to be sacrificed, and that's what I see."[35] I understand the idea that many people would be willing to make a sacrifice and even risk death, but I think that the priorities here are reversed. In any case, elderly people deciding to take risks for their own families is different than the government making the decision for them.

Around the same time Patrick said this, I was invited to speak to a group of Episcopal priests in New Hampshire to discuss the kind of pastoral work they did and whether they should keep their churches open. Their natural instinct was to do so, they said, "just as Jesus welcomed everyone," including the sick. Their motivation was different but the objective was the same. But I pointed out that, in the case of the coronavirus, the compassionate response was to avoid fostering the spread of the disease. If we really wanted to help our neighbors, we had to stay at home. Contact reduction was in fact *not* a selfish or even a timid thing to do. But it's an odd thing to persuade people that sitting on their couch is an act of generosity.

Yet people were more often willing to follow rules regarding physical distancing if they understood it was primarily to help others.[36] After all, humans are moral actors capable of transcending their own self-interest. One study evaluated how public health messaging could be more successful. What was more effective, telling people, "Follow these steps to avoid *getting* coronavirus" or telling them, "Follow these steps to avoid *spreading* coronavirus"? It turns out that the emphasis on the public threat of coronavirus is at least as effective as, and sometimes more so than, the emphasis on the personal threat. This is in keeping with other work that shows that people are motivated to get vaccinated not only out of self-interest but also out of concern for the common good.[37]

In the early part of my medical career as a hospice doctor, I often made house calls to patients who were near death, and I had the privilege of hearing their hopes and fears as they faced the end of their lives. One of the most consistent concerns raised again and again was their worry about how their death would affect their families. Seriously ill people who had only a few weeks or months left to live would make choices to limit or stop chemotherapy so that their illness would be less burdensome to their family. People would tell me they were worried about a loved one having to drive them daily for radiation treatment, or they would want to forgo particular drugs with difficult side effects, not to spare themselves pain, but to spare their loved ones. People on the brink of death would tell me that they were worried not so much about their own impending mortality but about their partners' grief.

As social critic Rebecca Solnit has argued, in the wake of major disasters, from earthquakes to hurricanes to bombings, people behave altruistically, working to care for those around them, not just family and friends but neighbors and complete strangers. Rather than the image of a selfish and violent and barbarous society, red in tooth and claw, people often (or even usually, I think) band together in disasters to confront their shared challenge. Solnit elicited this account from one man after a hurricane struck Halifax, Nova Scotia, in 2003: "Everybody woke up the next morning and everything was different.... There was no electricity, all the stores were closed, no one has access to media. The consequence was that everyone poured out into the street to bear witness.... It was a sense of happiness to see everybody even though we didn't know each other."[38] The people worked together to improvise a community kitchen and check on the elderly, and they formed new social relationships in the aftermath.

Of course, there is selfishness and violence as well. People take advantage of the chaos for their own ends, to settle old scores or just to take what they can. Solnit found such stories during Hurricane Katrina. And accounts of plagues over the centuries bear witness to this aspect of human nature—people abandon their friends, blame outsiders and burn them at the stake, or loot houses with the weakened victims still inside them. But anarchy and selfishness are more often the exception than the rule. In fact, in a phenomenon that Stanford psychologist Jamil Zaki has called "catastrophe compassion," survivors typically form communities of mutual aid and they experience greater solidarity.[39] This sense of union and the desire to do good can even result in individuals outside of the disaster zone rushing to the area with donations and volunteering to help.

An increased sense of shared identity is very common among those experiencing catastrophes, and this is a powerful source of cooperative behavior and goodwill. One of the ways this happens is that a widely shared peril erodes prior divisions, bringing large numbers of people into the category of "us." Everyone becomes part of the group confronting the problem. And the shared adversity creates what is possibly the most important division of all: those who are facing the same threat one is facing and those who are not. That activates an inborn tendency to look kindly upon members of one's own group, and this in turn elicits a natural desire to be good to them.

This in-group identity is further reinforced by another common practice we see during pandemics: individuals are more likely to discuss their shared adverse experiences with each other, including their fears, negative feelings, and sense of vulnerability. Ordinarily, people do not do this with those they do not know well, out of concern about imposing on others or out of fear that they might be judged adversely or even stigmatized. But when it is abundantly clear that everyone is in the same boat and facing the same fears, self-disclosure becomes less difficult. This in turn fosters trust and builds solidarity, and these deeper connections make mutual help easier.

A pandemic, unlike a more focused disaster such as a tornado, creates more spillover consequences from one individual or group to others. The virus observes no boundaries. Hence, the reckless lack of mask-wearing by some people may well have serious repercussions for more rule-abiding citizens. In Vermont, normally welcoming locals became incensed when a large group of unmasked motorcyclists from neighboring New York descended on a small town. This is the reason that one state governor deciding to relax physical-distancing rules affects us all. An analysis of twenty-two million mobile-phone users found that, when people lived in counties that were connected to other counties via social ties or geographic proximity, the stay-at-home rules passed by officials in those latter counties were very effective at keeping people at home in the former counties.[40]

These sorts of spillover effects can be managed effectively only with a spirit of cooperation. But this cooperation was, and remains, especially challenging in a large country like the United States when there is limited federal leadership and when uncoordinated approaches to control the virus across states are rife. In the absence of coordination, as one online commentator put it plainly, it feels like designating one end of the swimming pool for urinating and hoping for the best.

Altruistic behavior is usually associated with improved subjective well-being and overall mental health for the do-gooder (as long as it does not feel unduly burdensome). Volunteering is often associated with reduced depression and anxiety.[41] This connection between altruism and human psychology is especially critical in the time of COVID-19, when mental health is itself a concern, whether the harm originates from a fear of the virus or from social isolation.[42] And so altruism and cooperation provide an antidote for many of the negative emotions we reviewed in chapter 4.

$$\text{\textgreater\textgreater}\!\longrightarrow$$

One particular kind of altruism that got attention during the pandemic was the very real personal risk that health-care workers (as well as those in other less glamorous occupations, such as grocery clerks and bus drivers) assumed. As we have seen, this phenomenon has accompanied epidemics for thousands of years. During the plague of Athens, in 430 BCE, Thucydides observed:

> Neither were the physicians at first of any service, ignorant as they were of the proper way to treat it, but they died themselves the most thickly, as they visited the sick most often.[43]

During the Black Death in the fourteenth century, echoing Hemingway's twentieth-century lover, friar and historian Jean de Venette observed:

> In many places not two men remained alive out of twenty. The mortality was so great that, for a considerable period, more than 500 bodies a day were being taken in carts from the Hôtel-Dieu in Paris for burial in the cemetery of the Holy Innocents. The saintly sisters of the Hôtel-Dieu, not fearing death, worked sweetly and with great humility, setting aside considerations of earthly dignity. A great number of the sisters were called to a new life by death and now rest, it is piously believed, with Christ.[44]

Health-care workers themselves have been keenly aware of their situation during contagious outbreaks. One study combined results from fifty-nine investigations of doctors and nurses coping with epidemics of everything from SARS to MERS to Ebola to influenza to COVID-19.[45] Staff in contact with potentially infected patients reported greater levels of psychological distress and post-traumatic stress disorder. Risk factors for negative psychological outcomes included younger age, having small children, and an absence of practical support from the hospital. The lack of protective equipment was particularly stressful. Negative psychological outcomes were also more common if people were obliged to perform their duties rather than being allowed to volunteer (examples of which we saw earlier with the SARS-1 outbreak in 2003 at the Prince of Wales Hospital in Hong Kong and the SARS-2 outbreak in Italy). And the frequent, up-close contact with death can be, I know, very hard.[46]

In the middle of March 2020, as expected, reports began to appear of health-care personnel falling ill and dying of coronavirus they had acquired in the course of their duties.[47] In the United States—hauntingly—nearly six hundred health-care workers had died of

COVID-19 by June 2020.[48] In other countries, too, from China to Italy to Brazil, many nurses and doctors died as health-care systems became overwhelmed.[49] An international website started tracking such deaths. As of May 1, 2020, it recorded over one thousand names from sixty-four countries, from young medical students to doctors pressed into service post-retirement, ranging in age from twenty to ninety-nine.[50] The list makes for powerful reading. Here are just a few: Isaac Abadi, MD, eighty-four, professor of internal medicine and rheumatology, Hospital Universitario de Caracas, Caracas, Venezuela; Luigi Ablondi, sixty-six, epidemiologist, Cremona, Italy; Mamoona Rana, forty-eight, physician, North East London Foundation Trust, London, England; Alvin "Big Al" Sanders, seventy-four, plant operations, maintenance mechanic, Tulane Hospital, New Orleans, Louisiana; Ellyn Schreiner, sixty-eight, registered nurse, Crossroads Hospice, Dayton, Ohio; Susan Sisgundo, fifty, registered nurse, neonatal ICU, Bellevue Hospital, New York City; Arthur Turetsky, pulmonary and critical care physician, Bridgeport Hospital, Bridgeport, Connecticut; Ehsan Vafakhah, thirty-eight, anesthesiologist, Torbat-e Heydarieh, Iran; Liang Wudong, sixty-two, otorhinolaryngologist, Hubei Xinhau Hospital, Wuhan, China; and hundreds of others, some memorialized just by their first name and hospital.

Many of the infections and deaths in health-care professionals were a direct result of the absence of PPE, as these people were obliged to work unprotected. When I practiced medicine during the HIV epidemic in the 1990s, we took risks when we drew blood or cared for HIV-positive patients. Blood and other body fluids would sometimes splatter on us, and we did worry about infection. HIV was not as transmissible in a health care setting as SARS-2, but it was, at the time, deadlier. Still, this risk has always been part of doctoring. Patient care is a calling, not merely a job. But a key difference is that we had proper equipment to reduce our risk!

In the early days of the COVID-19 pandemic, however, American doctors, nurses, and EMTs were expected to take these risks without adequate equipment to protect themselves. The situation was so appalling that a collaborative group of software engineers and physicians from across the United States banded together and made a website, www.GetUsPPE.org, to coordinate requests for PPE and possible donors. As of May 2, 2020, there were 6,169 unique requests from all fifty states, with hospitals, outpatient facilities, and nursing homes accounting for the majority. Institutions and individual health-care workers were especially desperate for N95 respirator masks, which accounted for 74 percent of the requests.[51]

During the first wave of the epidemic, health-care workers filed 4,100 complaints with the Occupational Safety and Health Administration (OSHA); there were at least 275 "fatality investigations" related to the lack of PPE, and, as of June 30, at least 35 hospital workers had died after OSHA received safety complaints about their workplaces.[52] For instance, Barbara Birchenough, sixty-five, worked at the Clara Maass Medical Center in New Jersey as a nurse. On March 25, she texted her daughter, *The ICU nurses were making gowns out of garbage bags. Dad is going to pick up large garbage bags for me just in case.* Later that day, she texted again to note that she had a cough and a headache and that she had been exposed to six patients who had tested positive for COVID-19. *Please pray for all health care workers,* she texted her daughter, *we are running out of supplies.* By April 15, she was dead from the disease. There was something deeply unnerving about the calls put out from big hospitals, such as the Yale New Haven system and the Dartmouth-Hitchcock Medical Center, asking local citizens for any PPE they might have stashed at home and could spare. "No donation is too small," Dartmouth added.

»———————→

The imperative to be generous is hardwired in us, and indeed the survival of our species has depended on an exquisite balance between altruists and free riders, between the people who run into a burning building to save lives and the people who take advantage of others. Across time, humans evolved to live socially, and cooperative impulses won out. Evolutionarily speaking, however, when it comes to our response to collective threats, something even more fundamental than cooperation is going on. The very fact that we knew what to do when the pandemic struck partly reflects another extraordinary ability in our species: the capacity for teaching and learning.

Most animals can learn about their environment. A fish in the sea can learn that if it swims up to the light, it will find food there. This is known as "independent learning" or "individual learning." A smaller number of animal species (including apes, dolphins, and elephants) can learn by watching each other, by imitation and observation. This is known as "social learning." A person can put their hand in a fire and learn that it burns, *independently*. Or I can watch that person put their hand in the fire and learn not to do so, *socially*. I gain almost as much knowledge but pay none of the price. Another person might eat red berries in the woods and die, and if I watch what happens, I'll learn not to do so myself. Such social learning is incredibly efficient.

But people do something even more dramatic than simply learn from one another by copying. We affirmatively and consciously *teach* each other things. We set out to transmit information from one person to another. Teaching is at the root of the capacity for culture, the ability to accumulate useful knowledge, the ability to share it widely, and the ability to learn from the past. Such teaching is very rare in the animal kingdom. But it is universal in us.

The human ability to survive in a diversity of habitats, from the arctic tundra where people hunt seals to the African deserts where they build wells, has depended only a little on physiological adaptations, such as higher fat stores and shorter stature to conserve heat among humans living in the far north. Much more important, human survival in challenging environments for which our soft bodies alone would otherwise be unsuited hinges on our capacity for culture, a capacity that is itself ingrained within us and that has allowed us to invent astonishing things like kayaks and parkas. No other species depends quite so much on the ability to create and preserve cultural traditions.

One of my favorite examples of this is the transmission of warnings via myths or inscriptions regarding major natural disasters that occur much less frequently than a single human life span. The northeast coast of Japan is dotted with so-called tsunami stones, which are large flat rocks, some as tall as three meters, with inscribed warnings about where to build villages so as to avoid tsunamis (which can kill tens of thousands of people) or where to flee when they strike. In Aneyoshi, a stone was erected a century ago with the warning DO NOT BUILD YOUR HOMES BELOW THIS POINT! Why would people take the time and effort to warn distant descendants or strangers at some unknown future point? And why would those future people heed the warning of the past?

When a tsunami struck there in 2011, killing twenty-nine thousand people and totally destroying everything in its path, the water stopped just one hundred meters below the stone. The people in all eleven households built above the marker survived. "They knew the horror of tsunamis, so they erected that stone to warn us," said Tomishige Kimura, speaking of the former residents of his village. Sometimes, the ancient wisdom is transmitted by words not inscribed on a rock, as with the village named Namiwake, or "Wave's Edge." It is located on a spot more than six kilometers from the ocean, marking the devastating reach of a tsunami in 1611.[53] There is also the related phenomenon of designating the low-water marks in European rivers. The Elbe River in the Czech Republic is dotted with "hunger stones" commemorating historic droughts; they have inscriptions like IF YOU SEE ME, WEEP, from as far back as five hundred years.[54]

Similarly, the Aboriginal tribes of the Andaman and Nicobar Islands in the Indian Ocean have passed down an oral tradition for millennia that advises people, when the earth shakes and the sea recedes, to immediately run to particular locations on higher ground in the forest. These tribes all survived the great tsunami of 2005 that killed thousands of residents of more technologically advanced societies.[55]

One good definition of *culture* is that it is "information capable of affecting individuals' behavior that they acquire from other members of their species through teaching, imitation, and other forms of social transmission."[56] A key part of this definition is its interpersonal quality: culture is a property not of individuals, but of groups. Other scientists put more emphasis on material artifacts such as tools or art or medicines, but of course, cultural knowledge is also the antecedent of the creation of such objects.

Culture can evolve over time. Just like human genetic mutations can lead to improved disease resistance, chance events can lead to better ideas or tools. Superior inventions such as a Bronze Age sword can outperform weaker ones, such as a Stone Age ax. And bigger populations facilitate the preservation of discoveries. If an individual stumbles on a better technique for building a fire, finding water, tracking animals, or fabricating a vaccine, there must be someone around to observe, copy, and remember it. Consequently, bigger populations are better suited to social learning and to maximizing opportunities for valuable innovation. Furthermore, teachers and learners must work to keep a complex tradition alive; larger populations mean more pupils to learn from the best members of a society and also provide for the occasional pupils who can surpass their teacher.

Because of the evolution of culture, if you learned calculus in high school, you know (or at least, you knew) so much math that if you were moved back five hundred years, you would be the most knowledgeable mathematician on the planet. Simply by being born in the twentieth or twenty-first century, you have all of the science, art, and inventions made by all prior humans and recorded in diverse ways (in folklore, in books, online) available to learn. You have available to you (in most parts of the world) a deep understanding of the cosmos; domesticated plants and animals to provide food; electricity and modern medicine; highways and maps; and glass and plastics in addition to bronze, iron, and steel.

This is *cumulative culture*. Human beings endlessly contribute to the accumulated wealth of knowledge that belongs to humanity, and each generation is generally born into greater such wealth (of course, there are also periodic reversals when knowledge is lost, as happened after the collapse of the Roman Empire, leading to Europeans living in concrete dwellings for seven hundred years that they lacked the know-how to construct). A few animal species have limited forms of culture. But our type of cumulative culture, across generations and in its elaborate form, is unique.[57]

It was this cumulative culture that allowed us to teach each other things about how to cope with the pandemic when it first struck. Even if people had forgotten or did not know what to do, the knowledge was still available for rapid activation. And we had established mechanisms—universities, books, chat rooms—for sharing knowledge. Information spread at lightning speed around the world about properties of the virus, about how to control it, and about how to care for its victims. Dozens of studies were put online by Chinese scientists as early as January. And this was all in addition to books we could read about how our ancestors had coped with other devastating plagues in the past.

$$\text{➣}\longrightarrow$$

Our capacity for culture is what makes science possible, and that, in turn, allows us to develop pharmaceutical interventions to complement the nonpharmaceutical interventions that we also, even earlier in our history, learned to implement. In fact, a key objective of flattening the curve, to which so much effort was directed, was precisely to give us time to invent new treatments or a vaccine to forestall some of the deaths.

One of the most distinctive features of the COVID-19 pandemic is that it is striking us in a century when our knowledge of the human body and medicine—painstakingly accumulated over centuries with an accelerating pace in the past two hundred years—allows us to respond in a way wholly unavailable to our ancestors. We can scramble to develop drugs and vaccines to target the virus, adding pharmaceutical interventions to the nonpharmaceutical interventions that were the only tools of centuries past. As we saw in chapter 3, pharmaceutical interventions have been less important than NPIs in prior epidemics, but there is a real possibility that this will not be the case in our fight against COVID-19.

Literally within weeks of the virus leaping into our species, scientists began to develop pharmaceutical countermeasures. By May 2020, over one hundred different vaccines of an astonishing variety were already in the works, supported by university laboratories, pharmaceutical companies, and governments around the world.[58] Many of these had entered human trials.[59] However, to put the challenge in perspective, one of the fastest vaccines ever developed, the vaccine for the Ebola virus, approved in 2019, took five years.[60] The usual time frame is closer to ten years.[61]

Despite the history of time-consuming efforts to develop vaccines, there is a lot of optimism about the possibility of rapid progress in the case of SARS-2. One reason is that the biology of the coronavirus is less daunting than that of some other viruses, including even the plain old seasonal flu virus. Another reason for optimism has to do with the way the development of this vaccine is being approached. With a broad and rapid line of attack, we are increasing the probability of quick success by attacking the problem from many angles at once.

Speed is critical because, while a vaccine will be useful no matter when we get it, to really have a significant impact on the course of the pandemic, it has to emerge in much less time than it would take the worldwide population to reach herd immunity. Since that land-mark will probably occur after two or three waves of the pandemic, corresponding to roughly two or three years, it was unclear to me in the summer of 2020 whether the vaccine would arrive in time to make much of a difference. Even if it were accomplished in record time, many people would already have been infected. And it is hard not to flip-flop between optimism and pessimism, with many scientists arguing it is possible to rapidly develop a safe and effective vaccine and others expressing skepticism. There are good reasons on both sides of this division, and I have found myself holding both opinions at different times.

But the very prospect of developing a vaccine, regardless of the speed, depends on decades of prior painstaking work, extensive accumulation of technical knowledge, cooperation among scientists around the world, and the altruistic sacrifices of countless patients who volunteered for previous pharmaceutical trials. Generations of patients, scientists, and physicians struggled to accumulate and pre-serve the knowledge that humans might now be the beneficiaries of.

For many diseases, it has been impossible to develop a vaccine. Human bodies are incredibly complex, and it is difficult to forecast how we will respond to vaccines as a species, let alone how any particular individual will respond. For example, after four decades, we do not have a vaccine against HIV, nor do we have a vaccine against many of the viruses that cause the common cold, though the demand for both is huge. But still, there are so many different approaches to developing a vaccine for SARS-2 that the variety offers a sense both of the likelihood of success and of human ingenuity.

The first step toward a vaccine was taken early on, as physicians observed that humans made antibodies in the course of recovering from COVID-19; this confirmed that it was possible to elicit an effective immune response of particular kinds. Furthermore, the existence of COVID-19 survivors allowed us to immediately deploy a century-old technology that was successfully used during the 1918 flu pandemic: injecting patients with convalescent serum (the fluid part of the blood) extracted from recovered patients (as in the case of the Orthodox Jews from Brooklyn). Patients who have survived COVID-19 make antibodies to the virus; these antibodies circulate in their blood, and they can be extracted and transfused into seriously ill patients, where they help disable the virus. Early reports confirmed the effectiveness of this approach against SARS-2, but larger, formal trials are needed.[62] Regardless, the supply from donors is limited, and convalescent serum does not prevent the disease.

Vaccines seek to elicit a long-term, protective, natural immune response in people—but without the risk of actually giving them the disease.[63] The human body has an immune system with multiple components including circulating antibodies and specialized cells that also attack microbes. These interlocking systems naturally fight invading organisms and can remember the invaders our bodies have encountered before.

To understand vaccine development, it's helpful to understand what, exactly, the vaccine is trying to repel. The coronavirus infects cells by using the spike proteins on its surface to bind to proteins on the surface of human cells known as ACE2 receptors, especially those cells that line the respiratory tract, but other tissues as well, as we saw in chapter 1. After a sequence of steps, the virus enters the cell, where it takes over cellular machinery in order to reproduce itself, releasing more virus into the body, which can both harm the individual and spread to other people.

Ordinarily, as the body tries to repel the invader, certain specialized cells engulf the virus and then display parts of it to other cells, known as helper T cells—like a child showing you the food in his mouth so you can assess what he is eating. This subsequently assists so-called B cells in making antibodies against the virus and in initiating other bodily defenses. In the case of COVID-19, most patients produce sufficient antibodies within a few days to clear the infection. In parallel with this, a different kind of T cell (known as a cytotoxic T cell) can learn to identify and destroy human cells that are infected with the virus. Crucially, some of those immunological defenses record the nature of the invader, and this information stays in the system for a long time after the acute infection. These "records" are collectively known as "memory immunity."

In 2020, scientists tried to activate this natural system using a number of approaches. The oldest approach involved giving people something known as a "live attenuated virus." This is old technology.

Cowpox is a disease that usually infects cows, but it can sometimes infect humans too, giving them a mild disease that resembles small-pox (although smallpox is much deadlier). Noting the folk wisdom that milkmaids were immune to smallpox, on May 14, 1796, English physician Edward Jenner conducted an experiment. He scraped some pus from cowpox blisters on the hands of a milkmaid named Sarah Nelmes and then injected it into James Phipps, the eight-year-old son of his gardener (he conveniently overlooked the option of trying out the procedure on his own children).

Jenner inoculated Phipps in both arms, and Phipps developed a fever and felt sick. Two months later—crucially and almost unbelievably—Jenner injected Phipps on two occasions with mate-rial from *smallpox* blisters taken from another human. And Phipps did not develop smallpox. Jenner's unique contribution was not that he decided to infect people with cowpox but that he thought to prove, by these subsequent challenges, that the subjects were thereafter truly immune to smallpox.[64] Jenner called the procedure *vaccination*, after the Latin word for "cow," *vacca*.

Today, we still use variations of this idea, trying to artificially de-velop mild variants of an otherwise deadly virus. Weakened forms of a virus are created by allowing the virus to grow and infect animal or human cells in vitro repeatedly for hundreds of generations until it picks up mutations that make it less able to cause disease in humans. The trick is to do this just right so that the virus does not make someone ill but still elicits a good memory immune response. The vaccines for smallpox, chicken pox, rotavirus, measles, and mumps all work in this fashion. Indeed, because this approach so closely approximates a natural infection, live attenuated virus vaccines are among the most effective methods of vaccination known, usually conferring very good and long-term immunity. Moderately successful veterinary vaccines for various coronavirus species (used in pigs, cows, cats, and so on) have been of the live attenuated variety, which raises hope for similar success in humans.[65]

Closely related to this approach is the use of an inactivated virus. Here, instead of trying to create a new, mutated strain, researchers treat the virus with chemicals, heat, or some other process so as to render it unable to cause infection while remaining immunogenic. These kinds of vaccines often require booster shots. The vaccines for hepatitis A and the seasonal flu fall into this category. Already by April 2020, Chinese scientists at a company called Sinovac had begun testing a vaccine of this type.[66] They completed trials in monkeys very soon after the epidemic began, and human phase 1 trials started in Jiangsu Province north of Shanghai that month.[67]

A further approach is to use *parts* of the virus, specifically protein fragments, to elicit an immune response. A protein alone cannot infect a person with the virus it comes from, but it can trigger antibody production. The difficult part is triggering a strong enough response to fight the actual virus off. Many successful vaccines fall into this category, including those for shingles, HPV, hepatitis B, and meningococcus.

Still another approach, which has never previously been employed successfully, is to use nucleic acids from the virus instead of proteins—that is, to use fragments of DNA or RNA that resemble the genetic information of the virus. In the case of the DNA approach, the idea is to instruct the body to make a viral protein inside its own cells, as if it had been infected by the virus, and this in turn elicits the usual immune response. A variant of this approach is to add DNA that matches genes of a virus to an entirely different species of another, milder virus.[68] This altered virus is then injected into the body, once again initiating an immune response.

The RNA approach is similar to the foregoing. But it offers some additional benefits, namely, that human bodies could, in the case of coronavirus, come to recognize the RNA itself as alien and possibly learn to attack it directly in ways that are helpful in warding off future infections. Injecting people with DNA or RNA so that it might be taken up by their cells and produce viral proteins may sound scary, but this is actually a less toxic way of doing what the virus does to people anyway. The molecules being injected into our bodies are a far cry from the actual virus itself.

The first coronavirus vaccine developed in the United States was an RNA vaccine of the type just discussed, produced by Moderna Therapeutics on February 24, 2020, just forty-two days after the gene sequence of the virus was published in China.[69] The first participant in the first trial of this agent was dosed on March 16.[70] This is astonishing speed. On that date, there were just 4,609 known cases of the disease in the United States and only 95 known deaths. By May 19, 2020, scientists reported preliminary results from the trial, and the vaccine showed promise.[71]

There are still other approaches to making a vaccine beyond those noted above. And there is other technical know-how that goes into all these approaches, painstakingly accumulated over decades by human beings who learned from and collaborated with one another to make it possible for us to survive diseases that wiped out entire populations in previous eras.

A further wonderful example of the accumulation of cultural knowledge across time (and bequeathed to future humans like us!) is found in the development of something known as adjuvants, which are often used as components of vaccines. During the 1920s, a French veterinarian named Gaston Ramon made major contributions to the development of vaccines for diphtheria and tetanus (which were leading causes of death at that time) based on a method he developed that used formaldehyde to inactivate the deadly toxins those pathogens produced.[72] This chemical treatment rendered the toxins ineffective at causing illness but still able to elicit a protective immune response. A very similar process is still used in vaccines for these conditions manufactured today. Ramon was nominated for a Nobel Prize at least 155 times, the most of anyone who did not go on to win the prize (and, to my mind, his contributions were surely on par with many who did win).[73]

In the process of experimenting with this, however, Ramon noticed that horses being injected with diphtheria had a stronger immune response (that is, a higher level of antibodies in their blood) when they showed evidence of irritation at the injection site. The horses were being deliberately infected in part so scientists could harvest these antibodies and use them to treat humans for diphtheria (a treatment approach that earned a different scientist the Nobel Prize in 1901). A similar process is used nowadays to create antitoxins against snake venoms, since huge horses can easily survive small amounts of poison. This also resembles the convalescent serum treatment for COVID-19 noted earlier.

Ramon came to wonder whether he could deliberately foster such irritation, thereby somehow heightening the immune response of the horse, by adding irritating chemicals to the substance being injected. He experimented with tapioca starch and aluminum salts—substances still in use today in human vaccines. Around the same time, incidentally, British immunologist Alexander Glenny independently observed something similar: if humans were given a diphtheria vaccine that had aluminum salts, they seemed to develop stronger immunity.[74]

Over the years, researchers have tried adding all kinds of irritants, including seaweed proteins and even bread crumbs.[75] Scientists have continued to explore the biology of this, and pharmaceutical companies have developed more and more effective irritants. For example, the vaccine against the virus that causes shingles has a special mix, a witches' brew of fats and proteins from salmonella bacteria along with extracts from the Chilean soapbark tree. None of these substances has anything to do with the virus that causes shingles, but they nevertheless help the vaccine to work. Step by step, working together, we accumulate knowledge over time and share it widely.

By enhancing the immunogenicity of the actual active agent, adjuvants make it possible to give lower doses of the vaccine, which means more people can be vaccinated and at lower risk. Consequently, many of the vaccines under development for SARS-2 involve sophisticated plans to thoughtfully deploy such adjuvants. And companies with advanced technology of this type, including GlaxoSmithKline, have committed to making their proven adjuvant substances available for use with COVID-19 vaccines developed by other groups.[76]

The deployment of ingenious technologies to help us tackle the pandemic was breathtaking in still further ways. As part of their contribution, some scientists attempted to create genetically engineered animals that expressed the human form of the ACE2 protein (the one that the coronavirus latches onto), which would allow still other scientists to more rapidly assess the efficacy of the vaccines they were developing. A whole independent set of technological tools, building on years of science, was brought to bear simply to develop a tool that would be used to tackle the virus—like the technologies developed to make Styrofoam cups that are then merged with entirely different ones required to harvest and prepare coffee.

There is such an amazing amount of knowledge and work that comes together. It's beautiful and mind-boggling. It's hard even to estimate how many years of effort by scientists, doctors, engineers, and others collaborating—often simply by sharing knowledge—have been required to help our species cope with this new coronavirus.

Still, there are lots of unknown factors that will shape how virus research develops. A big one at the beginning of the vaccine development efforts is precisely how strong immunity to SARS-2 is and how long it will last (whether that immunity is prompted by natural infection or vaccination). There is no way to accelerate the acquisition of such knowledge. We simply have to wait for time to pass. Prior studies done with one of the species of coronavirus that causes the common cold and with SARS-1 have shown that the initial antibody response wanes with time, lasting about a year, but that people nevertheless still retained potent memory immunity (based on memory T cells).[77]

Another crucial factor is safety. The usual rate of serious complications for approved human vaccines is approximately one out of a million recipients. The seasonal flu vaccine might kill one out of ten million people to whom it is given, a figure that is clearly offset by the thousands of lives saved annually by the vaccine. But no matter how many lives a vaccine saves, safety is a serious issue, especially in some populations (like children) who face relatively less risk of getting or succumbing to COVID-19. Some vaccine candidates for other coronaviruses actually made the infections *worse* in animal testing, in part by worsening the natural way the animal's body attempted to fight off the infection.[78] Another risk is that if the wrong kind of immune response is elicited, the body might attack itself in what is known as an autoimmune reaction. This happened with the flu vaccine in 1976, when many patients developed a kind of paralysis (from which most people recovered) known as Guillain-Barré syndrome.[79]

Rushing to bring a vaccine to market could also cause other sorts of safety problems in its manufacturing. Infamously, this happened in the Cutter Incident during the early launch of the polio vaccine, in 1955. When the polio vaccine was made available, mass vaccination days were organized by local communities. More than 120,000 children received a batch of the vaccine in which the process of inactivating the live virus was incomplete. Within days, there were reports of children developing paralysis, and the mass immunization program was abandoned within a month. Investigation showed that two batches of the vaccine, manufactured by Cutter Laboratories, had the live virus, resulting in symptoms in forty thousand people, permanent paralysis in fifty-one, and death in five; and this does not include cases of the virus spreading to other children.[80] This episode was described as a perfect storm of sloppy company practices, greed, and lax federal oversight—leading to a tragedy for the victims.

It's hard to move with both speed and prudence. A survey in May 2020 revealed that 73 percent of Americans were confident a vaccine would be developed. Still, 64 percent said that scientists and companies should take the time they need to ensure that pharmaceuticals are safe.[81] But in the rush to develop a COVID-19 vaccine, some pharmaceutical companies are skipping some crucial steps, such as trials in animals, and neglecting needed preliminary work in small groups of humans, which might adversely affect the safety of the resulting vaccine. I am concerned that a vaccine that seems safe in trials will, in fact, be revealed to have problems when administered to millions of people. Any adverse reactions are sure to be breathlessly covered by the media, reducing public interest in vaccination when it is most needed.

Another unusual step taken to accelerate availability is the construction of manufacturing facilities even before a vaccine is proven effective or approved. Philanthropist Bill Gates has said that he would financially support, at tremendous cost, the construction of seven different factories using different manufacturing methods in advance of knowing which vaccines might work.[82] Similarly, drug companies have indicated that they are ramping up production capabilities even before knowing if their candidate vaccines are effective.

There is much to worry about in the development of a COVID-19 vaccine. This is a new virus for which new vaccine approaches are being attempted, requiring novel manufacturing procedures as well.

However, we could wind up with more than one type of vaccine in the end, and some types may be more or less suitable for particular populations, such as children, the elderly, the immunocompromised, and so on. No matter when a safe and effective vaccine appears, it will help to prevent deaths from COVID-19.

»———————→

The efforts to find drugs that can treat COVID-19 have proceeded with similar levels of international cooperation, speed, and ingenuity. Within a few months of the onset of the epidemic, dozens of chemicals were being repurposed or proposed as treatments for the virus. Clinical trials were launched, often reflecting partnerships between pharmaceutical companies, universities, governments, and even international bodies, such as the World Health Organization. For example, in March 2020, the WHO initiated what it called the Solidarity trial in ten countries, enrolling thousands of patients infected with SARS-2 to explore the potential utility of four existing antiviral medications previously used for other viruses.

The variety of drugs studied has been impressive, ranging from hydroxychloroquine (which we discussed in chapter 4) to an antiviral medication known as remdesivir that inhibits RNA synthesis, to other medications acting in biochemically ingenious ways (such as favipiravir or lopinavir), to various human monoclonal antibodies, to steroids, and many other drugs and approaches.

An important trial for remdesivir appeared in May 2020, and showed that the drug could decrease the amount of time patients spent in intensive care by a few days.[83] To be clear, this was a modest outcome and applied only to severely ill patients. The investigators were not able to show that the drug could prevent death (regardless of its impact on hospitalization duration). Nor were they trying to show that the drug could prevent progression to serious disease among the much larger population of infected-but-not-sick people.

But still, even if it did not ultimately prove to have those other beneficial effects, the drug could offer benefits for our society. Freeing up ICU capacity is a crucial objective.

On June 16, 2020, it was announced that a steroid known as dexamethasone could actually reduce mortality in hospitalized patients. The announcement came via a press release from Oxford University (a scientific paper followed a couple of weeks later), indicating how important the investigators felt it was to rapidly, publicly share the information.[84] Dexamethasone is very inexpensive and widely available; it was discovered in 1957 as part of work conducted by Philip Showalter Hench on the treatment of rheumatoid arthritis.

Steroids as a class (and dexamethasone in particular) *suppress* the immune system, which appeared to provide relief for patients whose lungs were ravaged by the overactive immune response that sometimes arises later in the course in severe cases of COVID-19. But patients still need functioning immune systems to fend off the virus itself. This is a tricky balance. The proper moment to use the drug may depend on the course of illness in particular patients. In the trial, a total of 2,104 patients were randomized to receive dexamethasone for ten days, and they were compared to 4,321 patients who got usual care. Overall, the drug reduced the twenty-eight-day mortality rate by 17 percent. This large reduction in mortality was welcome news. But the drug showed the greatest benefit among those sicker patients requiring intubation. Indeed, there was a hint in the findings that patients who did not require ventilators might have done slightly *worse* on the drug (perhaps by interfering with the ability of the immune system to ward off the virus). Developing, testing, and deploying drugs is not easy.

Still, the emergence of drugs such as remdesivir and dexamethasone, and others to come, vindicated the whole strategy of deploying the nonpharmaceutical interventions to flatten the curve. By buying time, we allowed ourselves the opportunity to use our capacities for teaching and learning in order to enhance our survival.

One classic study of immunity and symptoms related to corona-virus (involving the 229E strain, which causes the common cold), published in 1990, involved an odd twist.[85] Fifteen volunteers were deliberately infected with the virus. They all went on to develop cold symptoms, and the amount of circulating antibodies in their blood was periodically monitored for a year, at which point the levels were observed to be very low. But had their immunity waned completely, or did they have some memory immunity? There was only one way to know for sure. Nine of the fifteen agreed to be "challenged"; that is, they went back to the lab to be deliberately re-infected with the virus. Although their antibody titers were extremely low or undetect-able at the time, they indeed still had some immunity, since only six of the nine became re-infected with the virus and none developed symptoms.

Vaccine and drug development is usually extraordinarily labo-rious. The process costs around one billion dollars from start to finish, and it typically takes ten years. Even in late stages of drug development, when human trials have already begun and people are excited about the promise of the drug being evaluated, it is not uncommon for toxic side effects to emerge and the drug to be abandoned. Drugs are scrapped as much as 50 percent of the time even late in the process. When new chemical agents are developed (not existing drugs that are simply being evaluated for a new use, like dexamethasone), little is known about their safety, toxicity, and pharmacokinetics. The process is even more confusing, and in vitro studies with human cells or with animals are done to try to get a handle on those parameters.

This is where the human tendency to altruism and cooperation becomes really helpful. Testing drugs always requires volunteers willing to assume some risk by taking a new and incompletely under-stood pharmaceutical agent. Ultimately, there must be some "first in human" or "first human dose" studies, known as phase 1 studies. These studies are then followed by phase 2 studies, which involve slightly larger numbers of subjects and attempt to further explore the safety of the drug and, more important, get an initial sense of its efficacy. If this phase is promising, then phase 3 studies, involving a much larger number of people in a randomized controlled trial, are initiated with an eye to quantifying the *actual* efficacy of the drug. In such pivotal phase 3 studies, one group of people is randomly assigned to get the active agent and another, the control group, gets either a placebo or whatever constitutes the usual standard of care. During phase 3, scientists also assess the possible emergence of rare toxicities that could not have been observed in the small phase 1 and phase 2 studies. Finally, even after a drug is released, ongoing monitoring is necessary to make sure the pharmaceutical works in a broader, representative sample of people who get it and to check for even rarer toxicities (this is known as phase 4).

The rates of the outcomes of interest (illness, death, toxicities, and so on) are monitored and compared in the two groups in a phase 3 study (the treatment group and the control group). In the case of a vaccine trial, this can take as long as a year while researchers wait for patients to be exposed to, and possibly get, the disease naturally. If fewer patients in the group that got the vaccine go on to develop the illness than in the group that did not get the vaccine, this provides the evidence needed that the vaccine works to prevent infection.

But a problem emerged as COVID-19 vaccine trials got under way. In many parts of the world, such as in China, because of NPIs like mask-wearing and work closures, not enough people were getting sick to allow for efficient testing of the vaccines! Since the efficacy of the vaccine is assessed by comparing how many people who got the vaccine contracted COVID-19 compared to how many people who did not get the vaccine contracted it, a low incidence of COVID-19 means that very large numbers of subjects had to be enrolled. If no one gets exposed to the virus naturally, there is no way to test the vaccine. Ironically, success in deploying nonpharmaceutical interventions to control the disease made it harder to assess the efficacy of pharmaceutical interventions.

For this reason, and in order to accelerate phase 3 of a vaccine trial, some scientists proposed an alternative to simply waiting for people to get sick with COVID-19 naturally. They could do what Jenner did and what the 229E coronavirus study did: affirmatively *challenge* vaccinated people by deliberately infecting them with SARS-2.[86] Of course, giving volunteers the live SARS-2 virus risks disabling or killing them. But accelerating the development of a vaccine could significantly reduce death and disease in the population at large. And the risk to volunteers could be minimized if the volunteers were young and healthy, received excellent care, or were otherwise already at high risk of infection (meaning that they had something to gain and less to lose from participating). For instance, good candidates for such an experiment would be health-care workers or, conversely, people who were caring for chronically ill relatives and who wanted to avoid getting COVID-19 for fear of transmitting it to their loved ones.

Once again, given the altruism in our species, such volunteers emerged. Josh Morrison, who lives in Brooklyn, New York, leads a nonprofit that tries to make kidney donation easier. In 2011, he donated a kidney himself, accepting a small risk of death in order to save the life of a stranger. Helping to accelerate the development of a vaccine for COVID-19 by taking a similar risk, he reasoned, could save *thousands* of lives. After hearing about the possibility of challenge trials, he established COVID Challenge, "a hub for people to volunteer and to advocate for safe and rapid vaccine development." Over 1,550 people signed up by April 2020.[87] Journalist Conor Friedersdorf interviewed some of them.

One of them was Gavriel Kleinwaks, a twenty-three-year-old studying mechanical engineering at the University of Colorado at Boulder. As a Jew, she had always been moved by the Talmudic saying "He who saves a life saves the world entire." And so, she said, "I am lucky in a lot of…ways, including good health…I'm young. I don't get sick a lot. This seems like a way that I can share some of that luck. I empathize with other people. The pain of losing someone you care about is the same no matter who you are. Anything to reduce that amount of pain is something I should try to do." Furthermore, she continued, "There's the risk of participating in a human trial, but there's also the risk just walking around. It's not that I'm not afraid of the virus. I am. But the trial didn't seem like an enormous added risk." Another volunteer, Mabel Rosenheck, a thirty-five-year-old historian at Temple University in Philadelphia, explained her motivation as follows: "The risks that other people are taking on are so much bigger, in terms of doctors and nurses, grocery-store workers, people who are out there every day. The risk that I could take would be comparable but not even as strong, because I would have good medical care and people watching over me right from the start."

Still more cooperation is required, beyond what is needed to develop and test the pharmaceutical interventions. People still need to actually get the vaccine, for their own and everyone else's benefit. Evidence emerged in May 2020 that the percentage of individuals who said they would likely get a hypothetical vaccine declined in parallel with waning fear of the pathogen at that time.[88] Given the misinformation about vaccine risks, the baseline anti-vax movement, and, perhaps most important, the political polarization that has unfortunately come to characterize many aspects of our collective response to the COVID-19 pandemic, it could require significant effort to persuade everyone to get the vaccine.

For a vaccine to work on a collective level and confer herd immunity, people would have to once again cooperate. As we saw before, given the R_0 of SARS-2, roughly 67 percent of the population needs to be immunized. One poll in early May indicated that 72 percent of Americans would get a vaccine if one were available.[89] The willingness varied by race; 74 percent of whites were willing compared to 54 percent of blacks, despite the greater burden of the disease in black people. This perhaps reflects the shameful legacy of racist medical experimentation in our country that has left a residue of suspicion of the medical establishment in some African-Americans. Political affiliation also seemed to play a role, and 79 percent of Democrats expressed willingness to be vaccinated compared with 65 percent of Republicans. However, another national poll, conducted in May 2020, found that, among those planning to receive a vaccine when it became available, 93 percent said they would do it to protect themselves. Nearly as many, 88 percent, said they would do it protect their families. And 78 percent said they would get a vaccine because "I want to protect my community."[90] Such pro-social motivations are ever present.

$$\gg\!\!\longrightarrow$$

As we discussed in chapter 4, epidemics foster fear and anxiety. These psychological states, like the germs themselves, spread from person to person. But good ideas also spread from person to person. Even though the pathogen is exploiting our species' evolved and quite natural tendency to gather in groups, there are other parts of our evolved sociability that the pathogen does not change. We evolved one of the great mysteries of evolutionary biology—the capacity to make sacrifices for each other, to cooperate, and to teach each other. Darwin himself was perplexed by how it could be that such altruism arose, evolutionarily speaking. How could selfish creatures make sacrifices to help others? And yet humans do this all the time. These capacities for altruism, cooperation, and teaching, which are so fundamental to us, are ones that the virus does not destroy. And it is these capacities that will allow us to confront it. Even as we maintain physical distance, we can still band together to fight the virus.

In the 1990s during the HIV pandemic, about one-third of the hospice patients in Chicago where I worked were dying, brutally, of AIDS. A further third were dying of cancer, and the final third were dying of all remaining conditions combined. The AIDS patients were mostly young men. It was awful to witness their deaths. The first effective drug against the condition, azidothymidine, or AZT, had been released in 1987, but the disease was generally still a death sentence. AIDS activists such as Larry Kramer, the American play-wright who founded ACT-UP (AIDS Coalition to Unleash Power) and who died, at age eighty-four, on May 27, 2020, vigorously pres-sured the government to do more, to support more research, to take the disease more seriously. Among his targets was Dr. Anthony Fauci, who was already leading NIAID at that time and who was frequently made aware of Kramer's frustration. At the activist's death, Fauci was quoted as saying, "It was an extraordinary 33-year relationship. We loved each other."[91] Famously, as part of their Storm the NIH campaign, ACT-UP organized a march by hundreds of protesters to the bucolic campus of the National Institutes of Health, in Bethesda, Maryland, on May 21, 1990. Carrying signs reading RED TAPE KILLS US and chanting, "NIH, you can't hide, we charge you with geno-cide," they were met by two hundred riot police on horseback.[92] But they succeeded in changing the agenda for drug development.

In 1995, after much effort, a drug known as saquinavir (a protease inhibitor) was discovered. It became part of a three-drug regimen—composed of saquinavir, dideoxycytidine (ddC), and AZT—a combi-nation called highly active antiretroviral therapy, or HAART. In 1996, results from two pivotal randomized controlled trials, involv-ing a total of over twelve hundred volunteers and supported by NIAID, showed that HAART was extremely effective.[93] Though the side effects were considerable and the daily dosing regimens very complex, it was lifesaving. With HAART, many patients saw the viral load of HIV in their blood become undetectable. And over the course of just a few months in early 1997, our hospice program virtually stopped having to care for any HIV patients. Just like that.

HAART converted a disease that was uniformly fatal into one that could be managed for a very long time, almost like any other chronic illness. It was incredible. Following the success of HAART, in 2003, President George W. Bush launched PEPFAR, the President's Emergency Plan for AIDS Relief, the largest international single-disease initiative in history. It is credited with saving many millions of lives, mainly in sub-Saharan Africa.[94]

All this is what cooperation and teaching were able to accomplish. And this is how, in the end, we will defeat the virus. By connecting, volunteering, and learning, we can affirmatively work together to outlast the predations and limit the damage of such a tiny thing.

7.

Things Change

In Monza, he happened to pass a shop which was open, and had bread set out for sale. He asked for two loaves, so that he would not have to go hungry later on, whatever might happen. The baker signed to him not to come in, and held out a small dish filled with water and vinegar on the blade of a shovel, telling him to drop the money in there. Then he passed the two loaves over to Renzo one after the other, with a pair of tongs.

—Alessandro Manzoni, *The Betrothed* (1827)

In March 2020, as the lockdowns were coming into force in Europe, seismologist Thomas Lecocq of the Royal Observatory of Belgium noticed that the Earth was suddenly still.[1] Every day, as we humans operate our factories, drive our cars, even simply walk on our sidewalks, we rattle the planet. Incredibly, these rattles can be detected as if they were infinitesimal earthquakes. And they had stopped.

After Lecocq's initial observation, seismologists around the world began to share data. With the anthropogenic shaking of our planet subdued, they could even detect the rushing of rivers far away. The unexpected calm allowed them to use naturally occurring background vibrations, such as the crashing of distant ocean waves, to better understand the deformation of the Earth's stony crust. The coronavirus had changed the way the Earth moved.

Other signs pointed to a changing world as well. In the spring of 2020, many videos went viral of wild animals moving into cities. Herds of wild goats, crocodiles, leopards, even elephants wandered streets now devoid of traffic. Satellites high above our planet looked down and detected the disappearance of pollution as manufacturing ceased. In India, where over 1.2 million people die each year as a consequence of air pollution, residents of the city of Jalandhar looked up to see the Dhauladhar peaks of the Himalayas over one hundred twenty miles away. The mountains appeared in startling relief against incredibly clear blue skies for the first time the older residents could remember since they were little children.[2]

While the rest of the natural world began to heal, humans continued to suffer. We reshaped our way of life to slow the spread of the virus. Of course, our nonpharmaceutical interventions were able to postpone and mitigate the pandemic, not stop it. Once the virus had been established in our species, the outcome of the pandemic was inevitable almost no matter what we did. There would be many deaths. The epidemiology alone—an R_0 of about 3.0 and a CFR of 0.5 to 1.2 percent—dictated that.

In the summer of 2020, I kept trying to push these thoughts from my mind. But I could not think of a sound reason to justify optimism. As I reviewed the graphs showing sudden spikes in caseloads in states like Arizona, Texas, Florida, and California at the end of June, and as I listened to President Trump and Vice President Pence insisting that this escalation was merely a function of "too much" COVID testing, I felt despair.[3] In talking to my epidemiology colleagues, I detected a similar gloom. In his public pronouncements, Dr. Fauci, though he expressed "cautious optimism" for the rapid development of a vaccine, was clearly somber.

Other concerning signs appeared over the summer as well. We learned that in India, for unclear reasons, younger people died of SARS-2 in higher proportions than in other countries.[4] That country was being hit so hard that it resorted to using railway cars to create eight thousand more beds for COVID-19 patients in the capital.[5] At the same time, the virus reemerged in China and in other populous countries, such as South Korea, that had previously successfully controlled it. In Brazil, whose president, Jair Bolsonaro, was so dismissive of what he called a "little flu" that a federal judge had to order him to wear a mask, the virus ran loose. In fact, like Boris Johnson in England before him, Bolsonaro became infected himself.[6] Very preliminary hints also emerged from genetics laboratories around the world that the virus might possibly have certain variants that were worse for humans—more deadly or more infectious or both.[7] And, with the passage of time, information began to accumulate about the long-term morbidity associated with the virus; some patients would be debilitated for months after recovery.[8]

In other words, for all we have learned about the virus through the early stages of the pandemic, there is still colossal uncertainty about exactly how it will continue to change the shape of our society in the coming years. Nevertheless, it's quite clear that the virus has already changed our world and that it will continue to do so for some time.

Let's first establish a time frame. If we are able to make a safe and effective vaccine, distribute it rapidly and widely, and get large numbers of people to take it, we might shorten the duration of the pandemic. But even if we are able to clear all of the hurdles for such a rollout, the vaccine still might not arrive before we achieve herd immunity. That is, since we are likely to reach an attack rate of roughly 40 to 50 percent by 2022 no matter what we do, unless the vaccine becomes widely available in early 2021 (which would be the fastest a vaccine had ever been developed by far), it will not make much of a difference in the overall course of the pandemic (though even then, the vaccine would still be enormously valuable to protect uninfected people).

Either way, until 2022, Americans will live in an acutely changed world—they will be wearing masks, for example, and avoiding crowded places. I'll call this the *immediate pandemic period.* For a few years after we either reach herd immunity or have a widely distributed vaccine, people will still be recovering from the overall clinical, psychological, social, and economic shock of the pandemic and the adjustments it required, perhaps through 2024. I'll call this the *intermediate pandemic period.* Then, gradually, things will return to "normal"—albeit in a world with some persistent changes. Around 2024, the *post-pandemic period* will likely begin.

We cannot anticipate all the ways our lives will change, and in fifty years, we may not even remember which changes the pandemic catalyzed. For example, spittoons and public spitting were widespread in the United States up to the beginning of the twentieth century.[9] But they were both abandoned in part because of the 1918 influenza pandemic, when they were rightly seen as unsanitary. For a more recent example, I was well into adulthood before it became obvious to the world that flying in an enclosed space or waiting in a hospital for a medical procedure were not opportune moments to light up a cigarette. In hindsight, these unneighborly practices seem ridiculous. We do not go into restaurants and wonder why there are no spittoons, and the NO SMOKING signs on airplanes feel like an abstract formality. We have forgotten how the world used to be.

≫———→

Many personal attitudes and practices, at home and at work, have had to change due to the pandemic. The deadly virus on the loose, the isolation, and the slowed economy all worked together to foster more self-reliance. We have already seen this with respect to personal responsibility for some of the nonpharmaceutical interventions. But many other activities also called for greater independence, from home cooking, to home haircuts, to the performance of minor home repairs. Why risk having a plumber come over if he might have the virus? Why spend the money if you are out of work? People also took more personal responsibility for their medical care and were obliged to make more careful judgments about whether to seek professional treatment, given the risks of going to a health-care facility.[10]

Children may have benefited the most from the increased independence. In contrast to the pre-pandemic helicopter-parenting culture that kept so many young people from acquiring autonomy, a lot of parents seemed to have waved the white flag after a few weeks of homeschooling, abandoning any pretense of adult control. Teenagers assumed vampire hours, forgoing family dinners and wolfing down a couple of frozen burritos late at night after a hard day's shut-eye. One father reported that he felt like a criminal investigator every morning: "I'll find the wrappers from some snacks that [my son] had, there will be dishes in the sink from what he ate. Sometimes he'll leave the TV on to what he was watching. It's like having a raccoon that came through my house in the night."[11]

Odd eating patterns aside, many children and parents reported that children were spending much more meaningful time with family as well as more time playing outdoors or unsupervised. Lenore Skenazy, an advocate for independence among children, ran an "independence challenge" essay contest in the spring of 2020, and some of the responses illustrated the ways in which children thrived with less supervision. One eight-year-old girl gleefully reported that she rode her bicycle farther and faster than she was allowed; a ten-year-old enjoyed the independence of doing the laundry; a seven-year-old who had previously been afraid of the stove started cooking his own eggs. The budding chef said proudly: "I am getting independent in cooking. I can make omelet—veggie and plain. Hardest part is cracking the egg right over the pan. Because you can get burned the first time you do it."[12]

Around the country, people started baking from scratch or growing food to reduce expenses and unnecessary shopping trips. One eighty-six-year-old woman in my community began growing a garden for the first time in her life; she painstakingly carried soil to her plant beds, one bucket at a time. "I'm not only prepping my garden, but also prepping my body," she said, anticipating the possibility of falling ill with the virus.[13] For some families, home gardening was part of homeschooling. For others, it was a way for them to contribute—they gave the food they grew to local food pantries or to their neighbors in a palpable manifestation of the kind of neighborly cooperation we saw before.

Another change was how much time we spent at home. Millions of people began working from home, adjusting the rhythm of their lives and the allocation of space. Millions of others were at home due to unemployment. And the pandemic greatly amplified the challenges of those without homes too. Some of the advantages of urban life—the interconnected webs of cultural offerings, workspaces, coffee shops, public transportation—began to look like liabilities. People withdrew from public spaces. And there was some exodus from major cities to surrounding suburbs and rural areas, as we have seen.

In some ways, these new modes of living hark back to the past. For many centuries, a greater percentage of humans lived outside of cities than in them. As late as 1950, only 29.5 percent of the world's population lived in cities, and the percentage was 55.5 percent even for the developed world. In 2018, those numbers stood at 60.4 and 81.4 percent, respectively.[14] For thousands of years, most people lived on farms and more independently took care of their own affairs. In 2020, many households returned to some features of this model, taking care of their own families in their own homes.

In heterosexual relationships, there was evidence that, while the demands of homeschooling primarily fell to women, other household obligations came to be divided more equitably between men and women. On average, men increased the percentage of childcare and housework they did.[15] And in these new living arrangements, people were much more likely to spend time with kin than with strangers or even with nearby friends. Of course, our nation did not entirely return to a nineteenth-century lifestyle in 2020. But this localized, familial way of living was not so unusual for our species.

Of course, even when humans lived in a more agrarian way, plagues still occurred, fueled by the densely packed cities where conditions were far worse than those in cities today. And farm living alone does not protect people from plagues. In some ways, this again reminds us that the virus is very effective in exploiting features of how humans live, albeit especially how they have lived since the agricultural revolution. Neither farms nor cities were features of our more distant evolutionary past, prior to approximately ten thousand years ago; they arose as humans abandoned a hunter-gatherer way of life, which, with much smaller groups of people and more limited interactions between them, was much less prone to explosive pandemics.

Yet there is a wrinkle in this line of thinking that offers hope for our modern metropolises. Some of the most densely populated places on earth—Asian cities—have, to date, been very successful in containing the coronavirus pandemic. As we saw before, this paradox illustrates another important feature of our evolutionary heritage: the human capacity for cultural innovation and learning. Modern living environments may incubate more germs, but they can also incubate more ways to combat them.

»———▶

Despite our plentiful supply of clean water, Americans are surprisingly germy people. In one commercial survey of hygiene practices in eight countries, only Germans were worse handwashers. Perhaps study participants in countries like India are more familiar with the devastating burden of infection. But lack of knowledge is not the primary problem. Another survey, conducted for the American Microbiology Society, found big gaps between Americans' professed and observed handwashing practices (the researchers actually monitored behavior in public restrooms).[16] In other words, Americans know they should wash their hands. They just do not translate that knowledge into action. This is particularly true for men. A quantitative assessment of eighty-five scientific studies found that, compared to men, women were better handwashers (and they were 50 percent more likely to implement NPIs on all fronts, including masking).[17]

Handwashing is a special concern because our social custom of handshaking provides a mode of disease spread. While some pandemic lifestyle changes were hard for many Americans to adopt, this long-held habit of reaching for a person's hand disappeared overnight early in the pandemic, even before people began to stay six feet apart. Officials were quick to acknowledge the importance of giving up handshaking. Dr. Fauci declared that post-pandemic America would involve "compulsive handwashing" and "the end of handshaking." Dr. Gregory Poland, director of the Mayo Clinic Vaccine Research Group, called handshaking an "outdated custom" and pointed out that "many cultures have learned that you can greet one another without touching each other."[18]

While the origins of handshaking are unclear, the greeting has been around for thousands of years. Some theorize that a clasping of empty—and therefore weaponless—right hands originally expressed peaceful intent or symbolized a sacred oath and that the shaking action verified that both parties had no weapons hidden up their sleeves. Whatever the origin, the practice is ancient. One of the earliest recorded handshakes is depicted in a ninth century BCE stone relief showing Assyria's King Shalmaneser III clasping palms with a Babylonian ruler. Mentions of handshaking appear throughout ancient art and literature, from Homer's epics to engravings on Roman coins.[19]

The tradition may also be rooted in human evolution, playing a role in social chemosignaling. As one investigator noted, "Human overt olfactory sampling and investigation of unfamiliar individuals is largely a taboo"—meaning that people do not explicitly approach strangers and sniff them. But handshaking may be a mechanism to sample the odors of other people. This possibility is supported by experiments that show that people sniff their right hands more after shaking hands with a stranger of their own gender, as if to evaluate them.[20] And chimpanzees will sometimes link hands (by their palms or wrists, depending on group membership) with their grooming partners and raise them upward, using their free hands to groom. Group-specific handshakes in chimpanzees are intergenerational, passed down from mother to offspring, perhaps constituting a form of cultural learning.[21]

Even animals that do not have hands have evolved displays of friendly greeting. An aggressive dog will have an upright posture with a rigid tail, a flexed body, and a forward-facing head, but a friendly dog will get low to the ground, crouch, look upward, and wag its tail. Charles Darwin accounted for this by what he called the "principle of antithesis," which refers to emotional displays that are the opposite of displays that evolved to serve another purpose. In humans, too, friendly displays are the opposite of aggressive ones. We open our hands rather than clench them; we approach other people closely rather than keeping a wary distance; and we expose vulnerable parts of our bodies, such as our faces. Of course, the precise means by which we do this varies from culture to culture, but all cultures have conventions regarding which sorts of greetings are friendly and which are threatening.

Thankfully, handshake-heavy cultures were able to abandon this tendency with relative ease because humans are an intelligent species capable of quick learning when another evolutionary pressure—survival during an infectious outbreak—demands it. While the pandemic forced people worldwide to forgo more intimate greetings—like handshaking, kissing, hugging, and the Maori *hongi* (a traditional greeting involving pressing together of noses)—many of the world's cultures had been practicing no-contact greetings for quite some time.[22] For instance, *namaste*, a greeting usually accompanied by the *añjali mudrā* gesture, which involves pressing one's palms together with hands pointing upward and thumbs against the chest, is several thousand years old; it was written about in the *Rig Veda*, a Hindu religious text. The *wai*, involving a bow of the head and pressed hands in front of the chest, is used widely in Thailand. *"Mulibwanji"* ("Hello") said together with a clapping of cupped hands is a common greeting in Zambia. In Japan, bowing as a greeting was first introduced via China in the seventh century. The bow was originally a practice of the noble class, but it became popular among samurai warriors during the twelfth century and eventually reached the commoners after the Edo period, approximately five centuries later. Bowing had a chance to catch on in America, as it was practiced in Puritan communities during colonial times, generally as an obeisance—inferiors bowed to their superiors, men bowed to women. But during the Revolutionary War period, some considered bowing undemocratic, and handshaking rose in popularity. Similarly, handshaking was popularized by seventeenth- and eighteenth-century Quakers in order to replace more hierarchical greetings.

The COVID-19 pandemic is not the first time that health professionals have recommended against handshaking. In 1929, nurse Leila Given observed that "hands are agents of bacterial transfer" and suggested that Americans adopt an alternative form of greeting.[23] Even earlier, following the 1793 yellow fever epidemic in Philadelphia, "the old custom of shaking hands fell into such general disuse, that many shrank back with affright at even the offer of the hand."[24] Although the custom was clearly never extinguished in the United States, the avoidance of handshaking is much more widespread now and more broadly described by authorities as a proper means of infection control. Like so many other personal behaviors, our greetings reflect our biology, history, and culture. But a deadly pandemic can reshape them.

>》————→

One of the ironies at the start of the pandemic was that people had either too much privacy or too little. In some ways, intimacy increased because people were staying home with their families, but in other ways, intimacy decreased, and not just because people avoided body contact during greetings. For instance, wearing masks can be anonymizing. And many Americans began to die far from their loved ones, among strangers. Those changes in privacy would be familiar to survivors of past plagues, but the COVID-19 pandemic also heightened fundamental concerns about norms of privacy that are intertwined with twenty-first-century developments in technology.

Technologies of mass surveillance were already growing in sophistication and ubiquity, but in the spring of 2020, they found a new deployment when over a million students were monitored while taking exams. Many millions of university students were sent home suddenly at the beginning of March, and universities made a quick switch to online learning. At the University of Washington in Seattle, a message went out on the afternoon of Friday, March 6, 2020, that all classes (for more than forty thousand students) would shift online by that Monday.[25] Students around the country who went home to continue their schooling at a distance had to take exams while being remotely monitored by employees of commercial proctoring firms, strangers who observed their every move, monitored their faces, listened in on the conversations in their homes, and demanded that the students point the camera in different directions so they could make sure nothing was amiss. The students could not see the proctors' faces, however. This led to some uncomfortable test-taking situations. For instance, while taking an exam in her bedroom, Cheyenne Keating, a sophomore at the University of Florida, felt she needed to vomit, but bathroom breaks were not allowed.[26] She looked into the camera on her computer and asked the person who was proctoring her test if it was okay for her to throw up at her desk. After the proctor said, "Yes," she vomited into a nearby wicker basket and then cleaned up as best she could using a blanket within reach.

Automated computational methods were brought to bear to predict cheating. If the student looked off-screen for more than four seconds more than twice in a minute, this was deemed suspect. To make sure that the student did not switch with a different person to take a test, the software used facial-recognition capabilities linked to ID cards. Even the cadence of the student's typing could be monitored and compared to a baseline that was distinctive for each person. The proctoring companies retained rights to all the audio and video they captured as well as to other private information about the students, leading some professors to decry these methods as a move to turn their universities into surveillance tools.

The loss of privacy related to remote learning played out in other ways too. In one case, a fifth-grade boy who joined a video call with his school had BB guns on a wall behind him, and they were spotted by his teacher. She took a screenshot and notified the police, who then paid an unexpected visit to the family. The officer left after twenty minutes, having concluded that the family was not breaking any laws.[27] The principal said that having a gun, even a BB gun, visible in a virtual classroom was the same as bringing one into a real classroom.

Even people who were long past graduation had strangers peering into their rooms in other intimate ways. As professionals around the country were interviewed at home and work groups met online over videoconferences, viewers got uncommon opportunities to see into other people's lives. In March 2020, I did a TV interview with *Amanpour and Company* from my home office. Later I was contacted by a reporter who was curious about a particular painting by my artist daughter. She described the contents of my office as well as the home environments of many other experts (including one U.S. senator who did an interview in front of her kitchen refrigerator) in an article that appeared in the *Los Angeles Times*. Part of me was flattered, but a bigger part felt weird. I ended up buying a large piece of fabric emblazoned with a Yale logo to use as a backdrop.

Remote monitoring played out on a grander stage too. Many forms of big-data and internet-tracking technologies were used to monitor people to see if they were self-isolating, to detect interactions that placed people at risk of infection, and to follow the response of whole populations. SARS-2 became so prevalent and spread so fast that it proved almost impossible to contain with manual contact tracing, as we saw in chapter 3. However, a contact-tracing app that made the process faster, broader, and more efficient—that automatically kept track of all the people who had been near a sick person and then somehow immediately notified contacts of positive cases—could help achieve epidemic control. Technology experts pointed out that contact-tracing apps with access to intimate location data from cell phones would be especially useful to governments in the case of SARS-2 because of its capacity for asymptomatic transmission—but only if enough people used the apps. This would require persuading or ordering a very large fraction of the population to permit the government to have access to phone records.[28] Those in favor of implementing such widespread and automatic systems noted that many people might willingly surrender some of their privacy so as to acquire more freedom. They argued that citizens might prefer being tracked by the government when they left home to the alternative of being ordered by the government to stay home.

Governments around the world, including in China, Singapore, and Israel, implemented various techniques to exploit such data, as did various European consortia of NGOs and companies. Apple and Google worked together to enable technology that would make contact tracing with cell phones possible (on an opt-in basis), although their solution has been met with pushback from some states.[29] In Russia and China, such technology was supplemented by a broad network of cameras equipped with facial-recognition software that was also used in containment efforts. In South Korea, investigators combined multiple sources of information—including location data from smartphones, security-camera footage, and financial records from credit card companies (indicating where and when people had been in stores)—to do contact tracing. Many Americans also seemed willing to make such sacrifices of their privacy as well. Some even argued that the government should be allowed to monitor smartphone data, including Bluetooth and GPS signals.[30] The COVID-19 pandemic thus reignited debates about the balance between privacy and civil liberties that had arisen in the wake of the terrorist attacks of 9/11.

In practice, however, these apps do not offer enough epidemiological benefit to justify the privacy trade-off. Cell phone–based contact-tracing apps might not be that effective because GPS signals are not precise enough to indicate whether people have been within six feet of each other. Bluetooth signals can cut through walls and so might falsely indicate that people were close to each other when they were not. Huge amounts of time might be wasted on false leads. But even if the apps worked, there could be issues. I was reminded of Benjamin Franklin's adage (which he articulated in a rather different context) that "those who would give up essential liberty, to purchase a little temporary safety, deserve neither liberty nor safety." The erosion of liberty weakens a democracy.

Partly in order to address these concerns, my team in the Human Nature Lab released an app called Hunala in late May 2020. The app respects user privacy, is voluntary, and provides a useful tool for people to manage risk. Unlike most contact-tracing apps, which are *retrospective* and indicate to subjects whether they have previously been in contact with someone who was infected, our app is *prospective*, forecasting the user's risk of coming into contact with someone who has the virus. It works like a traffic app that collects data from many people about the location of traffic jams and then aggregates it to give anonymized information to other nearby users.

Our app maps social networks of individuals based on information they provide and invites people to report their symptoms as often as they wish. It then uses computational algorithms to notify people that their risk of contracting a respiratory disease is elevated because, for instance, twenty days earlier, their friends' friends' friends—who are strangers to them—reported that they had fevers. With a network of connections, it's possible to predict the chances that the virus could spread to them across a sequence of contacts.[31] No information about particular individuals is shared (just like a traffic app does not inform people who exactly has been stopped by the police on a highway). But being notified of elevated personal risk allows a person to take precautionary action, such as staying at home, the same way a driver might exit the highway when alerted that traffic is at a standstill a few miles ahead.

»———————→

Pandemics can impinge on other deeply personal matters. During historical plagues, religious fervor often increased as a means to cope with the seemingly indiscriminate deaths. Appeals to deities, whether out of fear or respect, were understandable responses to catastophe that seemed to lack a worldly explanation. However, especially when the plague involved a very high death toll, religious disillusionment was also common. How could a caring God cause or even allow such a calamity? Would belief even make a difference? During the plague of Athens in 430 BCE, Thucydides observed that for his fellow citizens, "Fear of gods...[did not] restrain them. . . . They judged it to be just the same whether they worshipped [the gods] or not, as they saw all alike perishing."[32]

In the immediate pandemic period, COVID-19 has affected personal religious behavior, particularly among highly religious people. For instance, in one representative national survey conducted in the spring of 2020, 55 percent of adults reported having prayed for an end to the spread of the virus, including 73 percent of Christians and 86 percent of people who ordinarily prayed daily. The less religious became somewhat more religious; 15 percent of people who seldom or never prayed and 24 percent of those without a religious affiliation also prayed for an end to the pandemic.[33] Another poll found that, while 78 percent of people reported that their personal faith or spirituality had remained unchanged during the pandemic, more people reported that their faith or spirituality had "gotten better" (20 percent) than those who said it had "gotten worse" (3 percent).[34] Still another poll conducted in late April showed that 24 percent of U.S. adults said that their faith had grown stronger because of the coronavirus, while only 2 percent said it had weakened.[35] Women and ethnic and racial minorities were relatively more likely to have increased faith.

Despite the net rise in religious sentiment, in-person religious attendance was generally prohibited in the early months of the pandemic as many churches, synagogues, mosques, and temples across the country shifted to online services. Places of worship posed special risks for the transmission of SARS-2 compared to other indoor locations, such as restaurants and stores, because people tended to be in closer proximity for longer periods, often singing with loud voices, which further enhanced the risk of transmission of the virus. Churches resemble bars and nightclubs in these regards. Many large outbreaks around the world involved religious groups or services, including a major outbreak in a Christian group in Korea in February 2020; several outbreaks in the United States, such as the deadly one involving a church choir in March that we discussed in chapter 2;

and an outbreak of more than one hundred cases in people who attended a religious service in Frankfurt, Germany, in May.[36]

By late April, 91 percent of adults who reported attending religious services at least monthly in 2019 said that their congregation had closed public religious services, usually in parallel with livestreaming the proceedings.[37] Overall, 40 percent of regular churchgoers substituted virtual worship for in-person services.[38] For many people, the virtual alternatives were just not the same, so some religious leaders adopted imaginative solutions to physical-distancing conditions, such as drive-in sermons in church parking lots or at drive-in theaters. Dozens of churches employed this strategy on Easter Sunday, April 12, 2020. That same week, a photo went viral of a priest in a clerical collar conducting an infant baptism by water pistol from six feet away. In May, several mosques held drive-through Eid events with live music and gift bags for children.[39]

Although virtual services were a disappointment to some worshippers, others saw them as an opportunity for spiritual growth. The Reverend Dr. Guy J. D. Collins, in a pastoral letter to the Episcopal congregation of St. Thomas Church in Hanover, New Hampshire, described a "theology of technology," linking the online worship services adopted in 2020 to a centuries-old tradition of spreading the word of God via technological innovations, many of which (such as the printing press) were considered heretical in their time. "Technology can be a barrier to worship," he noted. "However, it is more frequently a channel of grace. Visual technologies have, in particular, been essential to passing down the story of Christian faith when literacy was low and when few comprehended the official Latin language of medieval worship."

As the bans on public gatherings were lifted in the late spring of 2020, a dizzying variety of rules were imposed across states. In Texas, alternate pews had to be blocked off; from a public health point of view, this makes little sense because people in the same pew could still be close together and even alternate pews could still be closer than six feet apart. In Massachusetts, masks were required in places of worship. And in New York, services had to be limited to fewer than ten people.[40]

By May, the decision regarding whether places of worship could have in-person services had unfortunately become a political battleground. There were competing claims to, on the one hand, constitutional protection for religious liberty and the right to assemble, and, on the other hand, the guarantee that the government would "promote the general welfare" as promised in the preamble to the Constitution. In California, several churches took Governor Newsom to court over his stay-at-home orders, arguing that they violated religious freedom, but a federal appeals court allowed his order to stand. The Supreme Court eventually agreed with the appeals court, ruling that as long as government public health policies did not single out churches for special restrictions or benefits, the restrictions did not violate the First Amendment.[41]

Regardless of religious belief, many people, whether prompted by thoughts of mortality or by the homebound solitude, were introspective about what gave their lives meaning. I think this opportunity for personal reflection was another factor that played an important role in the massive protests for social justice that took place in June 2020. The pandemic also led people to reevaluate their social interactions, in many cases fostering more empathy and awareness of others. Stories emerged, for instance, of the bitterest of ex-spouses finding a way to communicate more humanely, even compassionately, in arranging the care of their children.[42] As we saw in chapter 6, disasters can bring out both the worst and the best in people, uniting them against a common enemy. I detected a heightened sense of morality animating people's lives as they stopped to consider their own values and more closely examine how to spend their limited time on earth.

»———→

COVID-19 prompted changes in long-standing modes of health-care practice in the immediate pandemic period, and many of these changes are very likely to last into the post-pandemic period. As we have seen, caring for people with the clinical disease itself affected treatment, including end-of-life care. And large numbers of patients with COVID-19 will suffer long-term effects such as pulmonary, renal, cardiac, or neurological damage.[13] That will lead to a bump in disability in the succeeding years, similar to what was seen after the polio and (as we will discuss below) 1918 flu pandemics. For this reason, hospital systems are preparing to care for large numbers of patients in newly established post-COVID clinics.

But so many of the usual ways of doing business in medicine were suspended during the pandemic, to no ill effect, that it is easy to wonder why we did business in the old way at all. For instance, in chapter 6, we saw some of the ways that vaccine development has been accelerated for COVID-19 (with companies rapidly conducting early trials or relying on widespread sharing of genetic sequence information). But there are others. Even though I no longer see patients, for years I have maintained two medical licenses, in Connecticut and Massachusetts. Due to the urgent need for doctors during the pandemic, rules about cross-state licensure were relaxed, so that I could have practiced in either state with a license from any state. The superfluous patchwork of state laws regarding medical licensure is likely to change after the pandemic recedes. It's a historical relic from a time when medical training was less standardized and some states were less vigilant in managing quacks.[44]

Other rules and procedures were rapidly modified by Medicare, Medicaid, and private insurers, most dramatically regarding reimbursement for face-to-face visits. Policymakers and experts had long advocated to allow medical care to be provided over the phone or the internet, and these practices were suddenly not just permissible, but actively encouraged. A large fraction of medical care was moved online to reduce congestion in health-care facilities. Obstetricians provided routine visits for normal pregnancies over the phone. Dermatologists diagnosed simple skin problems by having patients show them video or photographs. My brother Dr. Quan-Yang Duh, an endocrine surgeon in San Francisco, found that he could do most of his surgical follow-up appointments easily by video, though he still insisted on meeting patients in person before surgery. Psychotherapists moved online, with varying results. Internists handled many problems by taking a history from a patient via online video-conference and then providing advice or calling in a prescription. Many of the specialty referrals that primary care doctors handled could also be managed remotely. One of my colleagues at Yale, physician Patrick O'Connor, told me that, with respect to telemedicine, "more was accomplished in two weeks than in five years."

The pointlessness of much in-person medical practice was driven home by another physician colleague, Michael Barnett, who noted that a major reason that patients came in to see their doctors had nothing to do with good medical care or their clinical needs— many of these appointments were to satisfy insurance regulations that mandated visits for routine matters (like getting a prescription refill) that could easily be handled remotely. In fact, much of the work of primary care internists could be done without tethering them to the exam room, and the switch to telemedicine revived one of the hallmarks of good medical care: taking a careful medical history.

Many physicians were happy to see this confirmed. The outpatient medical clinic at the Brigham and Women's Hospital in Boston shifted to mostly online care in the spring of 2020 and found that only 5 percent of patients needed in-person visits.[45] In May 2020, after the peak of the first wave of the pandemic had passed, Yale New Haven Hospital announced that it planned to take advantage of a "strategic opportunity to expand in this area," and it set the target that by July 2020, "at least one third of ambulatory visits [would be] converted to telehealth."

The pandemic highlighted how much of medical care could be provided at home, especially when paired with devices such as home blood pressure cuffs, glucose monitors, and oximeters to gather basic information. The changes introduced to make this sort of care permissible during the peak of the crisis are unlikely to be abandoned as the pandemic wanes. The venues and the types of providers who offer various services will likely change over time, and certain medications, such as birth control pills and travel vaccines, may eventually be available without a doctor visit, in the same way that pharmacies now provide routine flu shots.

Policy experts such as physician and ethicist Zeke Emanuel have pointed out that a number of policy interventions could capitalize on these insights and improve the American health-care system.[46] For example, an online visit with a doctor should be reimbursed at the exact same rate as an office visit for the same problem; otherwise, the pricing structure will perversely incentivize doctors to have patients come to the office. Incredibly, many hospitals actually lost money or faced bankruptcy during the pandemic, despite being packed with patients and despite providing a critical lifesaving service to our nation, because reimbursement for caring for seriously ill people with infection is less than reimbursement for elective procedures for trivial problems. This made no sense before the pandemic and no sense afterward. Reimbursement policies are sure to change as the pandemic ebbs.

The pandemic also highlighted the long-standing issue of *iatrogenic* (doctor-caused) illnesses or injuries arising in the course of medical care. By some metrics, this problem is a leading killer in our society, resulting in as many as fifty thousand to a hundred thousand deaths every year, in and out of hospitals. Medical errors run the gamut from surgical mistakes (leaving a sponge in somebody's abdomen) to medication mistakes (prescribing Lasix instead of Losec, for instance—one is a diuretic and the other reduces stomach acid). People associate medical error with a surgeon accidentally removing the wrong kidney, but hospital-acquired (also known as *nosocomial*) infections happen vastly more often—things like urinary tract infections, surgical site infections, lung infections, bloodstream infections. The ugly truth is that most of these stem from preventable breakdowns in sterile procedures—that is, poor hygiene. Overall, possibly as many as 1 percent of patients admitted to American hospitals die from a medical mistake![47] Doctors know this. When I was doing my medical internship at the University of Pennsylvania Medical Center in Philadelphia in 1989, a senior doctor advised the new interns to think carefully before admitting patients: "Hospital admission is not a benign procedure," he cautioned.

Americans often equate more with better when it comes to medicine, but much data suggest otherwise. One way to assess the potentially deadly impact of health care is to look at what happens when doctors go on strike. This is a rare event, but an analysis of five strikes by physicians around the world, occurring between 1976 and 2003 and lasting between nine days and seventeen weeks, showed that, overall, mortality either stayed the same or *declined* during the strike.[48] Possible explanations for a decrease in mortality include a delay in elective surgeries (with their attendant risks) and a decline in medical errors and injuries.

What about other times when doctors are relatively unavailable? One study examined mortality among elderly Americans who had heart attacks or heart failure and were hospitalized during the two annual national cardiology conferences (when fewer cardiologists were available to treat them) over the course of ten years. The study found that, of the tens of thousands of patients with heart failure who were admitted during meeting days (when they were cared for by doctors who were *not* cardiologists), 17.5 percent died, and of the tens of thousands of heart-failure patients who were admitted during non-meeting days (when they were cared for by cardiologists), 24.8 percent died. Yes, cardiac patients died less when cardiologists did not get their hands on them. Cardiac mortality was not affected by patients being hospitalized during oncology, gastroenterology, or orthopedics conferences, when other specialists were away but cardiologists were present. Finally, the authors cleverly looked at gastrointestinal hemorrhage and hip fracture mortality and found these conditions were unaffected by hospitalization during the cardiology meetings.[49] It really did seem that the cardiologists and their interventions were causing some cardiac patients to die at a higher rate. As a doctor, I find this extraordinarily demoralizing.

During the spring of 2020, hospitals deferred elective surgeries and nonemergency visits for all conditions, both to protect patients from infection exposure and to preserve surge capacity for COVID-19 patients. And patients themselves chose to stay away for minor (or even serious) conditions. As a result, it's likely the deaths from medical mistakes or overtreatment declined. A colleague of mine, physician H. Gilbert Welch, argued that we were probably previously treating too many minor issues—everything from mammographic irregularities (which might disappear on their own) to mild heart attacks (which often have better outcomes without medical care than with it—since patients are often subjected to risky procedures driven more by the financial exigencies of the hospitals and specialists than by the patients' needs).[50] The pandemic forced hospitals to increase the threshold for admitting sick patients so as to keep beds available for COVID-19 patients. Similarly, doctors order billions of dollars of unnecessary tests and procedures every year, but the coronavirus pandemic provided a lesson in their lack of utility.

So, just like the pandemic gave us a glimpse of a world with less traffic, it also gave us a glimpse of a world with less medical injury. It's a lesson the medical system is likely to internalize, because health-care workers certainly do not want to harm people. Once detailed studies of the vast natural experiment provided by COVID-19 are analyzed in the post-pandemic period, thresholds for the treatment of a host of conditions are likely to be reconsidered.

Though iatrogenesis is real, I do not mean to impugn medical professionals, whose hard work and dedication are profound and whose efforts saved the lives of countless patients during the pandemic. Medicine is a profession, not merely an occupation. But as we saw before with the cases of staff being muzzled for revealing PPE shortages, health-care providers are increasingly seen as cogs in a vast bureaucratic machine. Yet, fundamentally, physicians and nurses are supposed to put the needs of their patients above their own. If your shift ends and the patient is sick and needs you, you do not just go home. You stay at the bedside. And indeed, physicians and nurses are supposed to take risks in times of infectious disease, as we have seen. Therefore, a final way in which the pandemic will change medicine is that the generation of doctors in training during the crisis are likely to have to confront their fears and think about their duty in a different way. I think training during a pandemic will heighten their sense of calling. For some trainees, no doubt, it will also affect the specialty they pick, with some of them finding infectious disease or public health more appealing. Contact with mortality has a way of driving a search for meaning among doctors just as it does for anyone else.

Perhaps most important, I think training during a large-scale, deadly infectious outbreak will affect the maturation of a whole cohort of trainees, enhancing their sense of purpose and responsibility. Some medical schools and nursing schools accelerated their graduations in the spring of 2020 so they could place more staff into the field immediately.[51] When I was a new intern in 1989 and struggling to find my identity as a physician, my father-in-law, James E. Zuckerman, an obstetrician-gynecologist trained in the previous generation, told me a story about his own internship in 1961 that has stuck with me. On his very first evening on call, he was taken aside by the senior neurosurgery resident who was supervising him and oriented to the heavy demands he would face that night at Cook County Hospital in Chicago, a famously extreme environment (perhaps the largest hospital in the world at the time, so huge that it had its own police precinct). In the event that the intern should face challenges during the night that he thought he could not handle, the senior resident advised, "Jimmy, tonight, you can call a doctor, or you can *be* a doctor."

$$\gg\!\!\longrightarrow$$

The pandemic also reshaped our economy in numerous ways over the immediate and intermediate terms, causing one of the largest global recessions in history.[52] On March 27, 2020, Congress passed the CARES (Coronavirus Aid, Relief, and Economic Security) Act, an almost inconceivably large trillion-dollar rescue package (we could establish a colony on Mars for less money). By the second-quarter report on the American economy released by the Commerce Department on July 30, 2020, the devastation was apparent. The American GDP fell by 9.5 percent and had an annualized decline of 32.9 percent—a drop so rapid and deep that it was unprecedented in our history. At this time as well, thirty million Americans were still receiving unemployment benefits after nineteen straight weeks of new unemployment claims exceeding one million. These losses would have been worse without the CARES Act and other measures. Nearly five years of growth in the American economy had been wiped away by the virus in just a few months.[53] Many feared that if high numbers of deaths resumed in the winter of 2020 and if people had to continue the nonpharmaceutical interventions, the overall economic impact of the COVID-19 pandemic could even surpass that of the Great Depression. As long as the virus poses a material threat to life, many people will not be willing to completely resume normal activities (such as eating out) or engage in many pre-pandemic purchasing behaviors. This reduced demand will unavoidably keep the United States in recession into the intermediate pandemic period.

Though the economy as a whole slowed down, some industries were forced into overdrive by the virus. We have seen how hospitals and health-care facilities had to rise to the occasion. But morgues, funeral homes, crematories, and cemeteries also had to cope with the sudden spike in deaths. In a funeral home in Brooklyn, a crematory oven broke down under the heavy volume of corpses. Joe Sherman, the owner of another funeral home in Brooklyn and who had been in the business for forty-three years, observed that the pandemic brought "so many more deaths than we could have ever imagined."[54]

The virus also increased demand for certain goods—testing equipment, hand sanitizer, drugs and vaccines, ventilators, and PPE. In March 2020, the forty-three men who worked at the Braskem petrochemical plant in Marcus Hook, Pennsylvania, volunteered to stay at the factory nonstop for twenty-eight days to produce huge amounts of the raw material required for the manufacture of PPE. They brought air mattresses and shaving kits to the factory, converted an office kitchen into a mess hall, and worked twelve-hour alternating shifts nonstop, day and night, without going home or allowing anyone else inside. Together they produced forty million pounds of polypropylene, enough to make five hundred million N95 masks. The front office pitched in too, paying the workers for all twenty-four hours of every day. The men worked with a renewed sense of purpose. One of their leaders, Joe Boyce, a twenty-seven-year veteran of the factory, said, when they finally exited a month later, "We were just happy to be able to help. We've been getting messages on social media from nurses, doctors, EMS workers, saying thank you for what we're doing. But we want to thank them for what they did and are continuing to do. That's what made the time we were in there go by quickly, just being able to support them."[55]

Many other companies responded by shifting production. Liquor distilleries began making hand sanitizer, and some gave it away for free to help out.[56] Sportswear manufacturers shifted from making T-shirts to making masks.[57] Working with GE and 3M, Ford Motor Company used repurposed car parts, such as fans and batteries, to make simplified ventilators.[58] Many of the industries that provide items to combat the virus itself will see elevated demand continue for the duration of the immediate pandemic period, though not with the same intensity as during the first wave.

While distilleries will probably switch back to making drinks, and Ford will go back to making cars, other structural changes in our economy could be long-lasting. Global supply chains could shrink, and there might be full repatriation of manufacturing facilities of certain industries—for instance, the pharmaceutical industry or high-tech machinery.[59] Before the pandemic, the emphasis was on just-in-time production, with parts being delivered just when they were needed in the manufacturing process. Maintaining stockpiles was expensive and inefficient. But in the post-pandemic period, the emphasis could shift to some extent to just-in-case supply chains. One model the pandemic may accelerate is flexible, automated small-batch factories that can manufacture goods to specification close to their needed consumption, which could even be cost-saving.

As people adopted physical distancing and prepared for shelter-in-place orders, the economy had to respond to the indirect effects of the virus as well. There was initial panic-buying of supplies, motivated in part by people seeking a sense of control. Filling a cart with canned goods, flour, cleaning products, and batteries made many people feel like they could influence what happened to them. Others, as we saw in chapter 5, responded to feelings of fear and insecurity by purchasing guns. Consumption of discretionary items declined as people postponed nonessential purchases. Indeed, the greater the degree of social contact people had with others who had contracted or died from the virus, the lower their consumption of nonessential items. For instance, a 10 percent increase in the number of COVID-19 cases in a person's social circle resulted in a 2 percent decrease in purchases of clothing and cosmetics by that person.[60]

Beer, wine, and liquor sales reached a historic high in March 2020 due to an unprecedented spike in demand.[61] Much of this was simply a shift away from purchases at bars and restaurants, since they were closed. Something similar helped explain the perplexing toilet-paper shortage in the early months of the pandemic. Though hoarding was a factor, it did not totally account for the shortage. Nor was there something about the virus that made people need more toilet paper, unlike hand sanitizer or cleaning products. Rather, since people no longer spent half their time at work, more bathroom breaks took place at home. But industrial paper products could not be easily diverted; unbeknownst to most people, the manufacture and distribution of toilet paper in the United States is bifurcated. Paper products for offices and factories form a totally different supply chain than those intended for home use, which meant that many grocery stores were short on toilet paper for months.[62]

After an initial drop-off in home deliveries in early March, companies like UPS and FedEx started showing huge demand, similar to the level seen around Christmas, beginning in April 2020, and they had to impose surcharges on deliveries.[63] Amazon had to hire one hundred thousand more warehouse workers to fill orders for all the people staying at home and give employees raises.[64] Grocery and takeout and other delivery services boomed as well.

While deliveries became more expensive, prices for goods and services less in demand dropped. Oil showed a stunning decline, which was reflected in the price of gas at the pump. For a brief period in April 2020, the price of oil fell below zero, meaning oil refineries had to pay their customers to take barrels of oil off their hands.[65] The price of clothes, cars, and airline tickets also dropped as demand dried up.[66] New vehicle sales declined by 40 percent, and General Motors and Ford shut down their factories.[67] But eggs and meat were scarcer, and their prices rose.

In an explosion of national solidarity and good marketing, many companies, especially those for which marginal costs were low, offered services for free. Companies providing online tools to stay connected, like file transfer or videoconferencing, gave away their products to facilitate working from home. A consortium of cable companies, including Comcast and Verizon, signed a pledge with the FCC to "keep Americans connected" in March 2020, not canceling internet service even if customers could not pay their bills.[68] U-Haul gave college students thirty days of free self-storage to help them cope with the disruptions in the middle of March 2020 when colleges shut down.[69]

The economy changed in still other ways. The demand for RVs in the summer of 2020 skyrocketed as people decided that this was a way to enjoy a vacation, isolated with their family, without the risk of airports or hotels.[70] RVs seemed a promising solution until the CDC, ever the killjoy, reminded the nation that "RV travel typically means staying at RV parks overnight and getting gas and supplies at other public places. These stops may put you and those with you in the RV in close contact with others."[71]

Businesses that rely on public gatherings, like restaurants and sports arenas, were greatly affected early in the pandemic, and this will persist through the intermediate pandemic period. By the end of March 2020 alone, 3 percent of restaurants had gone out of business outright, and 11 percent feared they would not make it through April.[72] Even once some reopened—in the late spring of 2020, in most places—restaurants were able to operate at only 50 percent capacity. About fifteen million people are employed in restaurants as waiters, cooks, and so on, and half of them saw their jobs disappear. The hotel industry was similarly devastated. Bookings completely evaporated. The same thing happened in the entertainment industry. And in the conference industry. And in the car-rental industry. And in the airline industry. An analysis using credit card transactions of a sample of sixty thousand small businesses around the nation showed that, at the nadir, in April 2020, 30 percent of small businesses had closed, and as of late May, 19 percent remained closed, their owners uncertain if they would reopen.[73] This is staggering.

A change in attitudes about risk prompted the emergence of

firms that offered, in business parlance, "safety as a service" or "safety as a value proposition." In hotels, airlines, restaurants, salons, gyms, and so on, "there will be a trade-off between price and safety in favor of safety," observed N. Chandrasekaran, the chief executive officer of the Tata conglomerate.[74] Indeed, hotels started to market themselves along these lines. "We take the greatest care to ensure your stay with us is safe, clean, and comfortable," the Warwick in New York City noted in a flyer it sent out. Gone were emphases on location, excitement, or dining. In their place were "enhanced preventative measures to comply with the latest guidance on cleaning and hygiene."

The ripple effects throughout the economy have been enormous, and we have seen only the beginning. Over the intermediate term, cities could be duller, as many small retail firms go out of business, leaving only large, well-capitalized chains to fill the urban landscape. As people continue to shift to working from home, employers could realize they need less office space, which means fewer custodians, building managers, rental agents, and so on. For some people, the reality of having to obey stay-at-home orders for a family of four in a two-bedroom apartment in the city might not be something they want to repeat, spurring them to look for housing in less urban areas, shifting demand in the enormous real estate industry. But even cities are finding ways to change their layouts during the pandemic; in May 2020, New York City closed over forty miles of streets to cars in order to facilitate outdoor recreation while respecting physical distancing, and cities around the country converted parking spaces to outdoor restaurants, as is typical in Europe.[75] Many of these changes will last beyond the immediate period.

New opportunities have arisen in multiple industries. Some indications suggest that the pace of patent applications is increasing, as inventors stuck at home have time to be more creative.[76] The pandemic could accelerate improvements in autonomous robotics.[77] Many robots able to clean surfaces with chemicals (or, even more efficiently, with UV light) have been deployed to help protect frontline custodial staff or hospital personnel entering rooms where COVID-19 patients were cared for. Other robots have been deployed to deliver groceries and food from restaurants around the country.[78] Contactless methods of payment have already proliferated, but we might see fully automated convenience stores, built like walk-in vending machines with automated tellers.

Working conditions will also change. Prior to the pandemic, less than half of shift workers in the United States had access to paid sick leave, so most workers still went to work when they were ill.[79] But the dynamics of a contagious disease make it abundantly clear why this is a bad idea, and hence many companies, from Apple to pizza-delivery firms, provided paid sick leave to hourly workers for the first time in their history. Not wanting to repeat the errors made by the meatpacking plants that we discussed in chapter 5, companies increased benefits to encourage their workers to stay home if sick with COVID-19. Such policies are likely to endure after the virus subsides, either because companies see the wisdom of it, legislation enforces it, or workers demand it.

The shift to working from home will also linger. In the post-pandemic period, workdays for many employees will get shorter or be better aligned with school days. Some companies have already eliminated in-office work, and others will follow. N. Chandrasekaran, of Tata, forecasts that most of the four hundred fifty thousand employees of Tata Consulting Services, one of the world's largest and most successful management consulting companies, will continue to work from home after the pandemic. Approximately a fifth of the company's employees (based in India, the United States, Britain, and elsewhere) worked from home before the pandemic, but he estimates that three-quarters will do so afterward. "The digital disruption is so significant that most of us cannot imagine the degree," Chandrasekaran said. "The pandemic has accelerated digital trends that will stick after it has gone."[80] Technology companies like

Twitter, Square, and Facebook announced that they were making working from home a permanent option for employees into the post-pandemic period. Ryan Smith, the CEO of Qualtrics, observed, "We have gone through a one-way door. We can't go back in part because some organizations have offered to let people work remotely permanently."[81]

Indeed, early studies revealed that the transition was unexpectedly smooth. Office workers' job satisfaction and engagement fell dramatically in the first two weeks of working from home in the United States when the lockdowns were initiated, one study found, but once people adjusted, eight weeks into the experience, their job satisfaction recovered rapidly. One employee observed, "I think it's weird how normal everything has become—the virtual meetings, the emails, everyone looking grungy." And a CEO noted that he felt this experience "put an end to the 'fly across the country for a one-hour meeting' expectation forever."[82]

Because of the positive experience, many of the changes are likely to be permanent. Why might the national experiment of working from home forced on the American economy have a better outcome than prior company-specific efforts that showed drops in productivity or increases in worker alienation? Two key factors are likely at play. First, this time, all the people in an organization, not just a small group that other employees might label as aberrant, are working from home, as are the workers in the companies they do business with. And second, people are motivated to band together to make the experience more functional. In the past, the subset of employees who worked from home would typically feel left out and be less able to contribute.

However, making such a change more permanent will pose other problems, ranging from the greater difficulty in onboarding new employees and fostering their enculturation to the norms of the firm to the loss of serendipitous (and often innovative) encounters in the physical space of the office. Moreover, working from home could undergo unappealing changes; surveillance of the kind we saw for exam proctoring might be attempted by some companies (for example, by using keystroke monitoring to make sure employees stay on task or tracking e-mails or calendars more stringently).

The economics and models of the education industry—and the workforce trained in educational institutions—are also undergoing change at multiple levels. Childcare and school reforms have been debated for centuries, of course, but sustained changes could be on the horizon. Access to childcare will be a real problem in the immediate pandemic period, lasting into the intermediate pandemic period. Childcare is an extraordinarily low-margin business, even as the costs of childcare are crushing to many American families, and the pandemic has heightened the precarious financial state of childcare centers and the two million providers who staff them (their 2017 median hourly wage was $10.72).[83] A July 2020 survey from the National Association for the Education of Young Children indicated that 40 percent of childcare centers (and half of minority-owned centers) would close permanently without significant public investment as a result of the COVID-19 pandemic.[84] Close to 90 percent of the centers open for business at the time of the survey had seen significant drops in enrollment as they faced increased expenses for PPE, extra cleanings, and other changes associated with infection prevention. Nearly three-quarters of the programs reported current or future furloughs, layoffs, and pay cuts. These ongoing pressures will likely prompt Americans to elect more officials who back imaginative solutions to this chronic problem for working parents and those who care for their children.

With respect to K-through-12 schooling, although much of the response to the 2020 remote-learning experience has been negative, we will likely see an increase in hybrid models of remote/in-person schooling, especially for high-school students and families who resist returning their children to bricks-and-mortar schools. We may also see a crisis in substitute teaching, a low-paying and often unrewarding job that schools find difficult to fill. An average of 6 percent of a child's K-through-12 experience is taught by substitute teachers, however, and with strict health criteria barring sick teachers from coming to school (as well as their fear and anxiety about being there), we can expect more teacher absences in the immediate and intermediate pandemic periods and fewer resources to deal with them.[85]

In the longer term and into the post-pandemic period, there is a real opportunity to revitalize a hundred-year-old model of K-through-12 delivery that cries out for innovation. Schools have always played a dual role of educating children and housing them while parents work, but this double duty reinforces a narrow view of what learning should look like, a view that is confined to particular buildings, schedules, and dates. Children's developmental needs are often ignored in this schema—for example, there are schools without outdoor spaces of any kind. To promote a more inclusive view of learning befitting twenty-first-century America, it may be time to think more imaginatively about how and where children learn best and then figure out how to make those learning opportunities possible in a way that protects parents' work hours.

Educators have noted these concerns for decades, but the post-pandemic pressures of disease prevention, new economic realities, and twenty-first-century technological advances may finally push school systems to explore radical innovation. Eventually, a smaller number of physical schools might come to function as primary learning hubs, with learning "spokes" found at home, in libraries, museums, community colleges, clubs, enrichment programs, and elsewhere in the community. We can expect that already heated debates about school consolidation and school choice will likely accelerate as public schools adapt to a more personalized approach to education and as a variety of adults (coaches, counselors, college students, homeschooling parents, online experts) assume greater responsibility for children's learning.

The disarray prompted by the pandemic is likely to provide an impetus for such a rethinking at higher levels of education too. When the pandemic struck, universities had already been trying to shift to provide more instruction online. Universities are sure to continue to offer both in-person and online experiences even after the virus subsides. Of course, unlike universities in Europe or at domestic community colleges, many four-year colleges in the United States are organized around the in-person residential experience. Personal interactions are seen as the key to emotional and intellectual growth. But a kind of residential-life arms race has proliferated in recent years, with huge investments in residential facilities (student centers and dormitories with fancy amenities) and lots of midlevel deans and administrators, contributing to higher tuitions and alarming debt loads. The availability of online instruction might shift the emphasis away from such expenditures or lead more applicants to question their value.

While I have spent my career advocating for the utility of face-to-face classroom instruction and on-campus living, the ability to teach classes asynchronously online, so that students do not all have to be in the same place at the same time, has its advantages. One likely outcome is that more professors will record lectures for online delivery, and students might be told to watch these lectures in preparation for class. Class time could then be reserved for more personal interactions or questions and discussion in what a colleague of mine, physicist and educational innovator Eric Mazur, calls a "flipped classroom," where the students do more talking than the teacher.

But a shift to online learning would change the value proposition for many colleges. Large universities would likely be able to offer an online alternative at low marginal cost. But hundreds of small colleges will face closure as students and families opt for online instruction at schools without an on-campus experience. As with the other cascade effects we have considered, this would put teachers and administrators and other employees out of work and would affect all the supporting businesses—from bars to bookshops—in countless small college towns across the country.

Even before the pandemic, there was already a shift toward moving academic (and other) conferences online, partly prompted by concern for the environment. It has been estimated that the 7.8 million researchers who travel to conferences each year produce the carbon emissions of a small nation.[86] Moving conferences online would also be more welcoming to scholars with disabilities, parents with young children, and those constrained by religious holidays.

One of the longest-lasting economic effects concerns the cohort of students graduating from college into a recession. They are unlikely to catch up economically; they are predicted to endure lower wages for at least twenty years.[87] Ironically, however, such people in prior recessions have reported *more* job satisfaction, even fifteen years later and even after accounting for occupation, income, and industry. When people begin their jobs during a recession, they appear to feel more fortunate about having a job in the first place.[88] Psychologist Adam Grant has argued that the ripple effects of the pandemic may even include more ethical leadership of our major corporations. Surviving tough times and beginning one's career when unemployment is high may reduce feelings of entitlement or narcissism, especially among men, and leaders of corporations may thereafter have a different sense of purpose.[89]

$$\gg\!\!\longrightarrow$$

After the intermediate pandemic period ends, in about 2024, there will still be normative, social, technological, and economic after-effects of the virus and our responses to it. Some are hard to foresee, but other changes are easier to predict. If history is a guide, it seems likely that consumption will come back with a vengeance. Periods of plague-driven austerity have often been followed by periods of liberal spending. This observation extends far back; Agnolo di Tura, a shoemaker and tax collector who chronicled the Black Death in 1348, noted:

> And then, when the pestilence abated, all who survived gave themselves over to pleasures: monks, priests, nuns, and lay men and women all enjoyed themselves, and none worried about spending and gambling. And everyone thought himself rich because he had escaped and regained the world, and no one knew how to allow himself to do nothing.[90]

If the Roaring Twenties following the 1918 pandemic are a guide, the increased religiosity and reflection of the immediate and intermediate pandemic periods could give way to increased expressions of risk-taking, intemperance, or joie de vivre in the post-pandemic period. The great appeal of cities will be apparent once again. People will relentlessly seek opportunities for social mixing on a larger scale in sporting events, concerts, and political rallies. And after a serious epidemic, people often feel not only a renewed sense of purpose but a renewed sense of possibility. The 1920s brought the widespread use of the radio, jazz, the Harlem Renaissance, and women's suffrage. Of course, the 1918 flu pandemic followed World War I and was itself more deadly. But we can expect to see similar technological, artistic, and even social innovations after the current pandemic—for instance, reflecting the ripple effects of larger numbers of people working from home.

The economic aftereffects of the COVID-19 pandemic will also likely be substantial. We have already considered the probable intermediate-term backlash against globalization, immigration, and urban living, but any such changes seem unlikely to persist past 2024, since the economic benefits of those long-term trends are so compelling. Other economic aftershocks of the pandemic may endure longer, however. A sustained recession could snowball into a true depression with effects of longer duration.

It can be difficult to differentiate the adverse economic impact of the virus itself from the adverse economic effects of the nonpharmaceutical interventions people implemented in response. Viruses can sicken and kill people and compromise the economy directly. And the precautions that people take in response, such as not spending their money or avoiding social interactions, can have adverse economic impacts on their own. A careful analysis of the 1918 pandemic in the United States that took advantage of variation in the timing of the arrival of the virus from place to place as well as variation in the timing of implementation of the nonpharmaceutical interventions as a kind of natural experiment concluded that it was the pandemic itself that depressed the economy, not the public health responses. Moreover, cities that implemented stricter nonpharmaceutical interventions and did so sooner in the course of the epidemic did not fare worse; indeed, their economies bounced back more rapidly after the pandemic was over. For example, reacting ten days earlier with respect to the arrival of the pandemic increased manufacturing employment by 5 percent after the epidemic ended.[91]

One analysis of the long-term economic impact of pandemics used painstakingly assembled data from twelve European pandemics, ranging from the first attack of the Black Death in the 1340s to the mild H1N1 pandemic of 2009, and including the European cholera epidemic of 1816, the 1918, 1957, and 1968 influenza pandemics, and various other outbreaks. In general, by killing *working-age* adults but leaving agricultural lands, buildings, mines, metals, and other capital assets relatively untouched, these deadly pandemics, on average, resulted in a rise in real wages and a long-term decline in interest rates.

After many people die in a serious pandemic, labor is generally scarce, relative to capital (this is in contrast to major wars, which typically result in destruction of capital in addition to human lives). Investment opportunities are depressed, given the excess of capital. Furthermore, people generally start saving more in the aftermath of a pandemic.[92] As a result, the natural rate of interest is reduced for nearly forty years overall, but especially over the first twenty, where, at the nadir, real interest rates are depressed by two percentage points. Real wages show the same pattern in reverse, remaining elevated for decades after a pandemic that kills many working-age adults, with wages peaking at 5 percent higher than they would have otherwise. To be clear, these effects are sustained across time, whereas the effects on the cohort of college students graduating into a recession (discussed earlier) relate to that one cohort facing the immediate shock.

The people enduring historical pandemics were often aware of some of these impacts. An account of the Black Death that devastated Rochester, England, in the middle of the fourteenth century, attributed to William de la Dene, noted a brewing class conflict:

> Such a shortage of workers ensued that the humble turned up their noses at employment, and could scarcely be persuaded to serve the eminent unless for triple wages. Instead, because of the doles handed out at funerals, those who once had to work now began to have time for idleness, thieving, and other outrages, and thus the poor and servile have been enriched and the rich impoverished. As a result, churchmen, knights, and other worthies have been forced to thresh their corn, plough the land, and perform every other unskilled task if they are to make their own bread.[93]

However, because the COVID-19 pandemic has largely spared working-age people and because it is not as intrinsically deadly as bubonic plague or smallpox, it is very unlikely to shift the balance of power between capital and labor quite as much as prior plagues have. Still, the pandemic may very well drive up wages via political pressure. The COVID-19 pandemic revealed the country's dependence on essential low-wage workers, and even without large numbers of working-age-adult deaths, laws that better enshrine worker protections in the United States in the post-pandemic period are likely. As we have seen, possible areas of improvement are paid sick and family leave, more flexible work schedules, and perhaps childcare subsidization. This is especially likely if there is sustained political activism beyond the initial burst of sympathy for grocery-store clerks, delivery drivers, and nursing-home staff. The process is unlikely to be smooth, but the COVID-19 pandemic struck, by coincidence, at a time when income inequality in the United States was already at a century-long high in ways that had increasingly come to be seen by many citizens as unsustainable.[94] Many Americans may also come to better appreciate essential but unglamorous jobs that keep their lives running and may be more sympathetic to wage demands.

The speed at which the United States recovers in relation to other countries could change its international standing. We have already seen a glimpse of this decline in stature as our leaders stumbled in their pandemic response to the first wave of the virus. As London-based journalist Tom McTague argued, "It is hard to escape the feeling that this is a uniquely humiliating moment for America. As citizens of the world the United States created, we are accustomed to listening to those who loathe America, admire America, and fear America (sometimes all at the same time). But feeling pity for America? That one is new."[95]

The loss of American economic power and the lack of U.S. leadership could create an opening for China to exert more influence, particularly in the developing world, where many countries desperately need help to cope with the virus and where American responses to the global pandemic have not been up to prior standards (though there will also be a backlash against China as the country of origin of the virus and especially given its initial lack of transparency). A lot of power will also flow to those countries that successfully develop a good vaccine or effective medicine.

The possible decrease in stature of the United States may

paradoxically both constrain and open up the future of young Americans. They may see themselves as more connected to the global community and more dependent on it too. Growing up during the pandemic will surely shape a generation of youth in other ways. Pandemics can leave a mark and change life trajectories in young adulthood, as we saw with the projected income trajectory forecast for new college graduates. The effects are different at younger ages. Our ten-year-old son, for example, mostly responded to the disruption in his life with insouciance, spending time outdoors and enjoying the suspension of school while he learned at home. But he also articulated that he was worried that my wife, Erika, and I would die. And the social isolation was very difficult for him.

For large numbers of American children in less privileged circumstances, the challenges have been far greater. Quite a few may have experienced the pandemic as a traumatic adverse childhood event—especially if their parents lost their jobs or their lives—and the memory of it will linger. Given that 45 percent of Americans already have at least one adverse childhood experience (for example, the death or mental illness of a parent), and 10 percent have three or more such experiences, the pandemic may amplify young people's already worrying mental-health trends, such as increasing behavioral problems and rising suicide rates.[96] We may see an epidemic of post-traumatic stress disorder in the post-pandemic period as these children grow older, especially if parents are unable to control their own anxiety and if traditional outlets for children's well-being (such as sports and free play) are curtailed. Some educators are concerned that the already well-documented diminishment of broader areas of curriculum—such as art, music, physical education, and social studies—will only accelerate as schools provide a more stripped-down educational experience in the immediate pandemic period; this could have ramifications for years. However, a minority of children may actually return to school with more resilience, and they could fare better in the long term too.

Pandemic disease can have effects even earlier in a child's life. Early-life exposures to the Spanish flu, either in utero or immediately after birth, had a lasting impact on morbidity, mortality, and socioeconomic status later in life. For example, Taiwanese children born in 1919 were shorter and experienced later growth spurts compared to their counterparts in adjacent birth cohorts.[97] For American children born between 1915 and 1923, prenatal exposure to the Spanish flu was linked to a more than 20 percent increase in the rate of ischemic heart disease after the age of sixty compared to those with little or no prenatal exposure.[98] Prenatal exposure to the Spanish flu has also been linked to lower educational attainment (a five-month decrease for those with infected mothers and a 4 to 5 percent lower likelihood of completing high school), annual income ($2,500 lower for sons of infected mothers), and higher rates of physical disability (8 percent greater likelihood of experiencing work-preventing disability).[99] Similar detrimental effects of in utero exposure to the Spanish flu have been seen in Brazilian and Swedish samples.[100]

Finally, our arts and literature will be imbued with pandemic-related symbolism. Already by the summer of 2020, I started to spot still lifes that featured face masks and other allusions to disease on Instagram accounts of artists. After severe epidemics in the past, the arts have gone in new directions. For instance, after 1918, romanticism faded and classicism made a comeback as artists, fashion designers, and architects attempted to "slough off" the excess of the turn of the century. The post-1918 decade was a time in which artists said, "We had nothing on the ancients, after all."[101] We were just as prone to catastrophe and death. In her essay "On Being Ill," Virginia Woolf complained that the culture had neglected an obvious source of inspiration: "Novels, one would have thought, would be devoted to influenza; epic poems to typhoid; odes to pneumonia.... But no."[102] The 1918 pandemic did indeed leave some mark on literature; for example, in chapter 6, we saw its influence on Hemingway's work.

The bubonic plague had a more dramatic effect on Western art, bringing to the fore—in shockingly graphic detail—the human preoccupations with death, pain, and sin. Pieter Bruegel the Elder's 1562 painting *The Triumph of Death* is a hellscape of bodily dysfunction—skeletons are seen pushing cartloads of skulls or decapitating, hanging, and drowning hapless victims while dogs pick at cadavers. Edvard Munch's *Self Portrait with the Spanish Flu* in 1919 is a startling evocation of the impact of the pandemic on the human body, with its echo of his more famous *The Scream of Nature* painting writ across his haunted, open-mouthed gaze. The young artist Keith Haring, who died of AIDS at age thirty-one, painted a famous poster in 1989 about the HIV pandemic with the legend SILENCE = DEATH in which three figures ape the "See no evil, hear no evil, speak no evil" proverb about the three wise monkeys.[103] It became an iconic image.

The pandemic may possibly reverse certain political and cultural trends that have, in my view, bedeviled our society in recent years. Early in the pandemic, I became worried that the thinning out of our intellectual life over the past twenty years would pose barriers for managing the spread of the virus. Political entrenchment and geographically segregated living patterns have made people less open to opposing ideas, and this has hampered addressing a variety of societal problems, from climate change to mass incarceration. I feared that, together with a number of other problematic, convergent features, this intellectual atrophy would make our response to the pandemic challenging.

First, there has been a progressive denigration of science. Science has come to be seen by too many as serving political ends. Many people have even abandoned the fundamental idea that it is possible to have an objective appreciation of the truth. For instance, right-wing politicians have not wanted to acknowledge findings from climate-science or gun-violence research, and left-wing politicians have wished to deny the role of genetics in human behavior. Rather than engage difficult topics head-on using our best efforts at objective research, many people have found it easier to ignore inconvenient truths and suppress the scientific inquiry that might reveal them.

Scientific literacy is low among the general public as well. A total of 38 percent of Americans believe that God created humans in their present form sometime in the last ten thousand years.[104] Over 25 percent of Americans believe that the sun revolves around the Earth. And 61 percent cannot correctly identify that the universe began with the big bang.[105] Substantial fractions of people reject the efficacy of vaccines, and some believe wild conspiracy theories, like the idea that airplane exhaust is used by the government to control the populace.

Distinct from the denigration of science, there has been a downgrading of expertise and a progressive anti-elitism in our society, fostered by extremists at both ends of the political continuum. Experts are seen as out-of-touch elites, and expert knowledge is seen as a kind of conspiracy aimed at obtaining resources for the privileged at the expense of the masses. But many people devote their lives to developing mastery in all areas of life. The sociologist Everett C. Hughes famously noted that one person's emergency is another person's ordinary employment.[106] When you have a big plumbing leak in your house, it's an unusual event and an emergency for you. But it is a routine event for the plumber who comes to repair it. This is how society has been organized since the time that people began to urbanize and develop specialties and means to barter knowledge as well as goods. When you are searching for a mechanic or a surgeon, you want expertise and competence.

The irony is that this tendency to disparage science and expertise coexists with a respect for scientists themselves. One national survey, conducted at the end of April 2020, found very large fractions of the population trusted scientists and physicians. For example, 88 percent of Americans reported that they had some or a lot of trust in the CDC; 96 percent had such trust in hospitals and doctors; and 93 percent had such trust in scientists and researchers.[107] What are we to make of the perplexing fact that so many aspects of our national response to the pandemic were nevertheless contentious? I think what is happening is that people trust science until it conflicts with their personal, religious, or ethical values. Most Americans (73 percent) believe that science generally has a positive effect on society. And 86 percent of Americans have a "great deal" or a "fair amount" of confidence in scientists to act in the public interest.[108] But many Americans also want scientists to stay in their labs and not actually influence policy, with 60 percent saying that scientists should "take an active role in policy debates" and 39 percent saying they should simply "focus on establishing sound scientific facts." This is divided along partisan lines—73 percent of Democrat-leaning citizens support scientists taking an active role compared with 43 percent of Republican-leaning citizens. Yet scientists have often been at the forefront of important issues, from nuclear war to disability rights.

In fact, Americans are evenly divided on whether scientific experts make better policy decisions than other people, with 45 percent saying they do and 48 percent saying they do not (7 percent say their decisions are usually *worse* than other people's). But there is a partisan divide here, too, with 54 percent of Democrats saying scientists' policy decisions are usually better and just 34 percent of Republicans feeling this way. And despite the overall confidence in science, many Americans also express suspicion: 63 percent say the scientific method "generally produces accurate conclusions," but 35 percent think it "can be used to produce any conclusion the researcher wants."

These unfortunate features of American culture, coupled with a centuries-old history of embracing visionaries, charlatans, and quacks, intersected with the especially politically polarized environment of 2020, making a bad situation worse.[109] A national survey conducted in April 2020 evaluated partisan divides across a range of public health behaviors recommended by experts. When asked if they were following a selection of recommendations, Democrats and Republicans reported different levels of participation; for avoiding contact with other people, the fraction was 75 percent versus 67 percent; avoiding crowded places was 79 percent versus 72 percent; and wearing a mask outside the home was 64 percent versus 50 percent.[110] None of these are remotely controversial from an epidemiological point of view.

Getting the evidence right as best we could at any given moment was crucial in controlling the epidemic. As Dr. Fauci noted in an interview where he tried to explain the importance of science in combating such threats, "Sooner or later, something that *really* is true will get confirmed, time after time after time. And something that in good faith was *thought* to be true, but isn't, when the scientific process repeats it over and over again, all of a sudden you realize, you know, 'There was something about that that wasn't quite right.' So as long as science is humble enough and open enough and transparent enough to accept the self-correction, it's a beautiful process."[111] But science cannot work as intended when scientific findings—say, about the utility of masks or vaccines—are interpreted as political statements.

Finally, there is the loss of the capacity for nuance in our public discourse. Problems and policies are framed—and seen—as black-and-white. The tolerance for shades of gray and complexity is low. That makes it difficult for scientists to communicate that we do not know exactly what will happen with this pandemic but that there is a range of options, each with a certain probability, and we should act accordingly. Neither blind confidence nor total panic is justified. In an era of sound bites, coping with the complexity of a pathogen that scientists are only beginning to understand has not been an easy task. When coupled with the underlying exponential growth seen in infectious disease outbreaks, which sneaks up on decision-makers, this often resulted in the public being behind the curve in its response.

Of course, as we saw, the desire for simplicity and certainty during a time of complexity, uncertainty, and danger can lead to lies and false reassurance by politicians and hucksters. Politicians around the country, including the president of the United States and others in the White House, promulgated information that was plainly scientifically false from the outset. Asymptomatic transmission *was* possible. Nonpharmaceutical interventions *did* save thousands of lives. COVID-19 *was* much more serious than the flu.

Despite all this, I think that one of the unexpected impacts of the COVID-19 pandemic may be that a society that feels besieged by the threat of the virus will increasingly treat scientific information, and not just scientists, seriously. We have seen this in other countries. Previously obscure doctors and physicians, and not just Dr. Fauci, have suddenly become household names as they calmly explained what was known about the epidemic.[112] It's possible this might be one of the lasting impacts of the COVID-19 pandemic: an increased respect for science and expertise, even when it leads to people taking actions they would rather avoid. Perhaps after the dust settles from the pandemic, and humanity moves on to other threats that require scientific understanding, such as climate change, the voice of experts might be given more weight.

After all, other large-scale threats have historically prompted scientific innovation, and we may see this again. The 1918 pandemic stimulated developments in microbiology and public health. The Manhattan Project during World War II contributed to tremendous advances in physics. The 1957 launch of Sputnik, the first Soviet satellite, prompted heavy American investment in space science and engineering. The "war on cancer" declared in 1971 had a similar impact (although it did not cure cancer, it advanced fundamental medical science). Perhaps the multitrillion-dollar hit to the American economy by the COVID-19 pandemic will make multibillion-dollar investments in science—from virology to medicine to epidemiology to data science—seem well worth it.

Plagues can also lead to long-term shifts in how we think about government and leaders. In medieval times, the manifest inability of rulers, priests, doctors, and others in positions of authority to control the course of plague led to a wholesale loss of faith in the corresponding institutions and a strong desire for new sources of authority. Some scholars have speculated that this set the stage for the rise of capitalism and even of the Reformation, since it became very clear that the priests had no way of stopping mortality from the plague. This may have spurred developments in empirical medicine too, since the doctors also were ineffective at stemming the tide of death.[113]

It is possible that the inability of our political institutions to fight the virus will have similar implications. We saw earlier that interest in collective state action will likely rise in the immediate and intermediate pandemic periods, but if the actions are incompetent, confidence in political institutions will fall. The incompetence of our government in confronting the pandemic (especially when compared to the responses of other countries) coupled with the essential necessity of strong collective action to combat epidemic disease may result in a shift in political preferences aimed at undoing the existing order.

Given the strong, coordinated state action that is required to achieve control of the virus, it is likely that the role of government itself will increase from the immediate pandemic period into the post-pandemic period. The worse the pandemic gets, the more people will expect from themselves, from others, and from the state.

8.

How Plagues End

> Dr. Rieux resolved to compile this chronicle … so
> that some memorial of the injustice and outrage
> done them might endure; and to state quite
> simply what we learn in time of pestilence: that
> there are more things to admire in men than
> to despise.
>
> —Albert Camus, *The Plague* (1947)

One day in 1902, an American second lieutenant named George
Marshall crossed a small river on the island of Mindoro, Philippines,
to pay a visit to a local leader and his three young daughters.
The attractive girls were apparently the main draw; they laughed
and made chitchat and sang "delightfully." Social visits were conducted in the mornings in those days to avoid the oppressive heat.
Marshall left. But later that same day, he had cause to return to
the village, where he joined in the funeral for the same four family
members who had welcomed him only hours before. They were
dead from a cholera outbreak that struck precipitously, eventually killing five hundred of the twelve hundred residents of the
hamlet.

Forty-six years later, now–secretary of state George C. Marshall (best known for the recovery act that got Europe on its feet after World War II) movingly recalled that grim experience when he was a young officer in opening remarks to an international conference of tropical disease experts held in Washington, DC. Like many Americans who felt optimistic and even triumphalist in the postwar period, Marshall envisioned a world where deaths like the ones in Mindoro would become relics of history. Vanquishing infectious disease was not an insoluble medical challenge, he declared, but "an international problem, and it should be solved by pooling of the genius and the resources of many nations."[1]

This optimism was widespread and sustained. Writing in 1963, physician and anthropologist T. Aidan Cockburn, an expert on ancient diseases who helped manage malaria outbreaks during the war, also noted his expectation that "within some measurable time, such as 100 years, all the major infections will have disappeared."[2] In 1978, Dr. Robert Petersdorf, an international leader in infectious diseases, opined in remarks to future infectious disease doctors, "Even with my great personal loyalty to [the specialty of] infectious disease, I cannot conceive a need for 309 more infectious-disease experts unless they spend their time culturing each other."[3]

Though these claims sound naive now, the period leading up to them and through the 1950s had seen astonishing advances. Many of the most devastating infectious diseases were in retreat, brought under control by a combination of factors that included rising wealth, better sanitation, improvements in food preparation, and— as a coup de grâce—the invention of antibiotics (as we saw in chapter 3). Penicillin, discovered in 1928, was a miraculous drug, and there were many more classes of antibiotics useful against bacteria of all kinds that followed. Vaccines became available in rapid succession for a wide range of diseases, from pertussis (1914) to tetanus (1924) to polio (1953) to measles (1963). The discovery of insecticides that killed mosquitoes and other insects that transmitted malaria and other diseases also raised hopes (in 1948, Paul Hermann Müller was awarded a Nobel Prize in Medicine for the discovery of DDT).

The identification of a path to eradicate smallpox a hundred and fifty years after Jenner's experiment with cowpox and our ultimate success against this scourge—it was declared eradicated in 1980—also fed this confidence. Of course, smallpox had the salient advantage of lacking any animal reservoir, so once gone from humans, it was forever gone. Polio nowadays offers a similar opportunity, and in 2016, after many years of coordinated international effort and support from the Bill and Melinda Gates Foundation, there were only forty-six cases of this disease worldwide.

The mid-twentieth-century optimism about disease eradication was quite broad. But smallpox and polio are exceptional cases, and these optimistic views were held despite the occurrence of the 1957 and 1968 influenza pandemics, an oversight that is hard to understand in hindsight. Of course, much of this triumphalism pertained to the rich Western countries. In the rest of the world, people still suffered and died from infectious diseases—in great numbers—in an ongoing reflection of global socioeconomic inequality.

In any case, it was not clear why human beings should be favored to win against microbes in an evolutionary arms race. Microbes have been around a lot longer than humans, are more numerous, do not mind dying, and can mutate rapidly, evading our defenses. How could we truly bring about their end? As molecular biologist Joshua Lederberg observed, "Pitted against microbial genes, we have mainly our wits."[4] And often, as we have seen, these wits are deployed not so much in the development of sophisticated pharmaceutical armaments but rather in the very basic implementation of the simplest of tools to fight our enemy—namely, staying six feet apart. While we can use our wits to win, perhaps, against a pathogen causing a particular outbreak, and while we can occasionally eliminate a pathogen like smallpox altogether, it is extremely doubtful we can win against all pathogens. Infectious disease care and control seem more realistic objectives than eradication.

As historian Frank Snowden has argued, the emergence of the worldwide pandemic of HIV in the 1980s was the death knell of this optimism.[5] Still, it took some time to reset expectations. In 1992, the U.S. government allocated just seventy-four million dollars for infectious disease surveillance.[6] In 1994, the CDC founded a new journal, *Emerging Infectious Diseases* (wherein many outstanding articles about COVID-19 would be rapidly published in 2020). In 1996, President Clinton issued a statement noting that emerging infectious diseases were "one of the most significant health and security challenges facing the global community" and, of course, the United States itself.[7] In 1998, the U.S. Department of Defense warned that "historians in the next millennium may find that the 20th century's greatest fallacy was the belief that infectious diseases were nearing elimination. The resultant complacency has actually increased the threat."[8] In 2000, the CIA labeled infectious disease as a serious threat.[9] Infectious diseases were seen as "nontraditional challenges," especially in the post–Cold War era, when military threats to the United States were seen as diminished. And in 2003, as we saw, President George W. Bush launched two global initiatives targeting HIV and malaria. From this perspective, nothing about the appearance of SARS-2 in 2019 should have been surprising. Rumors of the end of infectious disease were, we can say with confidence, greatly exaggerated.

Globalization, mass migrations, rapid airline links, the ever rising size of the human population, and humanity's increasing localization in huge and densely packed metropolises also contribute to the persistence of deadly infectious diseases. Outbreaks of novel pathogens reflect, among other things, changes in the way humans come into contact with animals. In fact, two of the biggest global challenges humans face—extreme weather events (like hurricanes and droughts) and periodic outbreaks of serious diseases—may be linked by climate change. People driven from their homes by changes in the weather or people clearing new land for cultivation may come into contact with animals (who may also be driven from *their* homes) in ways that increase the likelihood of emergence of new pathogens in our species. A review in 2008 documented that, contrary to the hopes from forty years earlier, there had been 335 new infectious diseases in the period between 1910 and 2004 and that their threat to global health was actually increasing.[10]

It's worth noting that all of our modern plagues are zoonoses coming to us from *wild* animals. Our other major contagious diseases (from smallpox to TB to measles) mostly came to us from animals that we domesticated beginning ten thousand years ago, which has allowed us time to coevolve with them, to some extent, and develop at least some genetic resistance. These were pathogens that once afflicted the wild ancestors of our cows, pigs, sheep, chickens, and camels. For example, measles appears to have entered our species quite recently, in the sixth century BCE, from a progenitor in cattle that caused a disease called rinderpest. This coincided not only with humans living so close to cattle (domesticated millennia earlier) but also with the establishment of large cities. These were crucial because once a pathogen arrives among us, in order for it to stay endemic and endure, it needs a population size above the "critical community size."[11] If a host population is too small, the pathogen runs out of runway and auto-eradicates. Human populations need to be large enough to sustain the ongoing transmission of any particular pathogen.

These observations help explain the perplexing fact that when Europeans came into contact with American Indians, the transfer of deadly germs was all one way (with the possible exception of syphilis).[12] There were no domesticated animals in the New World (other than the Peruvian llama), which meant humans there had no opportunity to evolve genetic resistance to particular diseases that originated in such animals before circulating among people.

And population sizes were smaller, even in the great civilizations of the Maya, Inca, and Aztec. Compared with Eurasia, the New World was thus a "microbial paradise," devoid of major endemic infectious disease.[13]

Domesticating animals (and bringing their pathogens up close to humans in a sustained way) had some other untoward consequences. Animal husbandry, and agriculture more broadly, contributed to the invention of cities in the first place by providing a steady supply of food. And this, in turn, fed into the development of far-flung trade routes and high-density habitation (which was often coupled with poor sanitation). These developments paved the way for the epidemics that began to afflict the great ancient civilizations, including Greece and Rome (in the second century CE, the city of Rome had over a million inhabitants!). But these agglomerations of humans and their institutions were able to survive major outbreaks too, for reasons we saw in chapter 6. So the picture is complex: devastating epidemics that brought societies low and fostered social changes but that were then brought under control by the power of humans living together, collaborating, and learning ways to fight off disease using even more ancient capacities with which our species was naturally equipped.

All of these features of how humans have lived for the past few thousand years are still with us. Why should infectious diseases not be? And like the serious plagues before it, the COVID-19 pandemic will be a historical watershed.

»——————→

How does SARS-2 compare with other major infectious killers in the United States over the past century? It's clear that, with the mere introduction of this new pathogen into our species, our life expectancy has been reduced. An exogenous force has changed our surroundings for the worse; like widespread contamination from a nuclear accident or a global change in temperature, it makes it harder for us to survive. Quantifying this overall impact is not straightforward. But we can start by simply counting deaths. It bears repeating that, given the intrinsic epidemiological parameters of SARS-2, had we done nothing to address it, the virus could easily have killed a million people in the United States during the first wave and countless more around the world—and may yet do so before the end.

From an individual perspective, it can be difficult to know what metric to use to evaluate risk of death—for instance, if we should look at it in absolute terms or relative to other causes. In chapter 2, we saw that even if a person gets the disease, unless he or she is older than seventy or eighty, the chance of death (the CFR) is still less than 1 percent. To be clear, this is pretty bad from the point of view of a doctor! To frame this a bit more, consider the chances of death once a person is hospitalized with COVID-19. Overall, the risk is on the order of 10 to 20 percent (though, again, this is very sensitive to age and illness severity). Roughly speaking, the risk of death of a forty-year-old patient who is hospitalized is about 2 to 4 percent. This is in the range of the risk of death of a seventy-year-old patient who is hospitalized for a heart attack in the United States.[14] In fact, at every age, being hospitalized with COVID-19 is meaningfully worse than being hospitalized for a heart attack in terms of the likelihood of making it out of the hospital alive.

But that is only the risk for people who actually come to be infected or come to be hospitalized. What about at baseline? Out of the U.S. population of 330 million, about 3 million people die each year, for a crude death rate of 9.1 people per thousand. If, for the sake of argument, we assume that over a year, the coronavirus pandemic causes one million deaths in the United States, the crude death rate would rise to 12.1 per thousand. The average person's absolute risk of dying from the virus would remain small—roughly a three-out-of-a-thousand chance (1,000,000 extra deaths from COVID-19 divided by 330,000,000 people). That seems low, but this level of mortality would still far surpass all of the threats to life an average person faced that year, making COVID-19 the number-one cause of death.

One careful analysis of weekly mortality data from Sweden compared deaths per day in 2020 to prior years, quantifying the excess deaths and assessing the impact on mortality across all ages (using the method invented by William Farr in the nineteenth century, discussed in chapter 2). It estimated that SARS-2 was a kind of shock that, if sustained, would lop off three years of life expectancy from men and two years from women. Even in a rich, functional society like Sweden, the pathogen would have that sort of intrinsic force.[15]

Another important subtlety here is that, while the great majority of deaths from COVID-19 occur in the elderly, this is true of virtually all other causes of death too. To fully understand the impact on mortality (leaving aside the further substantial impact on disability and morbidity) requires us to compare the virus's effect on overall risk of death after taking age into account. Being young and not fearing COVID-19 because fatal disease seems rare is a reasonable posture (though not caring about infecting others is not!). But it is crucial to recognize that, at baseline, young people do not face quantitatively large risks of death of any kind. Nevertheless, COVID-19 increases the baseline risk of death at *every* age. Most parents worry about all kinds of uncommon calamities happening to their children. But if we worry at all about our children drowning or being kidnapped, we should rationally be worrying more about COVID-19.

How can the extra risk of death from COVID-19, even if low in absolute terms, be put into perspective, given the background risk faced by people at any given age? Sophisticated demographic estimates offer a way of thinking about this. For instance, the death of 125,000 Americans over a three-month period corresponds to artificially "aging" everyone by 1.7 years, in terms of their risk of mortality. In other words, during such a period, a 20-year-old would have the risk of death faced by a 21.7-year-old in normal (non-pandemic) times, and a 60-year-old the risk of a 61.7-year-old. These numbers may again seem trivial, but in population terms, they are not. Of course, if the number of deaths from the pandemic during the period were larger because we did nothing to mitigate it, the numbers would rise proportionately. If half a million people were to die during the first six months or so of the pandemic, through October 2020, this would translate into roughly a 3.3-year increase such that, say, a 60-year-old would face the risk of a 63.3-year-old, and so on.[16]

How could such a large number of people dying over such a relatively short time result in these individual risks that might not seem like that big a deal? As we have seen repeatedly, SARS-2, while serious, is clearly not as deadly as the major plagues of prior centuries. And at baseline in the twenty-first century, especially in the more developed world, death is a relatively uncommon statistical event for most people in most years of their lives. Even an eighty-year-old man has a risk of death in the next year that is "only" 5 percent. And since the greatest risk for this pandemic is experienced by the elderly, it can be easier to lose sight of the mortality impact.

But most important to me as a public health expert and physician, these small changes in individual mortality scale up very quickly and become really alarming on a population level. A better way to come to terms with the worrisome nature of COVID-19 is to return to the population level and compare it to other epidemics using the metric of *years of life lost* relative to other threats. In this way, demographers Joshua Goldstein and Ronald Lee were able to generate the estimates shown in figure 16.[17] For comparison purposes, they assumed that one million people will die from COVID-19 by the time the pandemic is over in the United States after several waves (which is not inconceivable), but the numbers in the figure can just be adjusted linearly if the death count winds up being different (for instance, the bar would be half as large if "only" five hundred thousand people die).

After taking into account the size of the population, the age distribution, and other background causes of death, we can then compare the gravity of the COVID-19 pandemic to other threats. It's clear that COVID-19 is a serious disease, though not as bad as the Spanish flu and perhaps one-third as bad as the overall loss of life caused by HIV over its nearly thirty years of afflicting our nation.

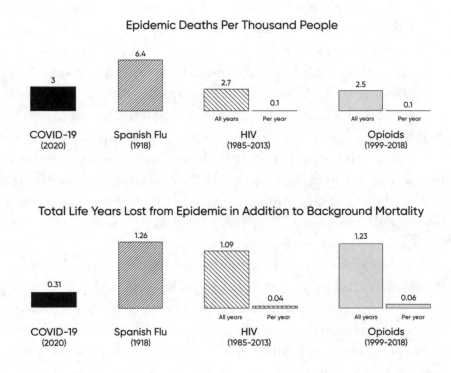

Figure 16: The mortality impact of COVID-19 in the United States can be quantitatively compared to that of other modern epidemics.

These sorts of demographic calculations also allow us to benchmark the financial benefits of saving lives against the financial costs of shutting down the American economy during the deployment of the NPIs. By using a standard benchmark of five hundred thousand dollars as the *economic* value of a year of life (or ten million dollars per life, regardless of age), we can estimate that one million coronavirus deaths (at the rough age distribution at which they occur) would be worth about six trillion dollars. Even at the highest end of a range of estimates of the consequences to our economy, including the expenditures by our government, we do not reach that sum. Strictly from an economic perspective, our response was commensurate to the threat posed by the pathogen. It's a bad virus.

》————→

A very basic way that pandemics have one sort of "end" has to do with the use of the nonpharmaceutical interventions to stop transmission of the pathogen. New cases can be driven to zero, as we saw in China and New Zealand. But this is not the real end of a plague. It's a kind of false bottom because the pathogen is still out there and will return the moment people resume normal interactions, as we saw with the recurrences in the United States over the summer of 2020.

After its dramatic initial appearance, SARS-2 will ultimately become endemic; it will regularly circulate among us at some low, steady level. This is connected to the second kind of end, which we have already considered: herd immunity. Here, the pathogen is still around, but it has a much more difficult time reestablishing itself. This resembles a well-vaccinated population for any infectious disease; there are only occasional, small outbreaks among nonimmune people.

By 2022 or so, we will reach this outcome naturally or via vaccination. Of course, if we do rapidly develop and distribute a safe and effective vaccine, we could reach herd immunity with fewer deaths. Based on the fundamental R_0 of SARS-2, as we saw in chapter 2, up to an estimated 60 to 67 percent of the population could be affected (or roughly two hundred million people in the United States). The necessary percentage could be lower, closer to 40 to 50 percent, given that social network structure means that different people spread the virus to different extents (as we also saw in chapter 2); or it could be higher, if the epidemic moves extremely fast and we overshoot the level required for herd immunity. Whatever the exact percentage, as a pathogen spreads, some people will die and others will recover and become immune, so eventually the virus will run out of places to go. This is the ordinary, natural way that, biologically speaking, epidemics *end*.

This is what we mean when we say that a pathogen is under control. But sometimes, plagues are so devastating that a society never recovers. It's very important to emphasize that, as bad as COVID-19 is, it's not remotely as bad as epidemics of bubonic plague, cholera, or smallpox that have killed much larger fractions of the population and that have had much larger and longer-lasting effects. Those types of plagues are even associated with the iconography of the Four Horsemen of the Apocalypse, Pestilence riding side by side with War, Famine, and Death. Those epidemics vindicated the adage that "too few of the living were left to bury the dead."

Such plagues end in utter destruction. Some American Indian tribes were almost entirely wiped out after European colonization, with less than 5 percent of the population left alive. Here is one indigenous account of a 1519 epidemic, possibly of measles or smallpox, that afflicted the Mayans in what is today Guatemala:

> The people could not in any way control the sickness. Great was the stench of the dead. After our fathers and grandfathers succumbed, half of the people fled to the fields. The dogs and the vultures devoured the bodies. The mortality was terrible.... So it was we became orphans.[18]

Even bubonic plague, which sometimes took whole cities, did not destroy whole societies. Thankfully, this sort of end is very rare.

»———→

As we saw earlier, from a Darwinian point of view, it does not suit a pathogen's interest to kill its victims, since it would rather that its hosts move around and transmit it to other people. Typically, with time, viruses become less lethal as a result of this preferential spread and survival of milder strains.

It is still too early to know how SARS-2 might mutate and over what time frame. Over the short term, it's possible that the virus could change to be either better or worse for us (in terms of its transmissibility, lethality, or both)—even though any long-term changes are likely to be positive for us, the unfortunate hosts. Many thousands of mutations in SARS-2 have already emerged naturally, but most of those mutations do not affect the action of the virus. As of the summer of 2020, there is not much evidence that the virus has mutated to be less severe, and there is some possible suggestion that one circulating variant might have become more transmissible (involving the D614G mutation in the spike protein).[19] Early clinical observations from China reported in the popular press suggested that the virus might have meaningfully different variants that caused primarily lung damage in some patients and primarily extrapulmonary damage (gut, nerves, kidney, or heart) in other patients.[20] It's possible that the variation in the lethality of outbreaks from place to place (for example, comparing New York City to Seattle) could relate, in some small part, to the virulence of the individual viral variants that were prevalent. But none of this was clear in the summer of 2020.

Still, one way a pandemic of a pathogen such as SARS-2 can come to an end is that the virus mutates over a period of years to get much milder. In fact, it's possible that the four coronavirus species that cause the common cold are distant echoes of long-ago pandemics, now domesticated through a combination of herd immunity and genetic change. There is some intriguing evidence that this might have happened with the OC43 coronavirus species that causes the common cold. By performing genetic analyses on a set of coronaviruses and developing a molecular clock based on fundamental mutation rates, scientists have been able to infer that OC43 arrived in our species from an animal reservoir sometime around 1890.[21]

At that time, a pandemic was raging that has traditionally been ascribed to a species of influenza. It is thought that the pandemic of 1889 began in May in Bukhara, a city in modern Uzbekistan. Known as the Russian flu (and also sometimes the Asiatic flu), it traveled quickly through Russia and then into Western Europe and the rest of the world. In the first week of December 1889, it struck its first major city, St. Petersburg, with a vengeance, infecting about half of the population (which is roughly the usual percentage for herd immunity for similar pathogens). From there, following rail links, it spread over the winter, attacking Berlin, Brussels, Paris, Vienna, Lisbon, Prague, and other major capitals and afflicting various leaders, such as the czar of Russia, the king of Belgium, and the emperor of Germany.

In a series of annual waves through 1892, it caused approximately 250,000 deaths in Europe. It reached the United States in the second week of January 1890.[22] In fact, the pandemic took only four months to circumnavigate the globe, peaking in the United States just seventy days after it had struck St. Petersburg (at that time, transatlantic travel by boat took about six days). The attack rates varied from 45 percent to 70 percent across geographic areas; the CFR ranged from 0.1 percent to 0.28 percent; and the R_0 was 2.1 (though the reproduction number varied substantially from one city to another).[23]

In London, the 1890 epidemic was anxiously anticipated by newspapers, which had reported outbreaks in other cities using the recently invented technology of the telegraph, publishing often gruesome and sometimes exaggerated accounts of the epidemic. Then, as with the COVID-19 pandemic now, some people thought that the epidemic was not real and that it had simply been "started by telegraph." Yet, as the anonymous author of an article about the epidemic in the British medical journal *The Lancet* noted, "The numbers crowding to the hospitals and dispensaries form sufficient answer to [that] suggestion," and the illness was indeed worse than "the common colds and catarrhs which are ordinarily rife at this season."[24] The author continued: "There is a growing tendency amongst the better educated to regard the epidemic as something almost too trivial for serious consideration, and this idea is often pushed to the extreme of thinking that it can be treated disdainfully by homely remedies, and by sufficient energy of self-control. But it is one thing to deny the reason for panic and another to urge the recklessness of unconcern."

The newspapers in New York City were indeed initially unconcerned. "It is not deadly, not even necessarily dangerous," the *Evening World* wrote before noting that "it will afford a grand opportunity for dealers to work off their surplus of bandanas." But as the death toll rose in the United States, people began to take it more seriously. As usual, New York City was hard hit, and during the first week of January 1890, the city reported a record number of deaths, 1,202. "On a Sixth Avenue Elevated train this morning, fully one-half of the passengers were coughing, sneezing, and applying handkerchiefs to noses and eyes, and many of them had their heads bundled up in scarves and mufflers," the *Evening World* later reported. "They were a dejected and forlorn appearing crowd."[25]

It has been speculated that this pandemic was the H3N8 strain of influenza or possibly the H2N2 strain.[26] But it's just possible that the 1889 pandemic was a major pandemic of a coronavirus. Clinicians' real-time observations of the disease when it struck London in 1890 described it as having three types of symptoms: "Sometimes pulmonary symptoms predominate; in other cases gastro-intestinal troubles appear to form the main feature; and in both there may be frequent complaints of racking pains in the head or in the limbs."[27]

Interestingly, COVID-19 similarly shows these three clinical types of illness, with emphasis on the respiratory, gastrointestinal, or musculoskeletal and neurological systems.[28] However, while patients with influenza clearly manifest respiratory symptoms (cough, sore throat, congestion, and shortness of breath), musculoskeletal symptoms (such as muscle pain or weakness, or headache) are not as typical, and neither are gastrointestinal symptoms (such as nausea, vomiting, or diarrhea).[29] Central nervous system symptoms were also more pronounced in that pandemic than is typical of influenza, according to historical accounts, and the OC43 coronavirus is known to be neuro-invasive.[30] Another suggestive piece of evidence comes from work that indicates that the Russian flu showed a disproportionate increase in case fatality with increasing age and possibly had a backward-L-shape mortality function, which is unusual for influenza (remember that influenza usually manifests a more pronounced U-shaped function).[31]

OC43 and a particular bovine coronavirus species show remarkable similarities in their genetic sequences and in the chemical structures of the proteins that elicit an immune response. And by comparison with other coronaviruses afflicting other animals, it is possible to reconstruct the evolution of OC43. The timing of separation from its bovine coronavirus cousin (that is, when the cold virus afflicting us went its separate way, evolutionarily speaking) proves to be approximately 1890, as noted above.

This jump to humans would have been similar to the ones made by SARS-1 and SARS-2. It turns out that, in the latter half of the nineteenth century, there was a highly infectious and deadly respiratory disease in cattle. Though we cannot be sure, it's possible that the cause of those outbreaks in cattle was the bovine coronavirus that OC43 so resembles. We do know that, from 1870 to 1890, many industrial countries engaged in major culling operations. During this slaughter, it seems quite possible that humans could have come into contact with cow respiratory secretions containing a coronavirus species, mutated from its bovine coronavirus ancestor, that would go on to cause the 1889 pandemic and eventually remain with us as OC43.

After being among us for a century, this virus would have further evolved to be a mild pathogen that just causes the common cold today. In addition to the virus evolving to be milder, our modus vivendi with it reflects an additional factor. Since OC43 is so widespread, most people are exposed to it in childhood and are spared any serious illness (remember the backward-L-shape mortality function). Thereafter, if they are exposed to OC43 again later in life, the virus just causes a mild cold, if it causes anything, because the hosts have some memory immunity. That's a stark difference from the situation now, as SARS-2 is having a field day in a wholly immunologically naive population. But it's possible that once we reach herd immunity in the coming years, people will simply be exposed to the SARS-2 virus as children, have a mild disease (most of the time!), get some immunity, and then avoid serious disease if they are re-exposed thereafter. Such a scenario is a quite possible eventual end to the story of SARS-2.

We are familiar with other viruses that follow this pattern, causing mild illness in children or in adults who have already experienced the illness as children, yet causing severe disease in adults who have not previously been exposed.[32] Chicken pox has a mortality rate twenty times higher in fifteen- to forty-four-year-olds than it does in five- to fourteen-year-olds, but it does not usually cause severe problems for adults who had it as children.[33] Similarly, Epstein-Barr virus causes mild disease in young children, but it can result in infectious mononucleosis in young adults and even be a risk factor for the onset of multiple sclerosis.[34] Other evidence suggests that if people are infected with this virus as children, they might simply have a minor upper respiratory disease, but if infected for the first time as adults, it might put them at risk for Hodgkin's lymphoma.[35]

>———→

The pathogens evolve to respond to *us*, but we, at a slower pace, also evolve to respond to *them*. Infectious diseases have been a part of our evolutionary history for so long that they have left a mark on our genes. For instance, humans have evolved genetic changes that have proven useful in coping with malaria beginning over one hundred thousand years ago, tuberculosis over nine thousand years ago, cholera and bubonic plague over six thousand years ago, and smallpox over three thousand years ago.[36]

Infectious pathogens (even if nonepidemic) have arguably been a crucial selective pressure throughout the evolution of our species.[37] The primary killers of human beings across evolutionary time are other human beings. Humans do not have any natural predators that substantially affect survival.[38] Except for our microscopic enemies.

The SARS-2 virus is a lot less lethal to people of reproductive age and can be combated with the lifesaving tools of modern medicine, so the impact on human evolution is surely going to be minimal. But, at least in theory, another way epidemics end is that hosts evolve to be resistant. And in fact, we may already have naturally occurring genetic variation in our species that affects the severity of COVID-19 in different populations, which would lay the groundwork for such evolution. Over generations, this can result in changes to the genetic makeup of the afflicted populations.

For example, over many millennia of exposure to malaria, populations in regions of the world where the pathogen is endemic evolved a number of genetic differences to help them survive, ranging from changes in levels of an enzyme known as G6PD to changes in hemoglobin structure. Northern Europeans, however, were not exposed to malaria and did not evolve such genetic resistance; they therefore fared very poorly when traveling to, or seeking to colonize, parts of the world where malaria was endemic. Conversely, for diseases like smallpox, where Europeans had some genetic resistance (and childhood immunity), the indigenous people they came into contact with or subjugated died in huge numbers.

For just one example from our ongoing catalog of such depressing cases, consider the steamboat *St. Peters,* which traveled up the Missouri River in 1837.[39] On or about April 29, when a deckhand was showing signs of smallpox, three indigenous women boarded the boat and subsequently returned to their village.[40] The disease then spread rapidly among the Mandan people, who had no natural genetic resistance. In July, the Mandan numbered about two thousand; by October, there were, by some accounts, fewer than thirty. Infectious diseases such as smallpox were arguably the most consequential factor in the decimation of indigenous populations throughout the colonization of the Americas beginning in the fifteenth century, compounding other horrors, including warfare and enslavement.[41] Sometimes, these diseases were even deliberately used as weapons.

An analysis spearheaded by anthropologist Ripan Malhi provides insight into the possible lingering genetic impact of such a catastrophe, illustrating how natural selection can work—brutally—on human beings.[42] The ancient Tsimshian people, an indigenous community in the Prince Rupert Harbor region of British Columbia, were well adapted to local diseases but not foreign maladies like smallpox. Working with the Tsimshian, Malhi compared the skeletal DNA of twenty-five individuals who resided in the region five hundred to six thousand years ago to the DNA of twenty-five living Tsimshian. There were differences in the prevalence of several immune-related genes between the two populations. For example, one variant called *HLA-DQA1* was found in nearly 100 percent of ancient individuals but in only 36 percent of modern ones, with an estimated genetic shift about a hundred and seventy-five years ago. That timing corresponded with widespread smallpox epidemics in the region. A demographic model based on the genetic analysis showed that, in keeping with historical accounts, nearly 80 percent of the community died in the decades after encountering the Europeans.[43]

Similar work has revealed analogous impacts for other deadly historical epidemics. For instance, one investigation identified convergent evolution between populations of Europeans and Roma (thought to have originated in India), both of which were afflicted by the Black Death, for certain genetic variants involved in the immune response to *Yersinia pestis*.[44] Along with the fact that these genetic variants are relatively absent in populations largely spared the bubonic plague, this suggests that the pestilence positively selected for this genetic shift. This is an active area of research, and more studies of ancient DNA are needed for a fuller picture.

While genetic adaptation to an epidemic can positively select for resistance traits that offer descendants protection from a pathogen, in some cases there can be negative consequences, such as an increased risk of genetic disease because of these mutations. The most well-known example is sickle cell anemia, a genetic variant causing deformed, sickle-shaped red blood cells that block small vessels and can lead to a host of problems, including premature death (although this is less frequent in modern times). Scientists were originally perplexed about why such a harmful genetic variant could be so widespread, but in 1954, geneticist Anthony C. Allison noticed a correspondence between the prevalence of sickle cell disease and malaria. This was a landmark study of human genetic susceptibility to infectious diseases.[45] It turns out that having the genetic variant that causes sickle-shaped red blood cells confers some immunity to malaria.

Other genetic disorders may have become prevalent in part due to the resistance they provided from infectious disease. Cystic fibrosis is an inherited disorder characterized by inflammation, tissue damage, and destruction of the lungs and pancreas, and until quite recently, it usually led to premature death. It is probable that the high cystic fibrosis rate among people of European descent reflects genetic advantages it confers with respect to infectious disease (scientists think the agent it protects against is most likely tuberculosis, but it could be cholera or typhoid).[46]

During the SARS-2 pandemic, geneticists have searched for genes that might be associated with susceptibility or resistance to the novel virus in hopes of identifying patients at special risk or of detecting possible drug targets by using natural human variation as a clue to what might be an effective pharmacological strategy. Initial work has identified certain genetic variants that occur unusually frequently in patients with the worst COVID-19 symptoms.[47] COVID-19 patients with blood type A are about 50 percent more likely to need oxygen therapy and be put on a ventilator than patients with other blood types, and individuals with blood type O experience a protective effect (ironically, this pattern is the reverse of that seen in the case of survival from cholera). A cluster of six genes on chromosome 3, including a gene known to affect ACE2 and a gene involved in the immune response to airway pathogens, is also associated with worse disease severity. So these findings all make physiological sense. And this cluster of genes, it turns out, was introduced to our species from Neanderthals and happens to be especially prevalent in Southeast Asia.[48]

These observations about human evolution indicate that, over very long periods, measured in thousands of years, pandemics end by reshaping our species. While such existing genetic variation, to the extent that it occurs, is not likely to play a material role in the course of the COVID-19 pandemic (with the possible exception of some small and localized populations who may happen to share a mutation that makes them less susceptible), it highlights the enduring ways that the social intersects with the biological when it comes to infectious diseases. Over very long time frames, we reach an uneasy genetic truce with pathogens.

»———————→

So we will reach herd immunity, or the pathogen will evolve to be less lethal, or (after a very long time) humans will evolve to be resistant. That is the biological end of the story. But pandemics are also sociological phenomena, driven by human beliefs and actions, and there is a *social* end to pandemics, too, when the fear, anxiety, and socioeconomic disruptions have either declined or simply come to be accepted as an ordinary fact of life.

In a narrow biomedical perspective on disease, historian Allan Brandt has argued, the social and environmental context is often discounted as a cause of, much less a treatment for, the condition. Physicians dispense "magic bullets" intended to restore sick patients to health.[49] But, as Brandt shows, infections and epidemics are about much more than the germs that cause them. Who gets sick, what we do about it, and even what counts as sickness in the first place are all culturally specified. This is true of all kinds of illness, from an influenza or cholera outbreak due to the dislocation of war to deaths from cigarettes or diabetes arising from the marketing of consumer products. The patient may be the one suffering from the disease, but the roots and meaning of ill health often lie in social circumstances.

If we see pandemics purely as a function of biological details—such as the mutations that make it possible for the pathogen to leave bats and spread in humans or the pharmacokinetics of drugs in our bodies—we may be lulled into thinking that there is nothing we can do to prevent or arrest such events. But if we see pandemics as sociological phenomena as well, we can more clearly recognize the role of human agency. And the more we see our own role in shaping the emergence and unfolding of pandemic disease, the more proactive and effective our responses can be.

It is tempting to call the SARS-2 pandemic a "perfect storm"—a virus mutated in just the right way in just the right host, acquired just the right epidemiological parameters (in terms of its R_0, CFR, latency period, and so on), had just the right serendipitous encounter with its very first victim, and appeared at just the right time, during the annual *chunyun* migration at a point in history when international air travel existed.

But the perfect-storm metaphor implies that the pandemic is an anomaly or that it is wholly unforeseeable. Yet pandemic disease is neither of those things. Even natural disasters, like the earthquake in Haiti in 2010, or actual storms, like Hurricane Katrina in 2005, that have devastating social and mortality consequences are not "perfect storms."[50] Earthquakes and hurricanes seem to be completely independent of human action, but long histories of economic inequality and the neglect of crucial infrastructure in both Haiti and New Orleans contributed substantially to the death toll.

Because pandemic disease is seen through human eyes, it is also stamped with powerful symbolic meanings that in turn affect how we react to what is happening. The COVID-19 pandemic assumed a significance that transcended mere host-pathogen biology. Masks, for example—the purpose of which is simply to physically stop the spread of virus-laden respiratory droplets—became one such symbol. Masks came to evoke issues of liberty and communality and were transformed into politicized emblems of either freedom (if one did not wear them) or virtue (if one did). This debate was culturally specific to America; in many other countries, people did not see wearing masks as a political act.

The controversy over what to call SARS-2 was another example of humans assigning arbitrary meaning to a seemingly factual issue. Calling it the "Chinese virus" highlighted the already fraught relationship between China and the United States and imbued this topic with ambient concerns about racism. Donald Trump and others exploited this when they called it the "Wuhan flu" and "Kung flu."[51] Yet, as we have seen, many pathogens used to be named for their origins. And the impact of symbolic meaning could even be found in a survey of American beer drinkers, 38 percent of whom would not buy Corona beer "under any circumstances" after the pandemic hit.[52]

We generate narratives of what is happening during a pandemic that are partially true and partially false, reflecting our hopes and fears. For instance, the existence of a routine and tolerated disease that people know as the seasonal flu complicated efforts to make sense of and cope with the much deadlier COVID-19. Since people's closest point of reference to the novel disease was a familiar illness that is viewed by many as a hassle but not a big risk—compounded further by the tendency to call any number of minor colds "the flu" and also by the great range of severity manifested by SARS-2—people struggled to respond optimally to the threat conceptually and, therefore, practically.[53]

This transformation of a biological entity into a social symbol greatly complicates efforts to control epidemics. This is most obvious in epidemics of sexually transmitted diseases in the past century. Such illnesses were used as rhetorical vehicles to demand the reform of sexual practices in soldiers in World War II and in gay men in the 1980s and 1990s, turning an infectious disease into a moral problem and thus taking the focus away from effective methods of controlling it. Such transformations have happened with other plagues, which have always assumed religious, moral, and political overtones, as we have seen.

Social variables and values also play a role when we think about *for whom* a pandemic has ended. For the elderly, the chronically ill, the poor, the imprisoned, and the socially marginalized, the SARS-2 pandemic might continue to be a threat biologically long after the majority of the population has moved on psychologically and practically and long after overall levels of the virus are low. This is another way that social attributes will structure when the pandemic ends. We saw this tension in the summer of 2020 as working parents and many child-development experts strongly advocated for children to return to (suitably prepared) schools in the fall while others, including representatives of teachers' unions and families with health risks, argued against it. These parties saw the disease differently and therefore acted differently.

This social construction of COVID-19 means that the end of the pandemic can also be socially defined. In other words, plagues can end when everyone believes they are over or when everyone is simply willing to tolerate more risk and live in a new way. If everyone willingly risks infection and resumes a semblance of normal life (or, implausibly, if everyone decides to employ physical distancing forever), then the epidemic can be said to have ended, even if the virus is still circulating. We got a glimpse of this phenomenon as well in the summer of 2020 as different states, tired of the lockdowns, acted as if the epidemic were over, even though, biologically speaking, it was not. It was wholly understandable that everyone was eager to leave the epidemic behind as quickly as possible. But the epidemiological reality did not submit to our desires. The pandemic was still claiming roughly a thousand lives per day, although Americans seemed inured to it. Many people, and not just self-interested politicians, seemed to believe the SARS-2 epidemic could end by fiat.

In early May 2020, as the United States began to ease the non-pharmaceutical interventions, Thomas Frieden, the former director of the CDC, observed, "We're reopening based on politics, ideology, and public pressure. And I think it's going to end badly."[54] It is one thing to determine what the epidemiology of the situation demands but conclude that the economics countermands it or that the public has had enough, but it is quite another thing to ignore the epidemiology and pretend that nothing bad is going to happen. In July 2020, states in the South faced overloaded hospitals and skyrocketing case counts, and again, many people seemed surprised. Houston recapitulated some of what happened in New York City. One day the governor of Texas, Greg Abbott, was saying that the situation was under control and then, three weeks later, like New York politicians before him, he was saying that "the worst is yet to come."[55]

A reasonable standard for states to reopen for business was fourteen days of declining coronavirus cases, but virtually no state waited to meet that standard before opening again. Americans had clearly lost the appetite for wide-scale physical-distancing measures, at least at the levels of overall mortality then present (higher levels might change their minds). As feared, the first round of reopenings after the initial lockdowns in the summer of 2020 did not go well in some large population centers. It's unclear what was gained from the haste, and the lack of effort to prepare the public for what could happen (and for the real trade-offs involved in opening schools and businesses during a time of contagious disease) was also dispiriting. The lack of scientific literacy, capacity for nuance, and honest leadership hurt us.

Another reason that the commitment to addressing the pandemic waned over the summer of 2020 was that the serious illnesses and deaths were still mostly happening offstage. While over one hundred thirty thousand people had died by the end of June, nearly half of them were in nursing homes, already isolated from the broader society, and most other people who died early on did so in hospitals that were overrun, so they often died alone. This meant that few people had personal experience with the impact of the virus. People sheltered separately, and those who died were not numerous enough or visible enough—except to their families—to highlight the threat, as we saw. Yet, as the pandemic continues to unfold in late 2020 and 2021, there will be more deaths, and as more people become personally familiar with the disease because they know someone who has died, attitudes will change.

Over the immediate pandemic period, in order to return to any semblance of normalcy, the United States will require much more widespread use of masks (and laws and policies mandating their use) and much more widespread testing (on the order of twenty to thirty million tests per day nationwide—as of July 2020, the country was performing only roughly eight hundred thousand per day). Basically, every worker who is in contact with other workers will need to be regularly tested. If the tests cost ten dollars each, the national expense would be about one and a half billion dollars per week, but that is still much cheaper than another massive economic shutdown. The virus is far too prevalent in most states in the United States to use contact tracing as an effective tool, though other sorts of electronic tools could help facilitate voluntary self-isolation. And reductions in social mixing via bans on gatherings will surely have to be present for a couple of years.[56] Employers will also have to redesign workplaces and business venues to allow for physical distancing. Schools will need to move teaching outside and make many other changes that will require national investment.

And a safe and effective vaccine, when it becomes available, has to be rapidly distributed and widely adopted, meaning that a massive public education campaign will be required (ideally with highly trusted and visible spokespeople, like actor Tom Hanks, who had COVID-19 himself, and others seen as credible on this subject, like Dr. Fauci). Ignorance about the benefits of vaccines, or resistance to them, will need to be vigorously fought back. This is an enormously complex issue, as some of those who have expressed resistance are not anti-vaxxers but instead harbor concerns about the extraordinary speed with which the vaccine is being developed and feel mistrust of the government, given its inept handling of the pandemic generally. Still others will want to wait some time to see if adverse effects emerge. We therefore need a major public education initiative to pave the way for vaccine adoption—if an effective one is even found—and there was not much evidence of such an effort as of the summer of 2020.

As Allan Brandt observed, "Many questions about the so-called end are determined not by medical or public health data but by sociopolitical processes."[57] I've seen this for years in the American response (or lack thereof) to mass shootings, automobile fatalities, rising suicide rates, and drug overdoses. After much anxiety and commentary, people eventually seem willing to accept an affliction that would have been intolerable at a prior point in time. The reverse can happen too; one of the reasons the mid-twentieth-century polio outbreaks generated so much public interest, despite relatively low levels of afflicted children, was that it seemed so culturally aberrant to see children dying of infection in an age of space travel. Americans saw polio as something that had to be wiped out. And so it was.

In some ways, our response to the pandemic can even be seen through the lens of psychiatrist Elisabeth Kübler-Ross's famous stages of coping with death.[58] Americans began with denial and anger, moved on to bargaining and depression, and will end with acceptance, which will mark the final, sociological end to the pandemic.

»———→

In 2015, Bill Gates released a popular TED Talk entitled "The Next Outbreak? We're Not Ready" in which he articulated the serious threat posed by pandemics; it has been viewed over thirty-six million times. The CDC has, for many years, maintained information on its websites and released many dozens of reports on pandemic preparedness, as have other governmental bodies. Our nation has countless well-trained epidemiologists, and many experts have been sounding the alarm.[59] Yet, because plagues are only in our distant collective memory, and few living individuals have personal memories of pandemics on the scale of COVID-19, these warnings were easy to ignore. Plus, as we have seen, epidemics are always accompanied by emotional contagions, like fear and denial.

And so Americans were caught unprepared—emotionally, politically, and practically. We did not even have the equipment needed, from PPE to tests to ventilators, to save our lives. But most of all, we did not have a collective understanding of the threat that we are facing. The COVID-19 pandemic awakened Americans to the importance of public health in the same way that 9/11 opened our eyes to the sophisticated threats to our national security, the great recession to the fragility of our financial system, and the election of various populist leaders around the world in the twenty-first century to the dangers of political extremism.

Respiratory pandemics recur. Figure 17 focuses just on those caused by influenza over the past three hundred years.[60] They have been appearing steadily all this time, every few decades. Especially bad respiratory pandemics occur every fifty to one hundred years. COVID-19 will not be the last pandemic. In fact, even as we were coping with the early phases of COVID-19, reports emerged in the summer of 2020 of a new influenza pathogen (of unclear severity) found during routine ongoing surveillance of pigs in China.[61] It is almost too horrible to contemplate having to face a pandemic from an entirely different class of pathogens overlapping with the ongoing coronavirus pandemic. But the threat is ever present.

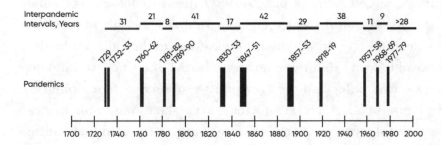

Figure 17: Serious influenza pandemics have recurred every few decades over the past three centuries.

For all the suffering it has caused, the COVID-19 pandemic also showed people new possibilities. The cessation of movement resulted in clean air and a reduction in carbon emissions on par with what is required (albeit in a more sustained way) to address climate change. Banding together to implement the various NPIs fostered a recognition of the importance of collective will and helped set the stage for political activism to address other long-standing problems in our society, from economic inequality to racial injustice to health-care inadequacies. The ability of the government to spend vast sums of money in the blink of an eye gave a palpable demonstration of the tremendous economic power it has to address a threat deemed important enough. The pandemic functioned as a kind of object lesson: *See? See what is possible?*

The COVID-19 pandemic also demonstrated in very material ways how interconnected we all are and how the commonweal is only as well off as its weakest members. In addition to raising important moral concerns, the existence of vulnerable groups in the United States and around the world who could serve as reservoirs of infection demonstrates the pragmatic utility of showing solidarity. When a deadly contagion is raging, it is in the interests of the strong to care for the weak. And effective disease containment, by definition, puts the needs of the collective ahead of the needs of individuals.

In *Prometheus Bound*, a play by Aeschylus, Prometheus is chained to a rock as punishment for giving humans the gift of fire (and thus technology). In addition, he gave us another gift: he made it impossible for us to foresee our own deaths. But since we still know that we can suffer and die (because we observe others doing so), this ignorance and uncertainty tends to make us miserable. We can use technology to predict the future, but this can also make things worse if the predictions are both accurate and dire. The chorus in the play asks Prometheus, "What cure did you discover for their misery?" And Prometheus responds, "I planted firmly in their hearts blind hopefulness." But blind hope is a fickle companion for our woe. It is not enough. Still, by forcing our gaze to the future, hope can serve another purpose: it can motivate us to prepare.

Microbes have shaped our evolutionary trajectory since the origin of our species. Epidemics have done so for many thousands of years. Like the myth of Apollo's arrows, they have been a part of our story all along. We have outlived them before, using the biological and social tools at our disposal. Life will return to normal. Plagues always end. And, like plagues, hope is an enduring part of the human condition.

Acknowledgments

I would like to thank my fantastic research assistants who worked to help me with this book: Eric Liu, Gina Markov, Drew Prinster, Caleb Rhodes, and Yiqi Yu.

Some of my own original research discussed in this book is based on efforts undertaken by the extraordinary personnel in the Human Nature Lab, including Mark McKnight and Wyatt Israel (who developed the Hunala app along with Alexi Christakis); Jacob Derechin, Eric Feltham, and Marcus Alexander, who have done work on COVID-19; and Maggie Traeger, who offered careful edits to the manuscript. My colleague Amin Karbasi also contributed to the app. I benefited from my collaboration with my Chinese colleagues, including Jayson Jia, Jianmin Jia, and others, in order to track SARS-2 in its early days in China. And Cavan Huang, a terrific graphic designer, developed all the figures.

I am very grateful to my dear friend Dan Gilbert for feedback on parts of the manuscript and to my other colleagues who read it, including Amy Cuddy, Paul Farmer, Jeff Flier, Steven Pinker, and William Nordhaus. My siblings, Quan-Yang Duh, Dimitri Christakis, and Anna-Katrina Christakis, also offered useful insights.

I am very grateful to several foundations that have generously supported my lab in recent years, including the Bill and Melinda Gates Foundation, the NOMIS Foundation, and the Robert Wood

Johnson Foundation, and also to the Tata Group for its long-term alliance with Yale University.

It is always a joy to work with my phenomenal longtime editor, Tracy Behar; she offers kind encouragement, wise advice, and perfect editing in an inimitable manner. I am also grateful to Ian Straus for his administrative help and to my agent, Richard Pine, for his confidence and careful reading as well. My copyeditor, Tracy Roe, was terrific. And I am especially appreciative to the cover designer, Julianna Lee, for making the book beautiful to look at.

My wife, Erika Christakis, a magnificent writer and brilliant thinker, read this book several times and vastly improved it. I am enormously grateful for the profound and loving life she has given me for thirty-three years, including with our beloved children, Sebastian, Lysander (who gave me much valuable feedback on this manuscript), Eleni, and Orien.

Notes

1. AN INFINITESIMAL THING

1   B. Westcott and S. Wang, "China's Wet Markets Are Not What Some People Think They Are," *CNN,* April 15, 2020.

2   F. Wu et al., "A New Coronavirus Associated with Human Respiratory Disease in China," *Nature* 2020; 579: 265–269; T. Mildenstein et al., "Exploitation of Bats for Bushmeat and Medicine," *Bats in the Anthropocene: Conservation of Bats in a Changing World,* Cham, Switzerland: Springer, 2016, pp. 325–375.

3   A.C.P. Wong et al., "Global Epidemiology of Bat Coronaviruses," *Viruses* 2019; 11: 174.

4   D. Ignatius, "How Did Covid-19 Begin? Its Initial Origin Story Is Shaky," *Washington Post,* April 2, 2020.

5   J.T. Areddy, "China Rules Out Animal Market and Lab as Coronavirus Origin," *Wall Street Journal,* May 26, 2020.

6   C. Huang et al., "Clinical Features of Patients Infected with 2019 Novel Coronavirus in Wuhan, China," *The Lancet* 2020; 395: 497–506.

7   X. Zhang et al., "Viral and Host Factors Related to the Clinical Outcome of COVID-19," *Nature* 2020; 583: 437–440.

8   Z. Wu and J.M. McGoogan, "Characteristics of and Important Lessons from the Coronavirus Disease 2019 (COVID-19) Outbreak in China," *JAMA* 2020; 323: 1239–1242; J.L. Zhou et al., "Raising Alarms: A Dialogue with the First Person to Report the Epidemic, Zhang Jixian," *XinhuaNet,* April 20, 2020.

9   W.W. Le and C.Z. Li, "Hubei Government Gives Zhang Dingyu and Zhang Jixian Great Merit Award," *XinhuaNet,* February 6, 2020.

10   D.L. Yang, "China's Early Warning System Didn't Work on COVID-19," *Washington Post,* February 24, 2020.

11   S.P. Zhang, "Huanan Seafood Market Is Closed Starting Today," *Beijing News,* January 1, 2020.

12   S.P. Zhang, "Patients of Unusual Pneumonia in Wuhan Transferred to an Infectious Disease Hospital, Residents near the Huanan Market Found Infected," *Beijing News,* January 2, 2020.

13   Anonymous, "China Detects Large Quantity of Novel Coronavirus at Wuhan Seafood Market," *XinhuaNet,* January 27, 2020.

14   J.Q. Gong, "The Whistle-Giver," *People Magazine* (China), March 10, 2020.

15   J.X. Qin et al., "The Whistleblower Li Wenliang: Truth Is the Most Important," *Caixin,* February 7, 2020.

16   D. Ji, "Third Session of 13th Hubei Provincial People's Congress Will Be Held on January 12th, 2020," *People's Daily Online,* November 29, 2019; P. Zhuang, "Chinese Laboratory That First Shared Coronavirus Genome with World Ordered Closed for 'Rectification,' Hindering Its COVID-19 Research," *South China Morning Post,* February 28, 2020.

17   T. Reals, "Chinese Doctor Was Warned to Keep Quiet After Sounding the Alarm on Coronavirus," *CBS News,* February 4, 2020.

18   K. Elmer, "Coronavirus: Wuhan Police Apologise to Family of Whistle-Blowing Doctor Li Wenliang," *South China Morning Post,* March 19, 2020.

19   C. Buckley, "Chinese Doctor, Silenced After Warning of Outbreak, Dies from Coronavirus," *New York Times,* February 6, 2020; Anonymous, "China Identifies 14 Hubei Frontline Workers, Including Li Wenliang, as Martyrs," *Global Times,* April 2, 2020.

20   C. Huang et al., "Clinical Features of Patients Infected with 2019 Novel Coronavirus in Wuhan, China," *The Lancet* 2020; 395: 497–506.

21   M.H. Wong, "3 Billion Journeys: World's Biggest Human Migration Begins in China," *CNN,* January 10, 2020.

22   J.S. Jia et al., "Population Flow Drives Spatio-Temporal Distribution of COVID-19 in China," *Nature* 2020; 582: 389–394.

23   C.Q. Zhou, "Xi Jinping Made an Important Directive Regarding Pneumonia Caused by the Novel Coronavirus: Citizens' Health and Safety Are Top One Priorities and the Spread of the Virus Must Be Controlled, Li Keqiang Made Further Arrangements," *XinhuaNet,* January 20, 2020.

24   Y. Wang, "Years after SARS, a More Confident China Faces a New Virus," *AP News,* January 22, 2020; J.C. Hernández, "The Test a Deadly Coronavirus Outbreak Poses to China's Leadership," *New York Times,* January 21, 2020.

25   E. Xie, "Build-Up to Coronavirus Lockdown: Inside China's Decision to Close Wuhan," *South China Morning Post,* April 2, 2020.

26   D.L. Yang, "China's Early Warning System Didn't Work on COVID-19," *Washington Post,* February 24, 2020.

27   S. Ankel, "A Construction Expert Broke Down How China Built an Emergency Hospital to Treat Wuhan Coronavirus Patients in Just 10 Days," *Business Insider,* February 5, 2020.

28   Anonymous, "As Xiangyang Railway Station Closes, All Cities in Hubei Are Now Placed in Lockdown," *Pengpai News,* January 29, 2020.

29   J.B. Zhu, "30 Provinces, Municipalities and Autonomous Regions Announce Highest-Level Public Health Emergency," *Sina News,* January 25, 2020.

30   P. Hessler, "Life on Lockdown in China: Forty-Five Days Avoiding the Coronavirus," *The New Yorker,* March 30, 2020.

31   Q.Y. Zhu, "Why Is China Able to Practice Closed-Off Community Management?" *China Daily,* April 7, 2020.

32   C. Cadell and S. Yu, "Wuhan People Keep Out: Chinese Villages Shun Outsiders as Virus Spreads," *Reuters,* January 28, 2020.

33   R. Zhong and P. Mozur, "To Tame Coronavirus, Mao-Style Social Control Blankets China," *New York Times,* February 20, 2020.

34   Y. Wang, "Must-See Instructions for Workplace Reopening! Does Your Workplace Implement These Eight Preventive Measures?" *State Council of the People's Republic of China,* February 22, 2020.

35   Anonymous, "March 31: Daily Briefing on Novel Coronavirus in China," *National Health Commission of the People's Republic of China,* March 31, 2020.

36   G. Cossley, "China Starts to Report Asymptomatic Coronavirus Cases," *Reuters,* April 1, 2020; W. Zheng, "Funeral Parlour Report Fans Fears over Wuhan Death Toll from Coronavirus," *South China Morning Post,* March 30, 2020.

37   S. Chen et al., "Wuhan to Test Whole City of 11 Million as New Cases Emerge," *Bloomberg,* May 12, 2020.

38   Anonymous, "First Travel-Related Case of 2019 Novel Coronavirus Detected in the United States," *CDC,* January 21, 2020; M.L. Holshue et al., "First Case of 2019 Novel Coronavirus in the United States," *New England Journal of Medicine* 2020; 382: 929–936.

39   T. Bedford et al., "Cryptic Transmission of SARS-CoV-2 in Washington State," *medRxiv,* April 16, 2020.

40   P. Robison et al., "Seattle's Patient Zero Spread Coronavirus despite Ebola-Style Lockdown," *Bloomberg Businessweek,* March 9, 2020.

41   M. Worobey et al., "The Emergence of SARS-CoV-2 in Europe and the US," *bioRxiv,* May 23, 2020.

42   J. Healy and S.F. Koveleski, "The Coronavirus's Rampage through a Suburban Nursing Home," *New York Times,* March 21, 2020; M. Baker et al., "Washington State Declares Emergency amid Coronavirus Death and Illness at Nursing Home," *New York Times,* February 29, 2020.

43   T.M. McMichael et al., "COVID-19 in a Long-Term Care Facility—King County, Washington, February 27–March 9, 2020," *CDC Morbidity and Mortality Weekly Report* 2020; 69: 339–342; T.M. McMichael et al., "Epidemiology of Covid-19 in a Long-Term Care Facility in King County, Washington," *New England Journal of Medicine* 2020; 382: 2005–2011; J. Healy and S.F. Kovaleski, "The Coronavirus's Rampage through a Suburban Nursing Home," *New York Times,* March 21, 2020.

44   T. Tully, "After Anonymous Tip, 17 Bodies Found at Nursing Home Hit by Virus," *New York Times,* April 15, 2020; H. Krueger, "Almost Every Day Has Brought a New Death from Coronavirus at the Soldiers' Home in Holyoke; 67 Have Died So Far," *Boston Globe,* April 27, 2020.

45   S. Moon, "A Seemingly Healthy Woman's Sudden Death Is Now the First Known US Coronavirus-Related Fatality," *CNN,* April 24, 2020; J. Hanna et al., "2 Californians Died of Coronavirus Weeks before Previously Known 1st US Death," *CNN,* April 22, 2020.

46   I. Ghinai et al., "First Known Person-to-Person Transmission of Severe Acute Respiratory Syndrome Coronavirus 2 (SARS-CoV-2) in the USA," *The Lancet* 2020; 395: 1137–1144; G. Kolata, "Why Are Some People So Much More Infectious Than Others?" *New York Times,* April 12, 2020.

47   M. Worobey et al., "The Emergence of SARS-CoV-2 in Europe and the US," *bioRxiv,* May 23, 2020.

48   S. Fink and M. Baker, "'It's Just Everywhere Already': How Delays in Testing Set Back the U.S. Coronavirus Response," *New York Times*, March 10, 2020.

49   Anonymous, "Coronavirus Disease 2019 (COVID-19) Situation Report—38," *WHO*, February 27, 2020.

50   Anonymous, "Coronavirus | United States," *Worldometer*, July 14, 2020.

51   D. Hull and H. Waller, "Americans Told to Avoid Cruises as Medical Team Boards Ship," *Bloomberg*, March 8, 2020.

52   L.F. Moriarty et al., "Public Health Responses to COVID-19 Outbreaks on Cruise Ships—Worldwide, February–March 2020," *CDC Morbidity and Mortality Weekly Report* 2020; 69: 347–352.

53   M. Hines and D. Oliver, "Coronavirus: More Than 1,000 Passengers Await Their Turn to Leave Grand Princess, Begin Quarantine," *USA Today*, March 11, 2020.

54   L.F. Moriarty et al., "Public Health Responses to COVID-19 Outbreaks on Cruise Ships—Worldwide, February–March 2020," *CDC Morbidity and Mortality Weekly Report* 2020; 69: 347–352.

55   S. Mallapaty, "What the Cruise-Ship Outbreaks Reveal about COVID-19," *Nature* 2020; 580: 18.

56   T.W. Russell et al., "Estimating the Infection and Case Fatality Ratio for Coronavirus Disease (COVID-19) Using Age-Adjusted Data from the Outbreak on the Diamond Princess Cruise Ship, February 2020," *Eurosurveillance* 2020; 25: pii=2000256.

57   A. Palmer, "Amazon Tells Seattle-Area Employees to Work from Home as Coronavirus Spreads," *CNBC*, March 5, 2020.

58   S. Mervosh et al., "See Which States and Cities Have Told Residents to Stay at Home," *New York Times*, April 20, 2020.

59   F. Wu et al., "A New Coronavirus Associated with Human Respiratory Disease in China," *Nature* 2020; 579: 265–269.

60   P. Zhuang, "Chinese Laboratory That First Shared Coronavirus Genome with World Ordered to Close for 'Rectification,' Hindering Its Covid-19 Research," *South China Morning Post*, February 28, 2020; J. Cohen, "Chinese Researchers Reveal Draft Genome of Virus Implicated in Wuhan Pneumonia Outbreak," *Science*, January 11, 2020.

61   P. Zhou et al., "A Pneumonia Outbreak Associated with a New Coronavirus of Probable Bat Origin," *Nature* 2020; 579: 270–273; T. Zhang et al., "Probable Pangolin Origin of SARS-CoV-2 Associated with the COVID-19 Outbreak," *Current Biology* 2020; 30: 1346–1351.; M.F. Boni et al., "Evolutionary Origins of the SARS-CoV-2 Sarbecovirus Lineage Responsible for the COVID-19 Pandemic," *Nature Microbiology*, July 28, 2020.

62   A. Rambaut, "Phylogenetic Analysis of nCoV-2019 Genomes," *Virological*, March 6, 2020.

63   T. Bedford et al., "Cryptic Transmission of SARS-CoV-2 in Washington State," *medRxiv*, April 16, 2020.

64   M. Worobey et al., "The Emergence of SARS-CoV-2 in Europe and the US," *bioRxiv*, May 23, 2020.

65   J.R. Fauver et al., "Coast-to-Coast Spread of SARS-CoV-2 during the Early Epidemic in the United States," *Cell* 2020; 181: 990–996.e5.

66   J. Goldstein and J. McKinley, "Coronavirus in N.Y.: Manhattan Woman Is First Confirmed Case in State," *New York Times*, March 1, 2020.

67   M. Worobey et al., "The Emergence of SARS-CoV-2 in Europe and the US," *bioRxiv*, May 23, 2020.

68   E. Lavezzo et al., "Suppression of COVID-19 Outbreak in the Municipality of Vo, Italy," *medRxiv*, April 18, 2020.

69   N. Gallón, "Bodies Are Being Left in the Streets in an Overwhelmed Ecuadorian City," *CNN*, April 3, 2020.

70   J.D. Almeida et al., "Virology: Coronaviruses," *Nature* 1968; 220: 650.

71   W.J. Guan et al., "Clinical Characteristics of Coronavirus Disease 2019 in China," *New England Journal of Medicine* 2020; 382: 1708–1720; C. Menni et al., "Real-Time Tracking of Self-Reported Symptoms to Predict Potential COVID 19," *Nature Medicine* 2020; 26: 1037–1040; A.B. Docherty et al., "Features of 16,749 Hospitalized UK Patients with COVID-19 Using the ISARIC WHO Clinical Characterization Protocol," *medRxiv*, April 28, 2020.

72   F. Hainey, "How Six People with Coronavirus Describe Suffering the Symptoms," *Manchester Evening News*, March 11, 2020.

73   J. Achenbach et al., "What It's Like to Be Infected with Coronavirus," *Washington Post*, March 22, 2020.

74   C. Goldman, "I Have the Coronavirus. So Far, It Hasn't Been That Bad for Me," *Washington Post*, February 28, 2020.

75   A. de Luca et al., "'An Anvil Sitting on My Chest': What It's Like to Have Covid-19," *New York Times*, May 7, 2020.

76   M. Bloom, "Chicagoan on What It's Like to Have Coronavirus: 'It Feels Like an Alien Has Taken Over Your Body,'" *Block Club Chicago*, May 14, 2020.

77   A. de Luca et al., "'An Anvil Sitting on My Chest': What It's Like to Have Covid-19," *New York Times*, May 7, 2020.

78   M. Longman, "What Coronavirus Feels Like, according to 5 Women," *Refinery29*, April 13, 2020; K.T. Vuong, "How Does It Feel to Have Coronavirus COVID-19?" *Mira*, May 13, 2020.

79   M. Wadman et al., "How Does Coronavirus Kill? Clinicians Trace a Ferocious Rampage through the Body, from Brain to Toes," *Science*, April 17, 2020.

80   S.A. Lauer et al., "The Incubation Period of Coronavirus Disease 2019 (COVID-19) from Publicly Reported Confirmed Cases: Estimation and Application," *Ann Intern Med*, May 5, 2020.

81   W.H. McNeill, *Plagues and Peoples*, New York: Doubleday/Anchor, 1976; A.W. Crosby, *The Columbian Exchange: Biological and Cultural Consequences of 1492*, Westport, CT: Greenwood Press, 1972.

82   E.N. Lorenz, "Predictability: Does the Flap of a Butterfly's Wings in Brazil Set Off a Tornado in Texas?" *American Association for the Advancement of Science, 139th Meeting*, December 29, 1972.

83   E.N. Lorenz, *The Essence of Chaos*, Seattle: University of Washington Press, 1993, p. 134.

84   E.N. Lorenz, "Deterministic Nonperiodic Flow," *Journal of the Atmospheric Sciences* 1963; 20: 130–141.

85   E.N. Lorenz, "The Predictability of Hydrodynamic Flow," *Transactions of the New York Academy of Sciences* 1963; 25: 409–432.

86   E.N. Lorenz, "The Butterfly Effect," *Premio Felice Pietro Chisesi e Caterina Tomassoni Acceptance Speech*, April, 2008.

2. AN OLD ENEMY RETURNS

1   M.S. Asher, *Dancing in the Wonder for 102 Years*, Seattle: Amazon, 2015, p. 7.

2   P. Dvorak, "At 107, This Artist Just Beat COVID-19. It Was the Second Pandemic She Survived," *Washington Post*, May 7, 2020; B. Harris, "Meet the 107-Year-Old Woman Who Survived the Coronavirus and the Spanish Flu," *Jerusalem Post*, May 7, 2020.

3   D.X. Liu et al., "Human Coronavirus-229E, -OC43, -NL63, and -HKU1," *Reference Module in Life Sciences*, May 7, 2020; H. Wein, "Understanding a Common Cold Virus," *NIH Research Matters*, April 13, 2009.

4   Anonymous, "Update 95—SARS: Chronology of a Serial Killer," *WHO*, July 4, 2003.

5   Anonymous, "Summary Table of SARS Cases by Country, 1 November 2002—7 August, 2003," *WHO*, August 15, 2003.

6   Anonymous, "SARS Outbreak Contained Worldwide," *WHO*, July 5, 2003.

7   Anonymous, "Summary Table of SARS Cases by Country, 1 November 2002—7 August, 2003," *WHO*, August 15, 2003.

8   E. Nakashima, "SARS Signals Missed in Hong Kong," *Washington Post*, May 20, 2003.

9   Anonymous, "Update 95—SARS: Chronology of a Serial Killer," *WHO*, July 4, 2003; E. Nakashima, "SARS Signals Missed in Hong Kong," *Washington Post*, May 20, 2003.

10   J.M. Nicholls et al., "Lung Pathology of Fatal Severe Acute Respiratory Syndrome," *The Lancet* 2003; 361: 1773–1776.

11   S. Law et al., "Severe Acute Respiratory Syndrome (SARS) and Coronavirus Disease—2019 (COVID-19): From Causes to Preventions in Hong Kong," *International Journal of Infectious Diseases* 2020; 94: 156–163.

12   T. Tsang et al., "Update: Outbreak of Severe Acute Respiratory Syndrome—Worldwide, 2003," *CDC Morbidity and Mortality Weekly Report* 2003; 52: 241–248.

13   E. Nakashima, "SARS Signals Missed in Hong Kong," *Washington Post*, May 20, 2003.

14   K. Fong, "SARS: The People Who Risked Their Lives to Stop the Virus," *BBC News Magazine*, August 16, 2013.

15   H. Feldmann et al., "WHO Environmental Health Team Reports on Amoy Gardens," *WHO*, May 16, 2003; I.T.S. Yu et al., "Evidence of Airborne Transmission of the Severe Acute Respiratory Syndrome Virus," *New England Journal of Medicine* 2004; 350: 1731–1739.

16   K.R. McKinney et al., "Environmental Transmission of SARS at Amoy Gardens," *Journal of Environmental Health* 2006; 68: 26–30.

17   K. Fong, "SARS: The People Who Risked Their Lives to Stop the Virus," *BBC News Magazine*, August 16, 2013.

18   Anonymous, "Update 95—SARS: Chronology of a Serial Killer," *WHO*, July 4, 2003.
19   B. Reilley et al., "SARS and Carlo Urbani," *New England Journal of Medicine* 2003; 348: 1951–1952.
20   C. Abraham, "How a Deadly Disease Came to Canada," *Globe and Mail*, March 29, 2003.
21   T. Tsang et al., "Update: Outbreak of Severe Acute Respiratory Syndrome—Worldwide, 2003," *CDC Morbidity and Mortality Weekly Report* 2003; 52: 241–248.
22   Y. Ye, *Biography of Zhong Nanshan*, Beijing: Writers Press, 2010, pp. 49–52.
23   Z. Shi and Z. Hu, "A Review of Studies on Animal Reservoirs of the SARS Coronavirus," *Virus Research* 2008; 133: 74–87.
24   A.C.P. Wong et al., "Global Epidemiology of Bat Coronaviruses," *Viruses* 2019; 11: 174.
25   S.J. Olsen et al., "Transmission of the Severe Acute Respiratory Syndrome on Aircraft," *New England Journal of Medicine* 2003; 349: 2416–2422.
26   D.M. Bell et al., "Public Health Interventions and SARS Spread—2003," *Emerging Infectious Diseases* 2004; 10: 1900–1906.
27   Anonymous, "Update 95—SARS: Chronology of a Serial Killer," *WHO*, July 4, 2003.
28   Anonymous, "Coronavirus Never Before Seen in Humans Is the Cause of SARS," *WHO*, April 16, 2003.
29   D.M. Bell et al., "Public Health Interventions and SARS Spread—2003," *Emerging Infectious Diseases* 2004; 10: 1900–1906.
30   K. Stadler et al., "SARS—Beginning to Understand a New Virus," *Nature Reviews Microbiology* 2003; 1: 209–218.
31   Anonymous, "Summary Table of SARS Cases by Country, 1 November 2002—7 August, 2003," *WHO*, August 15, 2003.
32   Anonymous, "Update 49—SARS Case Fatality Ratio, Incubation Period," *WHO*, May 7, 2003.
33   A. Forna et al., "Case Fatality Ratio Estimates for the 2013–2016 West African Ebola Epidemic: Application of Boosted Regression Trees for Imputation," *Clinical Infectious Diseases* 2020; 70: 2476–2483; N. Ndayimirije and M.K. Kindhauser, "Marburg Hemorrhagic Fever in Angola—Fighting Fear and a Lethal Pathogen," *New England Journal of Medicine* 2005; 352: 2155–2157.
34   D. Cyranoski, "Profile of a Killer: The Complex Biology Powering the Coronavirus Pandemic," *Nature*, May 4, 2020.
35   J. Howard, "Novel Coronavirus Can Be Spread by People Who Aren't Exhibiting Symptoms, CDC Director Says," *CNN*, February 13, 2020.
36   T. Subramaniam and V. Stracqualursi, "Fact Check: Georgia Governor Says We Only Just Learned People without Symptoms Could Spread Coronavirus. Experts Have Been Saying That for Months," *CNN*, April 3, 2020.
37   Z. Du et al., "Serial Interval of COVID-19 among Publicly Reported Confirmed Cases," *Emerging Infectious Diseases* 2020; 25: 1341–1343.
38   W. Xia et al., "Transmission of Corona Virus Disease 2019 during the Incubation Period May Lead to a Quarantine Loophole," *medRxiv*, March 8, 2020.
39   X. He et al., "Temporal Dynamics in Viral Shedding and Transmissibility of COVID-19," *Nature Medicine* 2020; 26: 672–675; S.M. Moghadas et al., "The Implications of Silent Transmission for the Control of COVID-19 Outbreaks," *Proceedings of the National Academy of Sciences*, July 6, 2020.
40   F.M. Guerra et al., "The Basic Reproduction Number (R_0) of Measles: A Systematic Review," *Lancet Infectious Diseases* 2017; 17: E420–E428; R. Gani and S. Leach, "Transmission Potential of Smallpox in Contemporary Populations," *Nature* 2001; 414: 748–751; A. Khan et al., "Estimating the Basic Reproductive Ratio for the Ebola Outbreak in Liberia and Sierra Leone," *Infectious Diseases of Poverty* 2015; 4: 13; M. Biggerstaff et al., "Estimates of the Reproduction Number for Seasonal, Pandemic, and Zoonotic Influenza: A Systematic Review of the Literature," *BMC Infectious Diseases* 2014; 14: 480.
41   M. Lipsitch et al., "Transmission Dynamics and Control of Severe Acute Respiratory Syndrome," *Science* 2003; 300: 1966–1970.
42   J.O. Lloyd-Smith et al., "Superspreading and the Effect of Individual Variation on Disease Emergence," *Nature* 2005; 438: 355–359; M. Small et al., "Super-Spreaders and the Rate of Transmission of the SARS Virus," *Physica D* 2006; 215: 146–158.
43   J. Riou and C.L. Althaus, "Pattern of Early Human-to-Human Transmission of Wuhan 2019 Novel Coronavirus (2019-nCoV), December 2019 to January 2020," *Eurosurveillance* 2020; 25: pii=2000058.
44   L.A. Meyers et al., "Network Theory and SARS: Predicting Outbreak Diversity," *Journal of Theoretical Biology* 2005; 232: 71–81.

45   O. Reich et al., "Modeling COVID-19 on a Network: Super-Spreaders, Testing, and Containment," *medRxiv*, May 5, 2020; A.L. Ziff and R.M. Ziff, "Fractal Kinetics of COVID-19 Pandemic," *medRxiv*, March 3, 2020.

46   G. Kolata, "Why Are Some People So Much More Infectious Than Others?" *New York Times*, April 12, 2020.

47   N.A. Christakis and J.H. Fowler, "Social Network Sensors for Early Detection of Contagious Outbreaks," *PLOS ONE* 2010; 5: e12948.

48   L. Hamner et al., "High SARS-CoV-2 Attack Rate Following Exposure at a Choir Practice—Skagit County, Washington, March 2020," *CDC Morbidity and Mortality Weekly Report* 2020; 69: 606–610.

49   S. Jang et al., "Cluster of Coronavirus Disease Associated with Fitness Dance Classes, South Korea," *Emerging Infectious Diseases* 2020; 26: 1917–1920.

50   E. Barry, "Days after a Funeral in a Georgia Town, Coronavirus 'Hit Like a Bomb,'" *New York Times*, March 30, 2020.

51   Anonymous, "Coronavirus Disease 2019 (COVID-19) Cases in MA as of March 26, 2020," *Massachusetts Department of Public Health*, March 26, 2020; F. Stockman and K. Barker, "How a Premier U.S. Drug Company Became a Virus 'Super Spreader,'" *New York Times*, April 12, 2020.

52   H. Qian et al., "Indoor Transmission of SARS-CoV-2," *medRxiv*, April 7, 2020.

53   Anonymous, "Middle East Respiratory Syndrome Coronavirus (MERS-CoV)," *WHO*, March 11, 2019.

54   M.S. Majumder et al., "Estimation of MERS-Coronavirus Reproductive Number and Case Fatality Rate for the Spring 2014 Saudi Arabia Outbreak: Insights from Publicly Available Data," *PLoS Current Outbreaks* 2014; 6.

55   M.S. Majumder and K.D. Mandl, "Early in the Epidemic: Impacts of Preprints on Global Discourse about COVID-19 Transmissibility," *Lancet Global Health* 2020; 8: e627.

56   R.E. Neustadt and H.V. Fineberg, *The Epidemic That Never Was: Policy-Making and the Swine Flu Scare*, New York: Vintage, 1983.

57   F.S. Dawood et al., "Estimated Global Mortality Associated with the First 12 Months of 2009 Pandemic Influenza A H1N1 Virus Circulation: A Modelling Study," *Lancet Infectious Diseases 2012;* 12: 687–695; J.K. Taubenberger and D.M. Morens, "1918 Influenza: The Mother of All Pandemics," *Emerging Infectious Diseases* 2006; 12: 15–22.

58   C. Reed et al., "Novel Framework for Assessing Epidemiological Effects of Influenza Epidemics and Pandemics," *Emerging Infectious Diseases* 2013; 19: 85–91.

59   W.P. Glezen, "Emerging Infections: Pandemic Influenza," *Epidemiologic Reviews* 1996; 18: 64–76.

60   A.D. Langmuir, "Epidemiology of Asian Influenza. With Special Emphasis on the United States," *American Review of Respiratory Disease* 1961; 83: 2–14.

61   C. Viboud et al., "Global Mortality Impact of the 1957–1959 Influenza Pandemic," *Journal of Infectious Diseases* 2016; 213: 738–745.

62   US Department of Health Education and Welfare Public Health Service, "Asian Influenza: 1957–1960," *Descriptive Brochure*, July 1960.

63   A.D. Langmuir, "Epidemiology of Asian Influenza. With Special Emphasis on the United States," *American Review of Respiratory Disease* 1961; 83: 2–14.

64   Anonymous, "Pneumonia and Influenza Mortality for 122 U.S. Cities," *CDC*, January 10, 2015.

65   J. Shaman and M. Kohn, "Absolute Humidity Modulates Influenza Survival, Transmission, and Seasonality," *Proceedings of the National Academy of Sciences* 2009; 106: 3243–3248.

66   R.A. Neher et al., "Potential Impact of Seasonal Forcing on a SARS-CoV-2 Pandemic," *Swiss Medical Weekly* 2020; 150: w20224; S.M. Kissler et al., "Projecting the Transmission Dynamics of SARS-CoV-2 through the Postpandemic Period," *Science* 2020; 368: 860–868.

67   W.P. Glezen, "Emerging Infections: Pandemic Influenza," *Epidemiological Reviews* 1996; 18: 64–76.

68   L. Zeldovich, "How America Brought the 1957 Influenza Pandemic to a Halt," *JSTOR Daily*, April 7, 2020.

69   N.P.A.S. Johnson and J. Mueller, "Updating the Accounts: Global Mortality of the 1918–1920 'Spanish' Influenza Pandemic," *Bulletin of the History of Medicine* 2002; 76: 105–115.

70   W.P. Glezen, "Emerging Infections: Pandemic Influenza," *Epidemiological Reviews* 1996; 18: 64–76.

71   L. Spinney, *Pale Rider: The Spanish Flu of 1918 and How It Changed the World*, New York: Public Affairs, 2017, pp. x–xi.

72   Ibid., p. 99.

73   J.K. Taubenberger et al., "Initial Genetic Characterization of the 1918 'Spanish' Influenza

Virus," *Science* 1997; 275: 1793–1796; T.M. Tumpey et al., "Characterization of the Reconstructed 1918 Spanish Influenza Pandemic Virus," *Science* 2005; 310: 77–80.

74   J.M. Barry, "The Site of Origin of the 1918 Influenza Pandemic and Its Public Health Implications," *Journal of Translational Medicine* 2004; 2: 3.

75   P.C. Wever and L. van Bergen, "Death from 1918 Pandemic Influenza during the First World War: A Perspective from Personal and Anecdotal Evidence," *Influenza and Other Respiratory Viruses* 2014; 8: 538–546; L. Spinney, *Pale Rider: The Spanish Flu of 1918 and How It Changed the World*, New York: Public Affairs, 2017, p. 38.

76   V.C. Vaughan, *A Doctor's Memories*, Indianapolis: Bobbs-Merrill, 1926, pp. 383–384.

77   S.E. Mamelund, "1918 Pandemic Morbidity: The First Wave Hits the Poor, the Second Wave Hits the Rich," *Influenza and Other Respiratory Viruses* 2018; 12: 307–313.

78   P. Toole, "The Flu Epidemic of 1918," *NYC Department of Records & Information Services*, March 1, 2018.

79   D. Barry and C. Dickerson, "The Killer Flu of 1918: A Philadelphia Story," *New York Times*, April 4, 2020.

80   G.H. Hirshberg, "Medical Science's Newest Discoveries about the 'Spanish Influenza,'" *Philadelphia Inquirer*, October 6, 1918; M. Wilson, "What New York Looked Like during the 1918 Flu Pandemic," *New York Times*, April 2, 2020.

81   Anonymous, "Drastic Steps Taken to Fight Influenza Here," *New York Times*, October 5, 1918.

82   A.M. Stein et al., "'Better Off in School': School Medical Inspection as a Public Health Strategy during the 1918–1919 Influenza Pandemic in the United States," *Public Health Reports* 2010; 125: 63–70.

83   H. Markel et al., "Non-Pharmaceutical Interventions Implemented by US Cities during the 1918–1919 Influenza Pandemic," *JAMA* 2007; 298: 644–654; F. Aimone, "The 1918 Influenza Pandemic in New York City: A Review of the Public Health Response," *Public Health Reports* 2010; 125: 71–79.

84   A.D. Langmuir, "William Farr: Founder of Modern Concepts of Surveillance," *International Journal of Epidemiology* 1976; 5: 13–18.

85   D.M. Weinberger, et al., "Estimating the Early Death Toll of COVID-19 in the United States," *medRxiv*, April 29, 2020.

86   D.M. Weinberger, et al., "Estimation of Excess Deaths Associated with the COVID-19 Pandemic in the United States, March to May 2020," *JAMA Internal Medicine*, July 1, 2020.

87   G. He et al., "The Short-Term Impacts of COVID-19 Lockdown on Urban Air Pollution in China," *Nature Sustainability*, July 7, 2020; R.K. Philip et al., "Reduction in Preterm Births during the COVID-19 Lockdown in Ireland: A Natural Experiment Allowing Analysis of Data from the Prior Two Decades," *medRxiv*, June 5, 2020; G. Hedermann et al., "Changes in Premature Birth Rates during the Danish Nationwide COVID-19 Lockdown: A Nationwide Register-Based Prevalence Proportion Study," *medRxiv*, May 23, 2020.

88   K.I. Bos et al., "A Draft Genome of *Yersinia pestis* from Victims of the Black Death," *Nature* 2011; 478: 506–510.

89   F.M. Snowden, *Epidemics and Society: From the Black Death to the Present*, New Haven, CT: Yale University Press, 2019, p. 48.

90   R.D. Perry and J.D. Fetherston, "*Yersinia pestis*—Etiologic Agent of Plague," *Clinical Microbiology Reviews* 1997; 10: 35–66.

91   B. Bramanti et al., "A Critical Review of Anthropological Studies on Skeletons from European Plague Pits of Different Epochs," *Scientific Reports* 2018; 8: 17655.

92   L. Mordechai et al., "The Justinianic Plague: An Inconsequential Pandemic?" *Proceedings of the National Academy of Sciences* 2019; 116: 25546–25554.

93   O.J. Benedictow, *The Black Death, 1346–1353: The Complete History*, Woodbridge, UK: Boydell & Brewer, 2004.

94   Anonymous, "Plague in the United States," *CDC Maps & Statistics*, November 25, 2019; N. Kwit, "Human Plague—United States, 2015," *CDC Morbidity and Mortality Weekly Report* 2015; 64: 918–919.

95   John of Ephesus, "John of Ephesus Describes the Justinianic Plague," ed. R. Pearse, *Roger Pearse blog*, May 10, 2017.

96   D. Defoe, *Journal of the Plague Year*, London: E. Nutt, 1722, p. 90.

97   K.E. Steinhauser et al., "Factors Considered Important at the End of Life by Patients, Family, Physicians, and Other Care Providers," *JAMA* 2000; 284: 2476–2482.

98   F.M. Snowden, *Epidemics and Society: From the Black Death to the Present*, New Haven, CT: Yale University Press, 2019, p. 70.

99   G. de Mussis, *Historia de morbo,* in *The Black Death,* trans. and ed. R. Horrox, Manchester: Manchester University Press, 1994, p. 22.

100   A. Cliff and M. Smallman-Raynor, "Containing the Spread of Epidemics," in *Oxford Textbook of Infectious Disease Control: A Geographical Analysis from Medieval Quarantine to Global Eradication,* Oxford: Oxford University Press, 2013.

101   O.J. Benedictow, *The Black Death 1346–1353: The Complete History,* Woodbridge, UK: Boydell & Brewer, 2004; O.J. Benedictow, *Plague in the Late Medieval Nordic Countries: Epidemiological Studies,* Oslo: Middelalderforlaget, 1992.

102   W.M. Bowsky, "The Impact of the Black Death upon Sienese Government and Society," *Speculum* 1964; 39: 1–34; N. Pūyān, "Plague, an Extraordinary Tragedy," *Open Access Library Journal* 2017; 4: e3643.

103   B. Bonaiuti, *Florentine Chronicle of Marchionne di Coppo Stefani,* ed. N. Rodolico, Città di Castello: Coi Tipi dell'editore S. Lapi, 1903, Rubric 643.

104   A.B. Appleby, "The Disappearance of Plague: A Continuing Puzzle," *Economic History Review* 1980; 33: 161–173; P. Slack, "The Disappearance of Plague: An Alternative View," *Economic History Review* 1981; 34: 469–476.

105   W.H. McNeil, *Plagues and Peoples,* London: Penguin, 1976; C.E. Rosenberg, *The Cholera Years: The United States in 1832, 1849, and 1866,* Chicago: University of Chicago Press, 1987; D.M. Oshinsky, *Polio: An American Story,* Oxford: Oxford University Press, 2005.

3. PULLING APART

1   L. Spinney, *Pale Rider: The Spanish Flu of 1918 and How It Changed the World,* New York: Public Affairs, 2017, p. 124.

2   T. McKeown and C.R. Lowe, *An Introduction to Social Medicine,* Oxford and Edinburgh: Blackwell Scientific Publications, 1966; T. McKeown and R.G. Brown, "Medical Evidence Related to English Population Changes in the Eighteenth Century," *Population Studies* 1955; 9: 119–141.

3   B. Pourbohloul et al., "Modeling Control Strategies of Respiratory Pathogens," *Emerging Infectious Diseases* 2005; 11: 1249–1256.

4   D. Cole and A. Main, "Top Infectious Disease Expert Doesn't Rule Out Supporting Temporary National Lockdown to Combat Coronavirus," *CNN,* March 15, 2020.

5   J. Kates et al., "Stay-at-Home Orders to Fight COVID-19 in the United States: The Risks of a Scattershot Approach," *KFF,* April 5, 2020.

6   K. Schaul et al., "Where Americans Are Still Staying at Home the Most," *Washington Post,* May 6, 2020.

7   COVID-19 Response Team, "Severe Outcomes among Patients with Coronavirus Disease 2019 (COVID-19)—United States, February 12–March 16, 2020," *CDC Morbidity and Mortality Weekly Report* 2020; 69: 343–346; Novel Coronavirus Pneumonia Emergency Response Epidemiology Team, "The Epidemiological Characteristics of an Outbreak of 2019 Novel Coronavirus Diseases (COVID-19)—China, 2020," *China CDC Weekly* 2020; 2: 1–10; G. Grasselli et al., "Critical Care Utilization for the COVID-19 Outbreak in Lombardy, Italy: Early Experience and Forecast during an Emergency Response," *JAMA* 2020; 323: 1545–1546.

8   Anonymous, "Hospital Beds (per 1,000 People)," *World Bank Data,* 2015.

9   L. Frias, "Thousands of Chinese Doctors Volunteered for the Frontline of the Coronavirus Outbreak. They Are Overwhelmed, Under-Equipped, Exhausted, and Even Dying," *Business Insider,* February 7, 2020.

10   M. Van Beusekom, "Doctors: COVID-19 Pushing Italian ICUs toward Collapse," *University of Minnesota Center for Infectious Disease Research and Policy,* March 16, 2020.

11   N. Winfield and C. Barry, "Italy's Health System at Limit in Virus-Struck Lombardy," *AP News,* March 2, 2020.

12   S. Hsiang et al., "The Effect of Large-Scale Anti-Contagion Policies on the COVID-19 Pandemic," *Nature,* June 8, 2020.

13   H.M. Krumholz, "Where Have All the Heart Attacks Gone?" *New York Times,* May 14, 2020.

14   Anonymous, "News Release," *US Department of Labor,* July 9, 2020.

15   K. Parker, et al., "About Half of Lower-Income Americans Report Household Job or Wage Loss Due to COVID-19," *Pew Research Center,* April 21, 2020.

16   H. Long and A. Van Dam, "Unemployment Rate Jumps to 14.7 Percent, the Worst since the Great Depression," *Washington Post,* May 8, 2020; D. Rushe, "US Job Losses Have Reached Great Depression Levels. Did It Have to Be That Way?" *The Guardian,* May 9, 2020.

17   J. Lippman, "Retail Meltdown Will Reshape Main St.: Popular Gelato Shop Won't Return, Could Be First of Many Downtown," *Valley News* (Lebanon, NH), May 8, 2020.

18   B. Casselman, "A Collapse that Wiped Out 5 Years of Growth, with No Bounce in Sight," *New York Times*, July 30, 2020.

19   Council of Economic Advisors, "An In-Depth Look at COVID-19's Early Effects on Consumer Spending and GDP," *White House*, April 29, 2020.

20   J. Dearen and M. Stobbe, "Trump Administration Buries Detailed CDC Advice on Reopening," *AP News*, May 7, 2020.

21   N. Qualls et al., "Community Mitigation Guidelines to Prevent Pandemic Influenza— United States, 2017," *CDC Morbidity and Mortality Weekly Report* 2017; 66: 1–34.

22   J. Rainey et al., "California Lessons from the 1918 Pandemic: San Francisco Dithered; Los Angeles Acted and Saved Lives," *Los Angeles Times*, April 10, 2020.

23   P. Gahr et al., "An Outbreak of Measles in an Under-Vaccinated Community," *Pediatrics* 2014; 134: e220–228.

24   H. Stewart et al., "Boris Johnson Orders UK Lockdown to Be Enforced by Police," *The Guardian*, March 23, 2020.

25   Y. Talmazan, "U.K.'s Boris Johnson Says Doctors Prepared to Announce His Death as He Fought COVID-19," *NBC News*, May 3, 2020.

26   T. Mulvihill, "Sweden's Divisive Lockdown Policy Could See It Excluded from Nordic 'Travel Bubble,'" *The Telegraph*, May 27, 2020.

27   J. Henley, "We Should Have Done More, Admits Architect of Sweden's Covid-19 Strategy," *The Guardian*, June 3, 2020.

28   D. Lazer et al., "The State of the Nation: A 50-State COVID-19 Survey," *Northeastern University*, April 20, 2020.

29   G.H. Weaver, "Droplet Infection and Its Prevention by the Face Mask," *Journal of Infectious Diseases* 1919; 24: 218–230.

30   World Health Organization, "Advice on the Use of Masks in the Context of COVID-19: Interim Guidance," *WHO*, April 6, 2020.

31   N.H.L. Leung et al., "Respiratory Virus Shedding in Exhaled Breath and Efficacy of Face Masks," *Nature Medicine* 2020; 26: 676–680; A. Davies et al., "Testing the Efficacy of Homemade Masks: Would They Protect in an Influenza Pandemic?" *Disaster Medicine and Public Health Preparedness* 2013, 7: 413–418; T. Jefferson et al., "Physical Interventions to Interrupt or Reduce the Spread of Respiratory Viruses: Systematic Review," *BMJ* 2008; 336: 77–80.

32   Y.L.A. Kwok et al., "Face Touching: A Frequent Habit That Has Implications for Hand Hygiene," *American Journal of Infection Control* 2015; 43: 112–114.

33   G. Seres et al., "Face Masks Increase Compliance with Physical Distancing Recommendations During the COVID-19 Pandemic," *Berlin Social Science Working Paper*, May 23, 2020.

34   J. Abaluck et al., "The Case for Universal Cloth Mask Adoption and Policies to Increase the Supply of Medical Masks for Health Workers," *Covid Economics*, April 6, 2020.

35   J. Howard et al., "Face Masks Against COVID-19: An Evidence Review," preprint July 10, 2020.

36   Anonymous, "Czech Video Inspires the World to Wear Face Masks during the Global Pandemic," *Czech Universities*, April 6, 2020; R. Tait, "Czechs Get to Work Making Masks after Government Decree," *The Guardian*, March 30, 2020.

37   D. Greene, "Police in Czech Republic Tell Nudists to Wear Face Masks," *NPR*, April 9, 2020.

38   L. Hensley, "Why Some People Still Refuse to Wear Masks," *Global News*, July 9, 2020.

39   J. Redmon et al., "Georgia Governor Extends Coronavirus Restriction While Encouraging Use of Face Masks," *Global News*, July 9, 2020.

40   S. Ryu et al., "Non-Pharmaceutical Measures for Pandemic Influenza in Non-Healthcare Settings—International Travel-Related Measures," *Emerging Infectious Diseases* 2020; 26: 961–966.

41   Monastery of Neuberg, *Monumenta Germaniae Historica — Scriptorum IX*, in *The Black Death*, trans. and ed. R. Horrox, Manchester: Manchester University Press, 2013, p. 59.

42   D. Lazer et al., "The State of the Nation: A 50-State COVID-19 Survey," *Northeastern University*, April 20, 2020.

43   M. Boyd et al., "Protecting an Island Nation from Extreme Pandemic Threats: Proof-of-Concept around Border Closure as an Intervention," *PLOS ONE* 2017; 12: e0178732.

44   D.F. Gudbjartsson et al., "Spread of SARS-CoV-2 in the Icelandic Population," *New England Journal of Medicine* 2020; 382: 2302–2315.

45   M. Boyd et al., "Protecting an Island Nation from Extreme Pandemic Threats: Proof-of-Concept around Border Closure as an Intervention," *PLOS ONE* 2017; 12: e0178732.

46   H. Yu, "Transmission Dynamics, Border Entry Screening, and School Holidays during the

2009 Influenza A (H1N1) Pandemic, China," *Emerging Infectious Diseases* 2012; 18: 758–766.

47   N. Ferguson et al., "Strategies for Mitigating an Influenza Pandemic," *Nature* 2006; 442: 448–452.

48   F. Stockman, "Told to Stay Home, Suspected Coronavirus Patient Attended Event with Dartmouth Students," *New York Times*, March 4, 2020.

49   S. Cohn and M. O'Brien, "Contact Tracing: How Physicians Used It 500 Years Ago to Control the Bubonic Plague," *The Conversation*, June 3, 2020.

50   A. Gratiolo, *Discorso di peste: Nel quale si contengono utilissime speculationi intorno alla natura, cagioni, e curatione della peste*, Venice: Girolamo Polo, 1576.

51   A. Boylston, "John Haygarth's 18th-Century 'Rules of Prevention' for Eradicating Smallpox," *Journal of the Royal Society of Medicine* 2014; 107: 494–499.

52   G.A. Soper, "The Curious Career of Typhoid Mary," *Bulletin of the New York Academy of Medicine* 1939; 15: 698.

53   G. Mooney, *Intrusive Interventions: Public Health, Domestic Space, and Infectious Disease Surveillance in England, 1840–1914*, Rochester, NY: University of Rochester Press, 2015.

54   A.M. Brandt, *No Magic Bullet: A Social History of Venereal Disease in the United States since 1880*, Oxford: Oxford University Press, 1987; G.W. Rutherford and J.M. Woo, "Contact Tracing and the Control of Human Immunodeficiency Virus Infection," *JAMA* 1988; 259: 3609–3610.

55   F. Fenner et al., *Smallpox and Its Eradication*, Geneva: World Health Organization, 1988, vol. 6; J.M. Hyman et al., "Modeling the Impact of Random Screening and Contact Tracing in Reducing the Spread of HIV," *Mathematical Biosciences* 2003; 181: 17–54; M. Begun et al., "Contact Tracing of Tuberculosis: A Systematic Review of Transmission Modelling Studies," *PLOS ONE* 2013, 8: e72470; K.T. Eames et al., "Assessing the Role of Contact Tracing in a Suspected H7N2 Influenza A Outbreak in Humans in Wales," *BMC Infectious Diseases* 2010; 10: 141; A. Pandey et al., "Strategies for Containing Ebola in West Africa," *Science* 2014; 346: 991–995; L. Ferretti et al., "Quantifying SARS-CoV-2 Transmission Suggests Epidemic Control with Digital Contact Tracing," *Science* 2020; 368: eabb6936.

56   P. Mozur et al., "In Coronavirus Fight, China Gives Citizens a Color Code, with Red Flags," *New York Times*, March 1, 2020.

57   L. Hamner et al., "High SARS-CoV-2 Attack Rate Following Exposure at a Choir Practice— Skagit County, Washington, March 2020," *CDC Morbidity and Mortality Weekly Report* 2020; 69: 606–610.

58   L.H. Sun et al., "A Plan to Defeat Coronavirus Finally Emerges, and It's Not from the White House," *Washington Post*, April 10, 2020.

59   Ibid.

60   D. Coffey, "Doctors Wonder What to Do When Recovered COVID-19 Patients Still Test Positive," *Medscape*, June 9, 2020.

61   T. Frieden, "Former CDC Head on Coronavirus Testing: What Went Wrong and How We Proceed," *USA Today*, March 31, 2020.

62   L.H. Sun et al., "A Plan to Defeat Coronavirus Finally Emerges, and It's Not from the White House," *Washington Post*, April 10, 2020.

63   E. Christakis, *The Importance of Being Little*, New York: Viking, 2015, p. 136.

64   Anonymous, "Map: Coronavirus and School Closures," *Education Week*, March 6, 2020.

65   National Center for Education Statistics, "Digest of Education Statistics: Table 105.20. Enrollment in Elementary, Secondary, and Degree-Granting Postsecondary Institutions," *U.S. Department of Education Institute of Education Sciences*, March 2019; National Center for Education Statistics, "Digest of Education Statistics: Table 105.40. Number of Teachers in Elementary and Secondary Schools," *U.S. Department of Education Institute of Education Sciences*, March 2019.

66   J. Couzin-Frankel, "Does Closing Schools Slow the Spread of Coronavirus? Past Outbreaks Provide Clues," *Science*, March 10, 2020.

67   W. Van Lancker and Z. Parolin, "COVID-19, School Closures, and Child Poverty: A Social Crisis in the Making," *Lancet Public Health* 2020; 5: e243–e244; J. Bayham and E.P. Fenichel, "Impact of School Closures for COVID-19 on the US Health-Care Workforce and Net Mortality: A Modelling Study," *Lancet Public Health* 2020; 5: e271–e278.

68   S.B. Nafisah et al., "School Closure during Novel Influenza: A Systematic Review," *Journal of Infection and Public Health* 2018; 11: 657–661; H. Rashid et al., "Evidence Compendium and Advice on Social Distancing and Other Related Measures for Response to an Influenza Pandemic," *Paediatric Respiratory Reviews* 2015; 16: 119–126; R.M. Viner et al., "School Clo-

sure and Management Practices during Coronavirus Outbreaks Including COVID-19: A Rapid Systematic Review," *Lancet Child & Adolescent Health* 2020; 4: 397–404.

69   S. Hsiang, et al., "The Effect of Large-Scale Anti-Contagion Policies on the COVID-19 Pandemic," *Nature,* July 8, 2020; S. Flaxman, et al., "Estimating the Effects of Non-Pharmaceutical Interventions on COVID-19 in Europe," *Nature,* June 8, 2020.

70   M. Talev, "Axios-Ipsos Poll: Americans Fear Return to School," *Axios,* July 14, 2020.

71   N. Ferguson et al., "Strategies for Mitigating an Influenza Pandemic," *Nature* 2006; 442: 448–452.

72   Ibid.; J. Zhang et al., "Changes in Contact Patterns Shape the Dynamics of the COVID-19 Outbreak in China," *Science* 2020; 368: 1481–1486.

73   H. Markel et al., "Non-Pharmaceutical Interventions Implemented by US Cities during the 1918–1919 Influenza Pandemic," *JAMA* 2007; 298: 644–654; M C J Bootsma and N.M. Ferguson, "The Effect of Public Health Measures on the 1918 Influenza Pandemic in US Cities," *Proceedings of the National Academy of Sciences* 2007; 104: 7588–7593.

74   Anonymous, "School Closures Begin as Japan Steps Up Coronavirus Fight," *Kyodo News,* May 2, 2020.

75   S. Kawano and M. Kakehashi, "Substantial Impact of School Closure on the Transmission Dynamics during the Pandemic Flu H1N1-2009 in Oita, Japan," *PLOS ONE* 2015; 10: e0144839.

76   J. Ang, "No Plans to Close Schools for Now, Says Education Minister Ong Ye Kung," *Straits Times,* February 14, 2020.

77   Anonymous, "Coronavirus: Italy to Close All Schools as Deaths Rise," *BBC,* March 4, 2020.

78   "Interim Guidance for Administrators of US K-12 Schools and Child Care Programs to Plan, Prepare, and Respond to Coronavirus Disease 2019 (COVID-19)," *CDC,* March 25, 2020.

79   H. Peele et al., "Map: Coronavirus and School Closures," *Education Week,* March 6, 2020.

80   N. Musumeci and G. Fonrouge, "NYC Parents, Teachers Worried about Coronavirus Spread in Public Schools," *New York Post,* March 13, 2020.

81   E. Christakis, "For Schools, the List of Obstacles Grows and Grows," *The Atlantic,* May 24, 2020; E. Christakis and N.A. Christakis, "Closing the Schools Is Not the Only Option," *The Atlantic,* March 16, 2020.

82   E. Jones et al., "Healthy Schools: Risk Reduction Strategies for Reopening Schools," *Harvard T.H. Chan School of Public Health Healthy Buildings Program,* June 2020; "COVID-19 Planning Considerations: Guidance for School Re-Entry," *American Academy of Pediatrics,* June 2020.

83   R. Louv, *Last Child in the Woods: Saving Our Children from Nature-Deficit Disorder,* Chapel Hill, NC: Algonquin Books, 2006; N.M. Wells et al., "The Effects of School Gardens on Children's Science Knowledge: A Randomized Controlled Trial of Low-Income Elementary Schools," *International Journal of Science Education* 2015; 37: 2858–2878; A. Faber Taylor and F.E. Kuo, "Children with Attention Deficits Concentrate Better after Walk in the Park," *Journal of Attention Disorders* 2009; 12: 402–409; M. Kuo et al., "Do Lessons in Nature Boost Subsequent Classroom Engagement? Refueling Students in Flight," *Frontiers in Psychology* 2018; 8: 2253.

84   J.D. Goodman, "How Delays and Unheeded Warnings Hindered New York's Virus Fight," *New York Times,* April 8, 2020.

85   B. Carey and J. Glanz, "Hidden Outbreaks Spread through U.S. Cities Far Earlier Than Americans Knew, Estimates Say," *New York Times,* April 23, 2020.

86   A.S. Gonzalez-Reiche et al., "Introductions and Early Spread of SARS-CoV-2 in the New York City Area," *Science,* May 29, 2020.

87   A.M. Cuomo, "Governor Cuomo Issues Statement Regarding Novel Coronavirus in New York," *Official Website of New York State,* March 1, 2020.

88   M.G. West, "First Case of Coronavirus Confirmed in New York State," *Wall Street Journal,* March 1, 2020; B. Carey and J. Glanz, "Hidden Outbreaks Spread through U.S. Cities Far Earlier Than Americans Knew, Estimates Say," *New York Times,* April 23, 2020.

89   J.D. Goodman, "How Delays and Unheeded Warnings Hindered New York's Virus Fight," *New York Times,* April 8, 2020.

90   A.M. Cuomo, "Governor Cuomo Issues Statement Regarding Novel Coronavirus in New York," *Official Website of New York State,* March 1, 2020.

91   M. Hohman and S. Stump, "New York's Coronavirus 'Patient Zero' Tells His Story for the First Time: 'Thankful That I'm Alive,'" *Today,* May 11, 2020.

92   J. Goldstein and J. McKinley, "Second Case of Coronavirus in N.Y. Sets Off Search for Others Exposed," *New York Times,* March 3, 2020; J. Millman, "Midtown Lawyer Positive for Coronavirus Is NY's 1st Case of Person-to-Person Spread," *NBC New York,* March 3, 2020.

93   L. Ferré-Sadurní et al., "N.Y. Creates 'Containment Zone' Limiting Large Gatherings in New Rochelle," *New York Times,* March 11, 2020.

94   J. Goldstein and M. Gold, "City Pleads for More Coronavirus Tests as Cases Rise in New York," *New York Times,* March 9, 2020.

95   J.D. Goodman, "How Delays and Unheeded Warnings Hindered New York's Virus Fight," *New York Times,* April 8, 2020.

96   C. Knoll, "New York in the Age of Coronavirus," *New York Times,* March 10, 2020.

97   W. Parnell and S. Shahrigian, "Mayor De Blasio Says Coronavirus Fears Shouldn't Keep New Yorkers Off Subways," *New York Daily News,* March 5, 2020.

98   E. Shapiro and M. Gold, "Thousands of Students in New York Face Shuttered Schools," *New York Times,* March 11, 2020.

99   A.L. Gordon, "NYC's Horace Mann School Closes as Student Tested for Virus," *Bloomberg,* March 9, 2020.

100   T. Winter, "Coronavirus Outbreak: NYC Teachers 'Furious' over De Blasio's Policy to Keep Schools Open," *NBC News,* March 15, 2020.

101   J.D. Goodman, "How Delays and Unheeded Warnings Hindered New York's Virus Fight," *New York Times,* April 8, 2020.

102   L. Stack, "St. Patrick's Day Parade Is Postponed in New York over Coronavirus Concerns," *New York Times,* March 11, 2020.

103   J.E. Bromwich et al., "De Blasio Declares State of Emergency in N.Y.C., and Large Gatherings Are Banned," *New York Times,* March 12, 2020; A.M. Cuomo, "During Novel Coronavirus Briefing, Governor Cuomo Announces New Mass Gatherings Regulations," *Official Website of New York State,* March 12, 2020.

104   J. Silverstein, "New York City to Close All Theaters and Shift Restaurants to Take-Out and Delivery Only Due to Coronavirus," *CBS News,* March 16, 2020.

105   A.M. Cuomo, "Governor Cuomo Signs the 'New York State on PAUSE' Executive Order," *Official Website of New York State,* March 20, 2020; A.M. Cuomo, "Video, Audio, Photos & Rush Transcript: Governor Cuomo Signs the 'New York State on Pause' Executive Order," *Official Website of New York State,* March 20, 2020; H. Cooper et al., "43 Coronavirus Deaths and Over 5,600 Cases in N.Y.C.," *New York Times,* March 20, 2020.

106   Anonymous, "'No Time to Be Lax': Cuomo Extends New York Shutdown, NJ Deaths Top 1,000," *NBC New York,* April 7, 2020.

107   J. McKinley, "New York City Region Is Now an Epicenter of the Coronavirus Pandemic," *New York Times,* March 22, 2020.

108   J. Marsh, "In One Day, 1,000 NYC Doctors and Nurses Enlist to Battle Coronavirus," *New York Post,* March 18, 2020.

109   L. Widdicombe, "The Coronavirus Pandemic Peaks in New York's Hospitals," *The New Yorker,* April 15, 2020.

110   M. Rothfeld et al., "13 Deaths in a Day: An 'Apocalyptic' Surge at a NYC Hospital," *New York Times,* March 25, 2020.

111   M. Myers, "The Army Corps of Engineers Has Two or Three Weeks to Get Thousands of New Hospital Beds Up and Running," *Military Times,* March 27, 2020.

112   J. McKinley, "New York City Region Is Now the Epicenter of the Coronavirus Pandemic," *New York Times,* March 22, 2020.

113   Ibid.

114   H. Cooper et al., "Coronavirus Hot Spots Emerging Near New York City," *New York Times,* April 5, 2020.

115   M. Bryant, "New York Veterinarians Give Ventilators to 'War Effort' against Coronavirus," *The Guardian,* April 2, 2020.

116   C. Campanile and K. Sheehy, "NY Issues Do-Not-Resuscitate Guideline for Cardiac Patients amid Coronavirus," *New York Post,* April 21, 2020.

117   L. Widdicombe, "The Coronavirus Pandemic Peaks in New York's Hospitals," *The New Yorker,* April 15, 2020.

118   M. Rothfeld et al., "13 Deaths in a Day: An 'Apocalyptic' Surge at a NYC Hospital," *New York Times,* March 25, 2020.

119   A. Feuer and A. Salcedo, "New York City Deploys 45 Mobile Morgues as Virus Strains Funeral Homes," *New York Times,* April 2, 2020.

120   "Research, Statistics, Data & Systems: National Health Expenditure Data; Historical," *Centers for Medicare & Medicaid Services,* December 17, 2019.

121   A. Correal and A. Jacobs, "'A Tragedy Is Unfolding': Inside New York's Virus Epicenter," *New York Times,* April 9, 2020.

122 J. Coven and A. Gupta, "Disparities in Mobility Responses to COVID-19," *NYU working paper,* May 15, 2020.

123 Ibid.

124 J.D. Goodman, "How Delays and Unheeded Warnings Hindered New York's Virus Fight," *New York Times,* April 8, 2020.

125 "New York Coronavirus Cases," *Worldometer,* March 31, 2020.

126 Anonymous, " 'No Time to Be Lax': Cuomo Extends New York Shutdown, NJ Deaths Top 1,000," *NBC New York,* April 7, 2020.

127 J.D. Goodman and M. Rothfeld, "1 in 5 New Yorkers May Have Had Covid-19, Antibody Tests Suggest," *New York Times,* April 23, 2020.; D. Stadlebauer et al., "Seroconversion of a City: Longitudinal Monitoring of SARS-CoV-2 Seroprevalence in New York City," *medRxiv,* June 29, 2020,

128 D. Carey and J. Glanz, "Travel from New York City Seeded Wave of U.S. Outbreaks," *New York Times,* May 7, 2020.

129 "State of the Restaurant Industry," *OpenTable.com,* July 18, 2020.

130 N. Musumeci and G. Fonrouge, "NYC Parents, Teachers Worried about Coronavirus Spread in Public Schools," *New York Post,* March 13, 2020.

131 John of Ephesus, "John of Ephesus Describes the Justinianic Plague," ed. Roger Pearse, *Roger Pearse blog,* May 10, 2017.

4. GRIEF, FEAR, AND LIES

1 G. Magallon, "Madera Woman Loses Mother and Will Miss Granddaughter's Birth Because of COVID-19," *ABC30 ActionNews,* April 10, 2020; S. Rust and C. Cole, "She Got Coronavirus at a Funeral and Died. Her Family Honored Her with a Drive-Up Service," *Los Angeles Times,* April 8, 2020.

2 C. Engelbrecht and C. Kim, "Zoom Shivas and Prayer Hotlines: Ultra-Orthodox Jewish Traditions Upended by Coronavirus," *New York Times,* April 16, 2020.

3 N.A. Christakis, *Death Foretold: Prophecy and Prognosis in Medical Care,* Chicago: University of Chicago Press, 1999; *Prognosis in Advanced Cancer,* ed. P. Glare and N.A. Christakis, Oxford: Oxford University Press, 2008.

4 K.E. Steinhauser et al., "Factors Considered Important at the End of Life by Patients, Family, Physicians, and Other Care Providers," *JAMA* 2000; 284: 2476–2482.

5 L. Widdicombe, "The Coronavirus Pandemic Peaks in New York's Hospitals," *The New Yorker,* April 15, 2020.

6 Petrarch, *Epistolae de rebus familiaribus et variae,* in *The Black Death,* trans. and ed. R. Horrox, Manchester: Manchester University Press, 2013, p. 248.

7 L. Spinney, *Pale Rider: The Spanish Flu of 1918 and How It Changed the World,* New York: Public Affairs, 2017, p. 31.

8 H. Warraich, *Modern Death: How Medicine Changed the End of Life,* New York: St. Martin's Press, 2017, pp. 43–45; S.H. Cross and H.J. Warraich, "Changes in the Place of Death in the United States," *New England Journal of Medicine* 2019; 381: 2369–2370.

9 Thucydides, *The History of the Peloponnesian War,* trans. Richard Crawley, London: Longmans, Green & Co., 1874, p. 132.

10 Marcus Aurelius, *Marcus Aurelius,* trans. C.R. Haines, Cambridge, MA: Harvard University Press, 1916, p. 235.

11 "Coronavirus Pandemic," *Gallup,* accessed May 24, 2020.

12 D. Lazer et al., "The State of the Nation: A 50-State COVID-19 Survey," *Northeastern University,* April 20, 2020.

13 A. McGinty et al., "Psychological Distress and Loneliness Reported by US Adults in 2018 and April 2020," *JAMA* 2020; 324: 93–94.

14 M. Brenan, "U.S. Adults Report Less Worry, More Happiness," *Gallup,* May 18, 2020, accessed May 24, 2020.

15 "Coronavirus Pandemic," *Gallup,* accessed May 24, 2020.

16 M. Brenan, "Targeted Quarantines Top U.S. Adults' Conditions for Normalcy," *Gallup,* May 11, 2020, accessed May 24, 2020.

17 F. Fu et al., "Dueling Biological and Social Contagions," *Scientific Reports* 2017; 7: 43634.

18 J.M. Epstein et al., "Couple Contagion Dynamics of Fear and Disease: Mathematical and Computational Explorations," *PLOS ONE* 2008; 3: e3955.

19 S. Taylor, *The Psychology of Pandemics,* Newcastle upon Tyne: Cambridge Scholars Publications, 2020.

20 K. King, "Daily Cheers Give Morale Boost to Medical Workers Fighting Coronavirus," *Wall Street Journal,* April 18, 2020; A. Mohdin, "Pots, Pans, Passion: Britons Clap Their Support for NHS Workers Again," *The Guardian,* April 2, 2020.

21 A. Finger, *Elegy for a Disease: A Personal and Cultural History of Polio,* New York: St. Martin's Press, 2006, p. 82.

22 J. Dwyer, "The Doctor Came to Save Lives. The Co-Op Board Told Him to Get Lost," *New York Times,* April 3, 2020.

23 E. Shugerman, "Coronavirus Heroes Are Getting Tossed from Their Homes by Scared Landlords," *Daily Beast,* June 23, 2020.

24 A. Gawande, "Amid the Coronavirus Crisis, a Regimen for Reëntry," *The New Yorker,* May 13, 2020.

25 G. Graziosi, "Doctor Loses Custody of Her Child over Coronavirus Fears," *The Independent,* April 13, 2020.

26 N.S. Deodhar et al., "Plague That Never Was: A Review of the Alleged Plague Outbreaks in India in 1994," *Journal of Public Health Policy* 1998; 19: 184–199.

27 H.V. Batra et al., "Isolation and Identification of *Yersinia pestis* Responsible for the Recent Plague Outbreaks in India," *Current Science* 1996; 71: 787–791.

28 D.V. Mavalankar, "Indian 'Plague' Epidemic: Unanswered Questions and Key Lessons," *Journal of the Royal Society of Medicine* 1995; 88: 547–551.

29 K.S. Jayaraman, "Indian Plague Poses Enigma to Investigators," *Nature* 1994; 371: 547; N.S. Deodhar et al., "Plague That Never Was: A Review of the Alleged Plague Outbreaks in India in 1994," *Journal of Public Health Policy* 1998; 19: 184–199; A.K. Dutt et al., "Surat Plague of 1994 Re-Examined," *Southeast Asian Journal of Tropical Medicine and Public Health* 2006; 37: 755–760.

30 H.V. Batra et al., "Isolation and Identification of *Yersinia pestis* Responsible for the Recent Plague Outbreaks in India," *Current Science* 1996; 71: 787–791; S.N. Shivaji et al., "Identification of *Yersinia pestis* as the Causative Organism of Plague in India as Determined by 16S rDNA Sequencing and RAPD-Based Genomic Fingerprinting," *FEMS Microbiology Letters* 2000; 189: 247–252.

31 N.A. Christakis and J.H. Fowler, *Connected: The Surprising Power of Our Social Networks and How They Shape Our Lives,* New York: Little, Brown, 2009.

32 J.F.C. Hecker, *The Epidemics of the Middle Ages,* trans. B.G. Babington, London: The Sydenham Society, 1844, pp. 87–88.

33 N.A. Christakis and J.H. Fowler, *Connected: The Surprising Power of Our Social Networks and How They Shape Our Lives,* New York: Little, Brown, 2009.

34 T.F. Jones et al., "Mass Psychogenic Illness Attributed to Toxic Exposure at a High School," *New England Journal of Medicine* 2000; 342: 96–100.

35 D. Holtz et al., "Interdependence and the Cost of Uncoordinated Responses to COVID-19," *MIT working paper,* May 22, 2020.

36 M.D. Lieberman, *Social: Why Our Brains Are Wired to Connect,* New York: Crown, 2013, p. 8.

37 Q. Jianhang and T. Shen, "Whistleblower Li Wenliang: There Should Be More Than One Voice in a Healthy Society," *Caixin,* February 6, 2020.

38 R. Judd, "ER Doctor Who Criticized Bellingham Hospital's Coronavirus Protections Has Been Fired," *Seattle Times,* March 27, 2020.

39 A. Gallegos, "Hospitals Muzzle Doctors and Nurses on PPE, COVID-19 Cases," *Medscape,* March 25, 2020.

40 E. Kincaid, "COVID-19 Daily: Physician Gag Orders," *Medscape,* March 25, 2020.

41 O. Carville et al., "Hospitals Tell Doctors They'll Be Fired If They Speak Out about Lack of Gear," *Bloomberg,* March 31, 2020.

42 S. Ramachandran and J. Palazzolo, "NYU Langone Tells ER Doctors to 'Think More Critically' about Who Gets Ventilators," *Wall Street Journal,* March 31, 2020.

43 M. Richtel, "Frightened Doctors Face Off with Hospitals over Rules on Protective Gear," *New York Times,* March 31, 2020.

44 L.H. Sun and J. Dawsey, "CDC Feels Pressure from Trump as Rift Grows over Coronavirus Response," *Washington Post,* July 9, 2020.

45 R. Ballhaus and S. Armour, "Health Chief's Early Missteps Set Back Coronavirus Response," *Wall Street Journal,* April 22, 2020.

46 L.H. Sun and J. Dawsey, "White House and CDC Remove Coronavirus Warnings about Choirs in Faith Guidance," *Washington Post,* May 28, 2020.

47 A. James et al., "High COVID-19 Attack Rate among Attendees at Events at a Church—Arkansas, March 2020," *CDC Morbidity and Mortality Weekly Report* 2020; 69: 632–635.

48 A. Liptak, "Supreme Court, in 5-4 Decision, Rejects Church's Challenge to Shutdown Order," *New York Times,* May 30, 2020.

49 E. Koop, "Surgeon General Koop: The Right, the Left, and the Center of the AIDS Storm," *Washington Post*, March 24, 1987.

50 L.M. Werner, "Reagan Officials Debate AIDS Education Policy," *New York Times*, January 24, 1987.

51 C. Friedersdorf, "Maybe Trump Isn't Lying," *The Atlantic*, May 19, 2020.

52 J. Margolin and J.G. Meek, "Intelligence Report Warned of Coronavirus Crisis as Early as November: Sources," *ABC News*, April 8, 2020.

53 P. Bump, "Yet Again, Trump Pledges That the Coronavirus Will Simply Go Away," *Washington Post*, April 28, 2020.

54 D.J. Trump, "Remarks by President Trump at a Turning Point Action Address to Young Americans," *White House*, June 23, 2020.

55 E. Samuels, "Fact-Checking Trump's Accelerated Timeline for a Coronavirus Vaccine," *Washington Post*, March 4, 2020.

56 N. Weiland, "Anyone Who Wants a Coronavirus Test Can Have One, Trump Says. Not Quite, Says His Administration," *New York Times*, March 7, 2020.

57 C. Paz, "All the President's Lies About the Coronavirus," *The Atlantic*, July 13, 2020.

58 I. Chotiner, "How to Talk to Coronavirus Skeptics," *The New Yorker*, March 23, 2020.

59 M. Segalov, "'The Parallels between Coronavirus and Climate Crisis Are Obvious,'" *The Guardian*, May 4, 2020.

60 J. Bertrand, *A Historical Relation of the Plague at Marseille in the Year 1720*, trans. Anne Plumptre, London: Billingsley, 1721.

61 D.D.P. Johnson and J.H. Fowler, "The Evolution of Overconfidence," *Nature* 2011; 477: 317–320.

62 D.B. Taylor, "George Floyd Protests: A Timeline," *New York Times*, July 10, 2020; W. Lowery, "Why Minneapolis Was the Breaking Point," *The Atlantic*, June 12, 2020.

63 D. Diamond, "Suddenly, Public Health Officials Say Social Justice Matters More Than Social Distance," *Politico*, June 4, 2020.

64 M. Bebinger et al., "New Coronavirus Hot Spots Emerge across South and in California, As Northeast Slows," *NPR*, June 5, 2020.

65 S. Pei et al., "Differential Effects of Intervention Timing on COVID-19 Spread in the United States," *medRxiv*, May 29, 2020.

66 A. Mitchell and J.B. Oliphant, "Americans Immersed in COVID-19 News; Most Think Media Are Doing Fairly Well Covering It," *Pew Research Center*, March 18, 2020.

67 D. Cyranoski, "Inside the Chinese Lab Poised to Study World's Most Dangerous Pathogens," *Nature*, February, 22, 2017.

68 A. Stevenson, "Senator Tom Cotton Repeats Fringe Theory of Coronavirus Origins," *New York Times*, February 17, 2020.

69 S.W. Mosher, "Don't Buy China's Story: The Coronavirus May Have Leaked from a Lab," *New York Post*, February 22, 2020.

70 A. Stevenson, "Senator Tom Cotton Repeats Fringe Theory of Coronavirus Origins," *New York Times*, February 17, 2020.

71 Anonymous, "Coronavirus: Trump Stands by China Lab Origin Theory for Virus," *BBC*, May 1, 2020.

72 K.G. Andersen et al., "The Proximal Origin of SARS-CoV-2," *Nature Medicine* 2020; 26: 450–455; P. Zhou et al, "A Pneumonia Outbreak Associated with a New Coronavirus of Probable Bat Origin," *Nature* 2020; 579: 270–273.

73 S. Andrew, "Nearly 30% in the US Believe a Coronavirus Theory That's Almost Certainly Not True," *CNN*, April 13, 2020; W. Ahmed et al., "COVID-19 and the 5G Conspiracy Theory: Social Network Analysis of Twitter Data," *Journal of Medical Internet Research* 2020; 22: e19458.

74 D. O'Sullivan et al., "Exclusive: She's Been Falsely Accused of Starting the Pandemic. Her Life Has Been Turned Upside Down," *CNN*, April 27, 2020.

75 L. Fair, "FTC, FDA Warn Companies Making Coronavirus Claims," *Federal Trade Commission*, March 9, 2020.

76 M. Shuman, "Judge Issues Restraining Order to 'Church' Selling Bleach as COVID-19 Cure," *CNN*, April 17, 2020.

77 A. Marantz, "Alex Jones's Bogus Coronavirus Cures," *The New Yorker*, April 6, 2020.

78 L. Fair, "FTC, FDA Warn Companies Making Coronavirus Claims," *Federal Trade Commission*, March 9, 2020.

79 Ibid.

80 D. Lazarus, "LA Animal Rights Advocate Peddled Pandemic Snake Oil, FTC Says," *Los Angeles Times*, April 30, 2020.

81 S. Jones, "As Coronavirus Panic Heats Up, So Do Sales of Snake Oil," *New York,* March 15, 2020.

82 D.D. Ashley and R. Quaresima, "Warning Letter," *United States Food and Drug Administration,* March 6, 2020.

83 K. Rogers, "Trump's Suggestion That Disinfectants Could Be Used to Treat Coronavirus Prompts Aggressive Pushback," *New York Times,* April 24, 2020.

84 L. Wade, "The Secret Life of Vintage Lysol Douche Ads," *Society Pages,* September 27, 2013.

85 M. Wang et al., "Remdesivir and Chloroquine Effectively Inhibit the Recently Emerged Novel Coronavirus (2019-nCoV) In Vitro," *Cell Research* 2020; 30: 269–271.

86 T. Nguyen, "How a Chance Twitter Thread Launched Trump's Favorite Coronavirus Drug," *Politico,* April 7, 2020.

87 J. Yazdany and A.H.J. Kim, "Use of Hydroxychloroquine and Chloroquine during the COVID-19 Pandemic: What Every Clinician Should Know," *Annals of Internal Medicine,* March 31, 2020.

88 J.C. Wong, "Hydroxychloroquine: How an Unproven Drug Became Trump's Coronavirus 'Miracle Cure,'" *The Guardian,* April 7, 2020.

89 D. Lazer et al., "The State of the Nation: A 50-State COVID-19 Survey," *Northeastern University,* April 20, 2020.

90 R. Savillo et al., "Over Three Days This Week, Fox News Promoted an Antimalarial Drug Treatment for Coronavirus Over 100 Times," *Media Matters for America,* April 6, 2020.

91 E. Edwards and V. Hillyard, "Man Dies After Taking Chloroquine in an Attempt to Prevent Coronavirus," *NBC News,* March 23, 2020.

92 N.J. Mercuro et al., "Risk of QT Interval Prolongation Associated with Use of Hydroxy-chloroquine with or without Concomitant Azithromycin among Hospitalized Patients Testing Positive for Coronavirus Disease 2019 (COVID-19)," *JAMA Cardiology,* May 1, 2020; J. Yazdany and A.H.J. Kim, "Use of Hydroxychloroquine and Chloroquine during the COVID-19 Pandemic: What Every Clinician Should Know," *Annals of Internal Medicine,* March 31, 2020.

93 K. Thomas and K. Sheikh, "Small Chloroquine Study Halted over Risk of Fatal Heart Complications," *New York Times,* April 12, 2020; J. Yazdany and A.H.J. Kim, "Use of Hydroxychloroquine and Chloroquine during the COVID-19 Pandemic: What Every Clinician Should Know," *Annals of Internal Medicine,* March 31, 2020.

94 K. Kupferschmidt, "Big Studies Dim Hopes for Hydroxychloroquine," *Science* 2020; 368: 1166 –1167; J. Geleris et al., "Observational Study of Hydroxychloroquine in Hospitalized Patients with COVID-19," *New England Journal of Medicine* 2020; 382: 2411–2418; E.S. Rosenberg et al., "Association of Treatment with Hydroxychloroquine or Azithromycin with In-Hospital Mortality in Patients with COVID-19 in New York State," *JAMA* 2020; 323: 2493–2502.

95 D.R. Boulware et al., "A Randomized Trial of Hydroxychloroquine as Post-Exposure Prophylaxis for COVID-19," *New England Journal of Medicine,* June 3, 2020; RECOVERY Collaborative Group, "Effect of Hydroxychloroquine in Hospitalized Patients with COVID-19: Preliminary Results from a Multi-Centre, Randomized, Controlled, Trial," *medRxiv,* July 15, 2020.; A.B. Cavalcanti et al., "Hydroxychloroquine With or Without Azithromycin in Mild-to-Moderate COVID-19," *New England Journal of Medicine, July 23,* 2020.

96 John of Ephesus, "John of Ephesus Describes the Justinianic Plague," ed. Roger Pearse, *Roger Pearse blog,* May 10, 2017.

97 L. Bode and E. Vraga, "Americans Are Fighting Coronavirus Misinformation on Social Media," *Washington Post,* May 7, 2020.

98 L. Singh et al., "A First Look at COVID-19 Information and Misinformation Sharing on Twitter," *arXiv,* April 1, 2020.

99 V.A. Young, "Nearly Half of the Twitter Accounts Discussing 'Re-Opening America' May Be Bots," *Carnegie Mellon University press release,* May 20, 2020.

100 W.J. Broad, "Putin's Long War against American Science," *New York Times,* April 13, 2020; M. Repnikova, "Does China's Propaganda Work?" *New York Times,* April 16, 2020.

101 N.F. Johnson et al., "The Online Competition between Pro- and Anti-Vaccination Views," *Nature* 2020; 582: 230–233; J.P. Onnela et al., "Polio Vaccine Hesitancy in the Networks and Neighborhoods of Malegaon, India," *Social Science and Medicine* 2016; 153: 99–106.

102 M. Baldwin, "Scientific Autonomy, Public Accountability, and the Rise of 'Peer Review' in the Cold War United States," *Isis,* 2018; 109: 538-558.

103 M.S. Majumder and K.D. Mandl, "Early in the Epidemic: Impacts of Preprints on Global Discourse about COVID-19 Transmissibility," *Lancet Global Health* 2020; 8: e627.

5. US AND THEM

1   S.K. Cohn, "The Black Death and the Burning of the Jews," *Past and Present* 2007; 196: 3–36.
2   Anonymous, "Examination of the Jews Captured in Savoy," in *Urkunden und Akten der Stadt Strassburg: Urkundenbuch der Stadt Strassburg,* in *The Black Death,* trans. and ed. R. Horrox, Manchester: Manchester University Press, 2013, p. 219.
3   J. Silver and D. Wilson, *Polio Voices: An Oral History from the American Polio Epidemics and Worldwide Eradication Efforts,* Westport, CT: Praeger, 2007, p. 22.
4   Ibid., p. 26.
5   D.J. Trump, "Remarks by President Trump to Reporters," *White House,* May 6, 2020.
6   K. Fukuda et al., "Naming Diseases: First Do No Harm," *Science* 2015; 348: 6235.
7   J.S. Jia et al., "Population Flow Drives Spatio-Temporal Distribution of COVID-19 in China," *Nature* 2020; 582: 389–394.
8   H. Yan et al., "What's Spreading Faster Than Coronavirus in the US? Racist Assaults and Ignorant Attacks against Asians," *CNN,* February 21, 2020; S. Tavernise and R.A. Oppel, "Spit On, Yelled At, Attacked: Chinese-Americans Fear for Their Safety," *New York Times,* June 2, 2020.
9   D.S. Lauderdale, "Birth Outcomes for Arabic-Named Women in California before and after September 11," *Demography* 2006; 43: 185–201.
10 E.I. Koch, "Senator Helms's Callousness toward AIDS Victims," *New York Times,* November 7, 1987.
11 R. Brackett, "Governor Says State Will Accept Florida Residents from Cruise Ship Stricken with Coronavirus," *Weather Channel,* April 1, 2020; M. Burke, K. Sanders, "Cruise Ship with Sick Passengers and Sister Ship Dock in Florida," *NBC News,* April 3, 2020.
12 D. Quan, "'Dreams Are Not Passports': Remote Arctic Village Residents Recount Bizarre Encounter with Quebec Couple Fleeing Coronavirus," *The Star* [Toronto], March 30, 2020.
13 Procopius, *History of the Wars,* trans. H.B. Dewing, Cambridge, MA: Harvard University Press, 1914, p. 453.
14 D. Haar, "Nobel Economist Shiller Says Crisis May Boost Income Equality," *Middletown Press* (CT), March 23, 2020.
15 Clement VI, *The Apostolic See and the Jews,* in *The Black Death,* trans. and ed. R. Horrox, Manchester: Manchester University Press, 2013, pp. 221–222.
16 J.L. Schwartzwald, *The Collapse and Recovery of Europe, AD 476–1648,* Jefferson, NC: McFarland, 2015, p. 123; D. Wood, *Clement VI: The Pontificate and Ideas of an Avignon Pope,* Cambridge: Cambridge University Press, 2003, p. 51.
17 A.K. Simon et al., "Evolution of the Immune System in Humans from Infancy to Old Age," *Proceedings of the Royal Society B* 2015; 282: 2014.3085.
18 L. Liu et al., "Global, Regional, and National Causes of Under-5 Mortality in 2000–15: An Updated Systematic Analysis with Implications for the Sustainable Development Goals," *The Lancet* 2016; 388: 3027–3035.
19 J.T. Wu et al., "Estimating Clinical Severity of COVID-19 from the Transmission Dynamics in Wuhan, China," *Nature Medicine* 2020; 26: 506–510.
20 WHO-China Joint Mission, "Report of the WHO-China Joint Mission on Coronavirus Disease 2019 (COVID-19)," *WHO,* February 16-24, 2020.
21 Q. Bi et al., "Epidemiology and Transmission of COVID-19 in 391 Cases and 1286 of Their Close Contacts in Shenzhen, China: A Retrospective Cohort Study," *Lancet Infectious Diseases,* May 5, 2020.
22 J.T. Wu et al., "Estimating Clinical Severity of COVID-19 from the Transmission Dynamics in Wuhan, China," *Nature Medicine* 2020; 26: 506–510; J. Zhang et al., "Changes in Contact Patterns Shape the Dynamics of the COVID-19 Outbreak in China," *Science* 2020; 368: 1481–1486.
23 L. Dong et al., "Possible Vertical Transmission of SARS-CoV-2 from an Infected Mother to Her Newborn," *JAMA* 2020; 323: 1846–1848.
24 J.T. Wu et al., "Estimating Clinical Severity of COVID-19 from the Transmission Dynamics in Wuhan, China," *Nature Medicine* 2020; 26: 506–510; H. Salje et al., "Estimating the Burden of SARS-CoV-2 in France," *Science* 2020; 369: 208–211; S. Riphagen et al., "Hyperinflammatory Shock in Children during COVID-19 Pandemic," *The Lancet* 2020; 395: 1607–1608.
25 WHO-China Joint Mission, "Report of the WHO-China Joint Mission on Coronavirus Disease 2019 (COVID-19)," *WHO,* February 16-24, 2020; J.T. Wu et al., "Estimating Clinical Severity of COVID-19 from the Transmission Dynamics in Wuhan, China," *Nature Medicine*

2020; 26: 506–510; W.J. Guan et al., "Clinical Characteristics of Coronavirus Disease 2019 in China," *New England Journal of Medicine* 2020; 382: 1708–1720; T.W. Russell et al., "Estimating the Infection and Case Fatality Ratio for Coronavirus Disease (COVID-19) Using Age-Adjusted Data from the Outbreak on the Diamond Princess Cruise Ship, February 2020," *Eurosurveillance* 2020; 25: pii=2000256.

26   Y.Y. Dong et al., "Epidemiological Characteristics of 2143 Pediatric Patients with 2019 Coronavirus Disease in China," *Pediatrics,* March 16, 2020; P. Belluck, "Children and Coronavirus: Research Finds Some Become Seriously Ill," *New York Times,* March 17, 2020.

27   CDC COVID-19 Response Team, "Severe Outcomes among Patients with Coronavirus Disease 2019 (COVID-19)—United States, February 12–March 16, 2020," *CDC Morbidity and Mortality Weekly Report* 2020; 69: 343–346.

28   A. Hauser et al., "Estimation of SARS-CoV-2 Mortality during the Early Stages of an Epidemic: A Modeling Study in Hubei, China, and Six Regions in Europe," *medRxiv,* July 12, 2020.

29   Severe Acute Respiratory Syndrome (SARS) Epidemiology Working Group, "Consensus Document on the Epidemiology of Severe Acute Respiratory Syndrome (SARS)," *WHO Department of Communicable Disease Surveillance and Response,* October 17, 2003.

30   M. Hoffmann et al., "SARS-CoV-2 Cell Entry Depends on ACE2 and TMPRSS2 and Is Blocked by a Clinically Proven Protease Inhibitor," *Cell* 2020; 181: 271–280; H. Zhang et al., "Angiotensin-Converting Enzyme 2 (ACE2) as a SARS-CoV-2 Receptor: Molecular Mechanisms and Potential Therapeutic Target," *Intensive Care Medicine* 2020; 46: 586–590; H. Gu et al., "Angiotensin-Converting Enzyme 2 Inhibits Lung Injury Induced by Respiratory Syncytial Virus," *Scientific Reports* 2016; 6: 19840; U. Bastolla, "The Differential Expression of the ACE2 Receptor across Ages and Gender Explains the Differential Lethality of SARS-CoV-2 and Suggests Possible Therapy," *arXiv,* May 3, 2020; L. Zhu et al., "Possible Causes for Decreased Susceptibility of Children to Coronavirus," *Pediatric Research,* April 8, 2020.

31   P. Verdecchia et al., "The Pivotal Link between ACE2 Deficiency and SARS-CoV-2 Infection," *European Journal of Internal Medicine* 2020; 76: 14–20.

32   E. Ciaglia et al., "COVID-19 Infection and Circulating ACE2 Levels: Protective Role in Women and Children," *Frontiers in Pediatrics* 2020; 8: 206.

33   A.K. Simon et al., "Evolution of the Immune System in Humans from Infancy to Old Age," *Proceedings of the Royal Society B* 2015; 282: 2014.3085.

34   L. Zhu et al., "Possible Causes for Decreased Susceptibility of Children to Coronavirus," *Pediatric Research,* April 8, 2020.

35   M.E. Rudolph et al., "Differences between Pediatric and Adult T Cell Responses to In Vitro Staphylococcal Enterotoxin B Stimulation," *Frontiers in Immunology* 2018; 9: 498; P. Mehta et al., "COVID-19: Consider Cytokine Storm Syndromes and Immunosuppression," *The Lancet* 2020; 395: 1033–1034.

36   L.E. Escobar et al., "BCG Vaccine Protection from Severe Coronavirus Disease 2019 (COVID-19)," *Proceedings of the National Academy of Sciences,* July 9, 2020.

37   M. Rawat et al., "COVID-19 in Newborns and Infants—Low Risk of Severe Disease: Silver Lining or Dark Cloud?" *American Journal of Perinatology* 2020; 37: 845–849; J. Wang and M.S. Zand, "Potential Mechanisms of Age Related Severity of COVID-19 Infection: Implications for Vaccine Development and Convalescent Serum Therapy," *University of Rochester Preprint,* March 21, 2020.; J. Mateus et al., "Selective and Cross-Reactive SARS-CoV-2 T-Cell Epitopes in Unexposed Humans," *Science,* August 4, 2020.

38   P. Brodin, "Why Is COVID-19 So Mild in Children?" *Acta Paediatrica* 2020; 109: 1082–1083.

39   S. Mallapaty, "How Do Children Spread the Coronavirus? The Science Still Isn't Clear," *Nature,* May 7, 2020; L. Rajmil, "Role of Children in the Transmission of the COVID-19 Pandemic: A Rapid Scoping Review," *BMJ Paediatrics Open* 2020; 4: e000722; D. Isaacs et al., "To What Extent Do Children Transmit SARS-CoV-2 Virus?" *Journal of Paediatrics and Child Health* 2020; 56: 978; X. Li et al., "The Role of Children in Transmission of SARS-CoV-2: A Rapid Review," *Journal of Global Health* 2020; 10: 011101; R.M. Viner et al., "Susceptibility to and Transmission of COVID-19 amongst Children and Adolescents Compared with Adults: A Systematic Review and Meta-Analysis," *medRxiv,* May 24, 2020.

40   K.M. Posfay-Barbe et al., "COVID-19 in Children and the Dynamics of Infection in Families," *Pediatrics* 2020; 146: e20201576; A. Fontanet et al., "SARS-CoV-2 Infection in Primary Schools in Northern France: A Retrospective Cohort Study in an Area of High Transmission," *medRxiv,* June 29, 2020.

41   G. Vogel and J. Couzin-Frankel, "Should Schools Reopen? Kids' Role in Pandemic Still a Mystery," *Science,* May 4, 2020.

42   L. Rosenbaum, "Facing COVID-19 in Italy—Ethics, Logistics, and Therapeutics on the Epidemic's Front Line," *New England Journal of Medicine* 2020; 382: 1873–1875; M. Vergano et al., "Clinical Ethics Recommendations for the Allocation of Intensive Care Treatments," *SIAARTI,* March 16, 2020; Y. Mounk, "The Extraordinary Decisions Facing Italian Doctors," *The Atlantic,* March 11, 2020.

43   S. Fink, *Five Days at Memorial,* New York: Crown, 2013.

44   L. Duda, "National Organ Allocation Policy: The Final Rule," *Ethics Journal of the American Medical Association* 2005; 7: 604–607; Anonymous, "How Organ Allocation Works," *U.S. Department of Health & Human Services,* n.d.

45   Anonymous, "NY Issues Do Not Resuscitate Guidelines for Cardiac Patients, Later Rescinds Them," *Journal of Emergency Medical Services,* April 22, 2020.

46   C. Huang et al., "Clinical Features of Patients Infected with 2019 Novel Coronavirus in Wuhan, China," *The Lancet* 2020; 395: 497–506.

47   C.M. Petrilli et al., "Factors Associated with Hospital Admission and Critical Illness among 5279 People with Coronavirus Disease 2019 in New York City: Prospective Cohort Study," *BMJ* 2020; 369: m1966.

48   E.J. Williamson et al., "Open SAFELY: Factors Associated with COVID-19 Death in 17 Million Patients," *Nature,* July 8, 2020.

49   S. Kadel and S. Kovats, "Sex Hormones Regulate Innate Immune Cells and Promote Sex Differences in Respiratory Virus Infection," *Frontiers in Immunology* 2018; 9: 1653.

50   P. Conti and A. Younes, "Coronavirus COV-19/SARS-CoV-2 Affects Women Less Than Men: Clinical Response to Viral Infection," *Journal of Biological Regulators and Homeostatic Agents* 2020; 34: 32253888; P. Pozzilli and A. Lenzi, "Commentary: Testosterone, a Key Hormone in the Context of COVID-19 Pandemic," *Metabolism* 2020; 108: 154252.

51   H. Schurz et al., "The X Chromosome and Sex-Specific Effects in Infectious Disease Susceptibility," *Human Genomics* 2019; 13: 2.

52   A. Maqbool, "Coronavirus: 'I Can't Wash My Hands—My Water Was Cut Off,'" *BBC News,* April 24, 2020.

53   T. Orsborn, "'We Just Can't Feed This Many,'" *San Antonio Express News,* April 9, 2020.

54   L. Zhou and K. Amaria, "The Current Hunger Crisis in the US, in Photos," *Vox,* May 9, 2020.

55   HUD, "2017 AHAR: Part 1—PIT Estimates of Homelessness in the U.S.," *HUD Exchange,* December 2017.

56   T. Baggett et al., "Prevalence of SARS-CoV-2 Infection in Residents of a Large Homeless Shelter in Boston," *JAMA* 2020; 323: 2191–2192.

57   Anonymous, "Coronavirus in the U.S.: Latest Map and Case Count," *New York Times,* July 17, 2020.

58   M. Huber, "Smithfield Workers Asked for Safety from COVID-19. Their Company Offered Cash," *Argus Leader* (Sioux Falls, SD), April 9, 2020.

59   K. Collins and M. Vazquez, "Trump Orders Meat Processing Plants to Stay Open," *CNN,* April 28, 2020.

60   Anonymous, "President Donald J. Trump Is Taking Action to Ensure the Safety of Our Nation's Food Supply Chain," *White House,* April 28, 2020.

61   J.W. Dyal et al., "COVID-19 among Workers in Meat and Poultry Processing Facilities—19 States, April 2020," *CDC Morbidity and Mortality Weekly Report* 2020; 69: 557–561.

62   L. Hamner et al., "High SARS-CoV-2 Attack Rate Following Exposure at a Choir Practice—Skagit County, Washington, March 2020," *CDC Morbidity and Mortality Weekly Report* 2020; 69: 606–610.

63   M.M. Harris et al., "Isolation of *Brucella suis* from Air of Slaughterhouse," *Public Health Rep* 1962; 77: 602–604; M.T. Osterholm, "A 1957 Outbreak of Legionnaires' Disease Associated with a Meat Packing Plant," *American Journal of Epidemiology* 1983; 117: 60–67.

64   M. Ferioli et al., "Protecting Healthcare Workers from SARS-CoV-2 Infection: Practical Indications," *European Respiratory Review* 2020; 29: 2000068.

65   M. Dorning et al., "Infections near U.S. Meat Plants Rise at Twice the National Rate," *Bloomberg,* May 12, 2020.

66   R.A. Oppel et al., "The Fullest Look Yet at the Racial Inequality of Coronavirus," *New York Times,* July 5, 2020; G.A. Millett et al., "Assessing Differential Impacts of COVID-19 on Black Communities," *Annals of Epidemiology* 2020; 47: 37–44.

67   APM Research Lab Staff, "The Color of Coronavirus: COVID-19 Deaths by Race and Ethnicity in the U.S.," *APM Research Lab,* July 8, 2020.

68   J. Absalom et al., *A Narrative of the Proceedings of the Black People, during the Late Awful Calamity*

in Philadelphia, in the Year 1793: And a Refutation of Some Censures, Thrown upon Them in Some Late Publications, Philadelphia: William W. Woodward, 1794, p. 15.

69   APM Research Lab Staff, "The Color of Coronavirus: COVID-19 Deaths by Race and Ethnicity in the U.S.," *APM Research Lab,* July 8, 2020.

70   C.W. Yancy, "COVID-19 and African Americans," *JAMA* 2020; 323: 1891–1892.

71   Anonymous, "COVID-19 Cases by IHS Area," *Indian Health Service,* July 17, 2020.

72   APM Research Lab Staff, "The Color of Coronavirus: COVID-19 Deaths by Race and Ethnicity in the U.S.," *APM Research Lab,* July 8, 2020.

73   D. Cohn and J.S. Passel, "A Record 64 Million Americans Live in Multigenerational Households," *Pew Research Center,* April 5, 2018.

74   P. Mozur, "China, Desperate to Stop Coronavirus, Turns Neighbor against Neighbor," *New York Times,* February 3, 2020; N. Gan, "Outcasts in Their Own Country, the People of Wuhan Are the Unwanted Faces of China's Coronavirus Outbreak," *CNN,* February 2, 2020.

75   R.D. Kirkcaldy et al., "COVID-19 and Post-Infection Immunity: Limited Evidence, Many Remaining Questions," *JAMA* 2020; 323: 2245–2246.

76   M.A. Hall and D.M. Studdert, "Privileges and Immunity Certification during the COVID-19 Pandemic," *JAMA* 2020; 323: 2243–2244.

77   Anonymous, "Immigrant and Refugee Health Frequently Asked Questions (FAQs)," *CDC,* March 29, 2012.

78   K. Olivarius, "Immunity, Capital, and Power in Antebellum New Orleans," *American Historical Review* 2019; 124: 425–455.

79   M. Myers, "Coronavirus Survivors Banned from Joining the Military," *Military Times,* May 6, 2020.

80   S.M. Nir, "They Beat the Virus. Now They Feel Like Outcasts," *New York Times,* May 20, 2020.

81   K. Collins and D. Yaffe-Bellany, "About 2 Millions Guns Were Sold in the US as Virus Fears Spread," *New York Times,* April 1, 2020.

82   T.L. Caputi et al., "Collateral Crises of Gun Preparation and the COVID-19 Pandemic: An Infodemiology Study," *JMIR Public Health Surveillance* 2020; 6: e19369.

83   A.M. Verdery et al., "Tracking the Reach of COVID-19 Kin Loss with a Bereavement Multiplier Applied to the United States," *Proceedings of the National Academy of Sciences* 2020; 117: 17695–17701.

6. BANDING TOGETHER

1   J. Tolentino, "What Mutual Aid Can Do during a Pandemic," *The New Yorker,* May 11, 2020.

2   S. Samuel, "How to Help People during the Pandemic, One Google Spreadsheet at a Time," *Vox,* April 16, 2020.

3   C. Milstein, "Collective Care Is Our Best Weapon against COVID-19," *Mutual Aid Disaster Relief,* June 6, 2020.

4   Anonymous, "Find Your Local Group," *Mutual Aid U.S.A,* May 6, 2020.

5   Anonymous, "Bay Area Mutual Aid and COVID-19 Resources," *94.1 KPFA,* n.d.

6   Anonymous, "Resources + Groups," *Mutual Aid NYC,* 2020.

7   D. Fallows, "Public Libraries' Novel Response to a Novel Virus," *The Atlantic,* March 31, 2020.

8   S. Zia, "As Coronavirus Impact Grows, Volunteer Network Tries to Help Health Care Workers Who Have 'Helped Us,'" *Stat News,* March 31, 2020.

9   S.S. Ali, "As Parents Fight on COVID-19 Front Lines, Volunteers Step In to Take Care of Their Families," *NBC News,* March 27, 2020.

10   Anonymous, "America's Hidden Common Ground on the Coronavirus Crisis," *Public Agenda,* April 3, 2020.

11   Anonymous, "Who Gives Most to Charity?" *Philanthropy Roundtable,* n.d.

12   Anonymous, "America's Hidden Common Ground on the Coronavirus Crisis," *Public Agenda,* April 3, 2020.

13   L. Rainie and A. Perrin, "The State of Americans' Trust in Each Other amid the COVID-19 Pandemic," *Pew Research Center,* April 6, 2020.

14   S.F. Beegel, "Love in the Time of Influenza: Hemingway and the 1918 Pandemic," in *War + Ink: New Perspectives on Ernest Hemingway's Early Life and Writings,* ed. S. Paul et al., Kent, OH: Kent State University Press, 2014, pp. 36–52.

15   L. Stack, "Hasidic Jews, Hit Hard by the Outbreak, Flock to Donate Plasma," *New York Times,* May 12, 2020.

16   T. Armus, "'Sorry, No Masks Allowed': Some Businesses Pledge to Keep Out Customers Who Cover Their Faces," *Washington Post,* May 28, 2020.

17   A. Finger, *Elegy for a Disease: A Personal and Cultural History of Polio,* New York: St. Martin's Press, 2006, p. 63.

18   H. Flor et al., "The Role of Spouse Reinforcement, Perceived Pain, and Activity Levels of Chronic Pain Patients," *Journal of Psychosomatic Research* 1987; 31: 251–259; S. Duschek et al., "Dispositional Empathy Is Associated with Experimental Pain Reduction during Provision of Social Support by Romantic Partners," *Scandinavian Journal of Pain* 2019; 20: 205–209; J. Younger et al., "Viewing Pictures of a Romantic Partner Reduces Experimental Pain: Involvement of Neural Reward Systems," *PLOS ONE* 2010; 5: c13309; K.J. Bourassa, "The Impact of Physical Proximity and Attachment Working Models on Cardiovascular Reactivity: Comparing Mental Activation and Romantic Partner Presence," *Psychophysiology* 2019; 56: e13324.

19   M. Slater, "'She Was Worth a Beating': Falling in Love through a Fence in a Concentration Camp," *The Yiddish Book Center's Wexler Oral History Project,* August 9, 2013.

20   V. Florian et al., "The Anxiety-Buffering Function of Close Relationships: Evidence That Relationship Commitment Acts as a Terror Management Mechanism," *Journal of Personality and Social Psychology* 2002; 82: 527–542.

21   W. Boston, "Two College Students Marry Quickly Before Escaping New York: 'The Only Way We Could Stay Together,'" *Wall Street Journal,* April 22, 2020.

22   G.L. White et al., "Passionate Love and the Misattribution of Arousal," *Journal of Personality and Social Psychology* 1981; 41: 56–62.

23   C. Cohan and S. Cole, "Life Course Transitions and Natural Disaster: Marriage, Birth, and Divorce Following Hurricane Hugo," *Journal of Family Psychology* 2002; 16: 14–25.

24   J. Lipman-Blumen, "A Crisis Framework Applied to Macrosociological Family Changes: Marriage, Divorce, and Occupational Trends Associated with World War II," *Journal of Marriage and Family* 1975; 37: 889–902.

25   C. Cohan and S. Cole, "Life Course Transitions and Natural Disaster: Marriage, Birth, and Divorce Following Hurricane Hugo," *Journal of Family Psychology* 2002; 16: 14–25.

26   S. South, "Economic Conditions and the Divorce Rate: A Time-Series Analysis of the Postwar United States," *Journal of Marriage and Family* 1985; 47: 31–41.

27   N. Raza, "What Single People Are Starting to Realize," *New York Times,* May 18, 2020.

28   Anonymous, "Domestic Violence Has Increased during Coronavirus Lockdowns," *The Economist,* April 23, 2020.

29   S. Zimmermann and S. Charles, "Chicago Domestic Violence Calls Up 18% in First Weeks of Coronavirus Shutdown," *Chicago Sun Times,* April 26, 2020.

30   A. Southall, "Why a Drop in Domestic Violence Reports Might Not Be a Good Sign," *New York Times,* April 17, 2020.

31   J. Ducharme, "COVID-19 Is Making America's Loneliness Epidemic Even Worse," *Time,* May 8, 2020.

32   A. Fetters, "The Boomerang Exes of Quarantine," *The Atlantic,* April 16, 2020.

33   H. Fisher, "How Coronavirus Is Changing the Dating Game for the Better," *New York Times,* May 7, 2020.

34   Anonymous, "Safer Sex and COVID-19," *NYC Health Department,* June 8, 2020.

35   A. Livingston, "Texas Lt. Gov. Dan Patrick Says a Failing Economy Is Worse Than Coronavirus," *Texas Tribune,* March 23, 2020.

36   J.J. Jordan et al., "Don't Get It or Don't Spread It? Comparing Self-Interested versus Prosocially Framed COVID-19 Prevention Messaging," *PsyArXiv,* May 14, 2020.

37   J.C. Hershey et al., "The Roles of Altruism, Free Riding, and Bandwagoning in Vaccination Decisions," *Organizational Behavior and Human Decision Processes* 1994; 59: 177–187; M. Li, "Stimulating Influenza Vaccination via Prosocial Motives," *PLOS ONE* 2016; 11: e0159780; J.T. Vietri, "Vaccinating to Help Ourselves and Others," *Medical Decision Making* 2012; 32: 447–458.

38   R. Solnit, *A Paradise Built in Hell,* New York: Viking, 2009, p. 4.

39   J. Zaki, "Catastrophe Compassion: Understanding and Extending Prosociality under Crisis," *Trends in Cognitive Sciences* 2020; 24: 587–589.

40   D. Holtz, et al., "Interdependence and the Cost of Uncoordinated Responses to COVID-19," *MIT Working Paper,* May 22, 2020.

41   S. Feigin et al., "Theories of Human Altruism: A Systematic Review," *Journal of Psychiatry*

and Brain Function 2014; 1: 5; T.D. Windsor et al., "Volunteering and Psychological Well-Being among Young-Old Adults: How Much Is Too Much?" *Gerontologist* 2008; 48: 59–70; C. Schwartz et al., "Altruistic Social Interest Behaviors Are Associated with Better Mental Health," *Psychosomatic Medicine* 2003; 65: 778–785; T. Fujiwara, "The Role of Altruistic Behavior in Generalized Anxiety Disorder and Major Depression among Adults in the United States," *Journal of Affective Disorders* 2007; 101: 219–225; M.A. Musick and J. Wilson, "Volunteering and Depression: The Role of Psychological and Social Resources in Different Age Groups," *Social Science & Medicine* 2003; 56: 259–269; S.L. Brown et al., "Coping with Spousal Loss: Potential Buffering Effects of Self-Reported Helping Behavior," *Personality and Social Psychology Bulletin* 2008; 34: 849–861; H.L. Schacter and G. Margolin, "When It Feels Good to Give: Depressive Symptoms, Daily Prosocial Behavior, and Adolescent Mood," *Emotion* 2019; 19: 923; K.J. Shillington et al., "Kindness as an Intervention for Student Social Interaction Anxiety, Affect, and Mood: The KISS of Kindness Study," *International Journal of Applied Positive Psychology*, May 14, 2020.

42   Anonymous, "Mental Health and Psychosocial Considerations during the COVID-19 Outbreak," *WHO*, March 18, 2020; Y. Feng et al., "When Altruists Cannot Help: The Influence of Altruism on the Mental Health of University Students during the COVID-19 Pandemic," *Globalization and Health* 2020; 16: 61.

43   Thucydides, *The History of the Peloponnesian War*, trans. Richard Crawley, London: Longmans, Green & Co., 1874, 2.47.4.

44   J. de Venette, *The Chronicle of Jean de Venette*, in *The Black Death*, trans. and ed. R. Horrox, Manchester: Manchester University Press, 2013, pp. 55–56.

45   S. Kisely, "Occurrence, Prevention, and Management of the Psychological Effects of Emerging Virus Outbreaks on Healthcare Workers: Rapid Review and Meta-Analysis," *BMJ* 2020; 369: m1642.

46   J. Hoffman, " 'I Can't Turn My Brain Off': PTSD and Burnout Threaten Medical Workers," *New York Times*, May 16, 2020.

47   K. Weise, "Two Emergency Room Doctors Are in Critical Condition with Coronavirus," *New York Times*, March 15, 2020.

48   C. Jewett et al., "Nearly 600—and Counting—US Health Care Workers Have Died of COVID-19," *The Guardian*, June 6, 2020.

49   M. Zhan et al., "Death from COVID-19 of 23 Health Care Workers in China," *New England Journal of Medicine* 2020; 382: 2267–2268; L. Magalhaes et al., "Brazil's Nurses Are Dying as COVID-19 Overwhelms Hospitals," *Wall Street Journal*, May 19, 2020.

50   Anonymous, "In Memoriam: Healthcare Workers Who Have Died of COVID-19," *Medscape*, May 1, 2020.

51   S. Gondi et al., "Personal Protective Equipment Needs in the USA during the COVID-19 Pandemic," *New England Journal of Medicine* 2020; 395: e90.

52   C. Jewett et al., "Workers Filed More Than 4,100 Complaints about Protective Gear. Some Died," *Kaiser Health News*, June 30, 2020.

53   M. Fackler, "Tsunami Warnings, Written in Stone," *New York Times*, April 20, 2011.

54   C. Domonoske, "Drought in Central Europe Reveals Cautionary 'Hunger Stones' in Czech Republic," *NPR*, August 24, 2018.

55   S. Bhuamik, "Tsunami Folklore 'Saved Islanders,' " *BBC News*, January 20, 2005.

56   P.J. Richerson and R. Boyd, *Not by Genes Alone: How Culture Transformed Human Evolution*, Chicago: University of Chicago Press, 2005, p. 5.

57   J. Henrich and C. Tennie, "Cultural Evolution in Chimpanzees and Humans," in *Chimpanzees and Human Evolution*, ed. M. Muller, R.W. Wrangham, and D.R. Pilbeam, Cambridge, MA: Harvard University Press, 2017.

58   T.T. Le et al., "The COVID-19 Vaccine Development Landscape," *Nature Reviews Drug Discovery* 2020; 19: 305–306; Anonymous, "Draft Landscape of COVID-19 Candidate Vaccines," *WHO*, April 20, 2020.

59   Anonymous, "China Has 5 Vaccine Candidates in Human Trials, with More Coming," *Bloomberg*, May 15, 2020.

60   Anonymous, "First FDA-Approved Vaccine for the Prevention of Ebola Virus Disease, Marking a Critical Milestone in Public Health Preparedness and Response," *United States Food and Drug Administration*, December 19, 2019.

61   E.S. Pronker et al., "Risk in Vaccine Research and Development Quantified," *PLOS ONE* 2013; 8: e57755.

62   K. Duan et al., "Effectiveness of Convalescent Plasma Therapy in Severe COVID-19 Patients," *Proceedings of the National Academy of Sciences* 2020; 117: 9490–9496; V.N. Pimenoff et al., "A

Systematic Review of Convalescent Plasma Treatment for COVID-19," *medRxiv,* June 8, 2020; L. Li et al., *"Effect of Convalescent Plasma Therapy on Time to Clinical Improvement in Patients with Severe and Life-Threatening COVID-19: A Randomized Clinical Trial," JAMA,* June 3, 2020.

63   D. Lowe, "Coronavirus Vaccine Prospects," *Science Translational Medicine,* April 15, 2020.

64   D.R. Hopkins, *The Greatest Killer: Smallpox in History, with a New Introduction,* Chicago: University of Chicago Press, 2002, p. 80.

65   L.J. Saif, "Animal Coronavirus Vaccines: Lessons for SARS," *Developments in Biologicals* 2004; 119: 129–140.

66   H. Wang et al., "Development of an Inactivated Vaccine Candidate, BBIBP-CorV, with Potent Protection against SARS-CoV-2," *Cell* 2020; 182: 1–9.

67   Q. Gao et al., "Development of an Inactivated Vaccine Candidate for SARS-CoV-2," *Science* 2020; 369: 77–81; J. Cohen, "Covid-19 Vaccine Protects Monkeys from New Coronavirus, Chinese Biotech Reports," *Science,* April 23, 2020.

68   F.C. Zhu et al., "Safety, Tolerability, and Immunogenicity of a Recombinant Adenovirus Type-5 Vectored COVID-19 Vaccine: A Dose-Escalation, Open-Label, Non-Randomised, First-in-Human Trial," *The Lancet* 2020; 395: 13–19.

69   Anonymous, "Moderna Ships mRNA Vaccine against Novel Coronavirus (mRNA-1273) for Phase 1 Study," *Moderna,* February 24, 2020.

70   Anonymous, "Moderna Announces Positive Interim Phase 1 Data for Its mRNA Vaccine (mRNA-1273) against Novel Coronavirus," *Moderna,* May 18, 2020.

71   Anonymous, "Moderna Reports Positive Data from Phase I COVID-19 Vaccine Trial," *Moderna,* May 19, 2020.

72   G. Ramon, "Combined (Active-Passive) Prophylaxis and Treatment of Diphtheria and Tetanus," *JAMA* 1940; 114: 2366–2368.

73   D. Butler, "Close but No Nobel: The Scientists Who Never Won," *Nature,* October 11, 2016.

74   A.T. Glenny and H.J. Südmersen, "Notes on the Production of Immunity to Diphtheria Toxin," *Epidemiology and Infection* 2009; 20: 176–220.

75   G. Ott and G.V. Nest, "Development of Vaccine Adjuvants: A Historical Perspective," in *Vaccine Adjuvants and Delivery Systems,* ed. M. Singh, New York: Wiley and Sons, 2007, pp. 1–31; R.R. Shah et al., "Overview of Vaccine Adjuvants: Introduction, History, and Current Status," in *Vaccine Adjuvants: Methods and Protocols,* ed. C.B. Fox, New York: Springer Science, 2017, pp. 1–13.

76   T.T. Le et al., "The COVID-19 Vaccine Development Landscape," *Nature Reviews Drug Discovery* 2020; 19: 305–306.

77   K.A. Callow et al., "The Time Course of the Immune Response to Experimental Coronavirus Infection of Man," *Epidemiology and Infection* 1990; 105: 435–446; L.P. Wu et al., "Duration of Antibody Responses after Severe Acute Respiratory Syndrome," *Emerging Infectious Diseases* 2007; 13: 1562–1564.

78   S. Jiang, "Don't Rush to Deploy COVID-19 Vaccines and Drugs without Sufficient Safety Guarantees," *Nature,* March 16, 2020.

79   H.C. Lehmann et al., "Guillain-Barré Syndrome after Exposure to Influenza Virus," *Lancet Infectious Diseases* 2010; 10: 643–651.

80   P.A. Offit, *The Cutter Incident: How America's First Polio Vaccine Led to a Growing Vaccine Crisis,* New Haven, CT: Yale University Press, 2007.

81   C.L. Thigpen and C. Funk, "Most Americans Expect a COVID-19 Vaccine within a Year; 72% Say They Would Get Vaccinated," *Pew Research Center,* May 21, 2020.

82   I.A. Hamilton, "Bill Gates Is Helping Fund New Factories for 7 Potential Coronavirus Vaccines, Even Though It Will Waste Billions of Dollars," *Business Insider,* April 3, 2020.

83   J.H. Beigel et al., "Remdesivir for the Treatment of COVID-19—Preliminary Report," *New England Journal of Medicine,* May 22, 2020.

84   Anonymous, "Low-Cost Dexamethasone Reduces Death by Up to One Third in Hospitalized Patients with Severe Respiratory Complications of COVID-19," *University of Oxford,* June 16, 2020; P. Horby et al., "Effect of Dexamethasone in Hospitalized Patients with Covid-19: Preliminary Report," *medRxiv,* June 22, 2020.

85   K.A. Callow et al., "The Time Course of the Immune Response to Experimental Coronavirus Infection of Man," *Epidemiology and Infection* 1990; 105: 435–446.

86   N. Eyal et al., "Human Challenge Studies to Accelerate Coronavirus Vaccine Licensure," *Journal of Infectious Diseases* 2020; 221: 1752–1756.

87   C. Friedersdorf, "Let Volunteers Take the COVID Challenge," *The Atlantic,* April 21, 2020.

88   Anonymous, "Expectations for a COVID-19 Vaccine," *AP-NORC Center,* May 2020.

89   C.L. Thigpen and C. Funk, "Most Americans Expect a COVID-19 Vaccine within a Year; 72% Say They Would Get Vaccinated," *Pew Research Center,* May 21, 2020.

90 Anonymous, "Expectations for a COVID-19 Vaccine," *AP-NORC Center,* May 2020.

91 D.G. McNeil Jr., " 'We Loved Each Other': Fauci Recalls Larry Kramer, Friend and Nemesis," *New York Times,* May 27, 2020.

92 D. Bernard, "Three Decades before Coronavirus, Anthony Fauci Took Heat from AIDS Protestors," *Washington Post,* May 20, 2020.

93 S.M. Hammer et al., "A Controlled Trial of Two Nucleoside Analogues plus Indinavir in Persons with Human Immunodeficiency Virus Infection and CD4 Cell Counts of 200 per Cubic Millimeter or Less," *New England Journal of Medicine* 1997; 337: 725–733; R.M. Gulick et al., "Treatment with Indinavir, Zidovudine, and Lamivudine in Adults with Human Immunodeficiency Virus Infection and Prior Antiretroviral Therapy," *New England Journal of Medicine* 1997; 337: 734–739.

94 A.S. Fauci and R.W. Eisinger, "PEPFAR—15 Years and Counting the Lives Saved," *New England Journal of Medicine* 2018; 378: 314–316.

7. THINGS CHANGE

1 E. Gibney, "Coronavirus Lockdowns Have Changed the Way Earth Moves," *Nature,* March 31, 2020.; T. Lecocq et al., "Global Quieting of High-Frequency Seismic Noise Due to COVID-19 Pandemic Lockdown Measures," *Science,* July 23, 2020.

2 L. Boyle, "Himalayas Seen for First Time in Decades from 125 Miles Away after Pollution Drop," *The Independent,* April 8, 2020; India State-Level Disease Burden Initiative Air Pollution Collaborators, "The Impact of Air Pollution on Deaths, Disease Burden, and Life Expectancy across the States of India: The Global Burden of Disease Study 2017," *Lancet Planetary Health* 2019; 3: e26–e39.

3 M. Vasquez, "Trump Now Says He Wasn't Kidding When He Told Officials to Slow Down Coronavirus Testing, Contradicting Staff," *CNN,* June 23, 2020.

4 S.S. Dutta, "People under 45 Make Up Higher Percentage of COVID-19 Deaths in India Compared to US, China," *New Indian Express,* May 1, 2020.

5 Anonymous, "Coronavirus: India to Use 500 Train Carriages as Wards in Delhi," *BBC,* June 14, 2020.

6 J.T. Lewis and L. Magalhaes, "Brazilian Court Rules President Bolsonaro Must Wear Mask in Public," *Wall Street Journal,* June 23, 2020.

7 L. Zhang et al., "Mutated Coronavirus Shows Significant Boost in Infectivity," *Scripps Research,* June 12, 2020.

8 P. Belluck, "Here's What Recovery from COVID-19 Looks Like for Many Survivors," *New York Times,* July 1, 2020.

9 S. Chapman, "Great Expectorations! The Decline of Public Spitting: Lessons for Passive Smoking?" *BMJ* 1995; 311: 1685.

10 P.H. Lai et al., "Characteristics Associated with Out-of-Hospital Cardiac Arrests and Resuscitations during the Novel Coronavirus Disease 2019 Pandemic in New York City," *JAMA Cardiology,* June 19, 2020.

11 N. Friedman, "Locked-Down Teens Stay Up All Night, Sleep All Day," *Wall Street Journal,* May 22, 2020.

12 L. Skenazy, "COVID Surprise: Kids Are Doing All the Stuff Helicopter Parents Used to Do for Them," *Big Think,* April 30, 2020.

13 N. Doyle-Burr, "Norwich Rallies Together to Grow Gardens as Part of COVID-19 Response," *Valley News* (Lebanon, NH), May 18, 2020.

14 United Nations, Department of Economic and Social Affairs, Population Division, *World Urbanization Prospects: The 2018 Revision (ST/ESA/SER.A/420),* New York: United Nations, 2019.

15 Anonymous, "Men Pick Up (Some) of the Slack at Home: New National Survey on the Pandemic at Home," *Council on Contemporary Families,* May 20, 2020.

16 Anonymous, "A Survey of Handwashing Behavior (Trended)," *Harris Interactive,* August 2010.

17 K.R. Moran and S.Y. Del Valle, "A Meta-Analysis of the Association between Gender and Protective Behaviors in Response to Respiratory Epidemics and Pandemics," *PLOS ONE* 2016; 11: e0164541.

18 J. Scipioni, "White House Advisor Dr. Fauci Says Handshaking Needs to Stop Even When Pandemic Ends—Other Experts Agree," *CNBC,* April 9, 2020.

19 E. Andrews, "The History of the Handshake," *History,* August 9, 2016.

20   I. Frumin et al., "A Social Chemosignaling Function for Human Handshaking," *eLife* 2015; 4: e05154.

21   S. Pappas, "Chimp 'Secret Handshakes' May Be Cultural," *Scientific American*, August 29, 2012.

22   N. Strochlic, "Why Do We Touch Strangers So Much? A History of the Handshake Offers Clues," *National Geographic*, March 12, 2020; S. Fitzgerald, "6 Ways People around the World Say Hello—Without Touching," *National Geographic*, March 23, 2020.

23   N. Strochlic, "Why Do We Touch Strangers So Much? A History of the Handshake Offers Clues," *National Geographic*, March 12, 2020.

24   S. Roberts, "Let's (Not) Shake on It," *New York Times*, May 2, 2020.

25   A. Witze, "Universities Will Never Be the Same after the Coronavirus Crisis," *Nature*, June 1, 2020.

26   D. Harwell, "Mass School Closures in the Wake of the Coronavirus Are Driving a New Wave of Student Surveillance," *Washington Post*, April 1, 2020.

27   C. Papst, "Police Search Baltimore County House over BB Guns in Virtual Class," *FOX45 News*, June 10, 2020.

28   L. Ferretti et al., "Quantifying SARS-CoV-2 Transmission Suggests Epidemic Control with Digital Contact Tracing," *Science* 2020; 368: eabb6936.

29   Anonymous, "Apple and Google Partner on COVID-19 Contact Tracing Technology," *Apple*, April 10, 2020; S. Overly and M. Ravindranath, "Google and Apple's Rules for Virus Tracking Apps Sow Division among States," *Politico*, June 11, 2020.

30   M. Giglio, "The Pandemic's Cost to Privacy," *The Atlantic*, April 22, 2020.

31   N.A. Christakis and J.H. Fowler, "Social Network Sensors for Early Detection of Contagious Outbreaks," *PLOS ONE* 2010; 5: e12948.

32   Thucydides, *The History of the Peloponnesian War*, trans. Richard Crawley, London: Longmans, Green & Co., 1874, p. 133.

33   Anonymous, "Most Americans Say Coronavirus Outbreak Has Impacted Their Lives," *Pew Research Center*, March 30, 2020; F. Newport, "Religion and the COVID-19 Virus in the U.S," *Gallup*, April 6, 2020.

34   F. Newport, "Religion and the COVID-19 Virus in the U.S," *Gallup*, April 6, 2020.

35   C. Gecewicz, "Few Americans Say Their House of Worship Is Open, but a Quarter Say Their Faith Has Grown amid Pandemic," *Pew Research Center*, April 30, 2020.

36   Anonymous, "Coronavirus: South Korea Church Leader Apologises for Virus Spread," *BBC*, March 2, 2020; W. Boston, "More Than 100 in Germany Found to Be Infected with Coronavirus after Church's Services," *Wall Street Journal*, May 24, 2020; L. Hamner et al., "High SARS-CoV-2 Attack Rate Following Exposure at a Choir Practice—Skagit County, Washington, March 2020," *CDC Morbidity and Mortality Weekly Report* 2020; 69: 606–610.

37   C. Gecewicz, "Few Americans Say Their House of Worship Is Open, but a Quarter Say Their Faith Has Grown Amid Pandemic," *Pew Research Center*, April 30, 2020.

38   Anonymous, "Most Americans Say Coronavirus Outbreak Has Impacted Their Lives," *Pew Research Center*, March 30, 2020.

39   J. Abdalla, "Michigan Muslims Find New Ways to Celebrate Eid amid a Pandemic," *Al Jazeera*, May 22, 2020.

40   I. Lovett and R. Elliott, "America's Churches Weigh Coronavirus Danger against the Need to Worship," *Wall Street Journal*, May 28, 2020.

41   A. Liptak, "Supreme Court, in 5-4 Decision, Rejects Church's Challenge to Shutdown Order," *New York Times*, May 30, 2020.

42   P. Drexler, "For Divorced Parents, a Time to Work Together," *Wall Street Journal*, April 25, 2020.

43   J. Couzin-Frankel, "From 'Brain Fog' to Heart Damage, COVID-19's Lingering Problems Alarm Scientists," *Science*, July 31, 2020; Y. Lu et al., "Cerebral Micro-Structural Changes in COVID-19 Patients—An MRI-Based 3-Month Follow-Up Study," *EClinicalMedicine*, August 3, 2020.

44   R. Kocher, "Doctors without State Borders: Practicing across State Lines," *Health Affairs*, February 18, 2014.

45   M.L. Barnett, "After the Pandemic: Visiting the Doctor Will Never Be the Same. And That's Fine," *Washington Post*, May 11, 2020.

46   E.J. Emanuel and A.S. Navathe, "Will 2020 Be the Year That Medicine Was Saved?" *New York Times*, April 14, 2020.

47   D.C. Classen et al., "'Global Trigger Tool' Shows That Adverse Events in Hospitals May Be Ten Times Greater Than Previously Measured," *Health Affairs* 2011; 30: 581–589.

48 S.A. Cunningham et al., "Doctors' Strikes and Mortality: A Review," *Social Science and Medicine* 2008; 67: 1784–1788.

49 A.B. Jena et al., "Mortality and Treatment Patterns among Patients Hospitalized with Acute Cardiovascular Conditions during Dates of National Cardiology Meetings," *JAMA Internal Medicine* 2015; 175: 237–244.

50 H.G. Welch and V. Prasad, "The Unexpected Side Effects of COVID-19," *CNN,* May 27, 2020.

51 E. Goldberg, "Early Graduation Could Send Medical Students to Virus Front Lines," *New York Times,* March 26, 2020.

52 J. Lu, "World Bank: Recession Is the Deepest in Decades," *NPR,* June 12, 2020.

53 B. Casselman, "A Collapse That Wiped Out 5 Years of Growth, with No Bounce in Sight," *New York Times,* July 30, 2020.

54 R. Sanchez, "'So Many More Deaths Than We Could Have Ever Imagined.' This Is How America's Largest City Deals with Its Dead," *CNN,* May 3, 2020.

55 M. Flynn, "They Lived in a Factory for 28 Days to Make Millions of Pounds of Raw PPE Materials to Help Fight Coronavirus," *Washington Post,* April 23, 2020.

56 S. Lewis, "Distilleries Are Making Hand Sanitizer and Giving It Out for Free to Combat Coronavirus," *CBS News,* March 14, 2020.

57 L. Darmiento, "How the L.A. Apparel Industry Became Mask Makers," *Los Angeles Times,* June 22, 2020.

58 D. Robinson, "The Companies Repurposing Manufacturing to Make Key Medical Kit during COVID-19 Pandemic," *NS Medical Devices,* April 1, 2020.

59 C. Edwards, "Onshoring in the Post-Coronavirus Future: Local Goods for Local People," *Engineering and Technology,* May 18, 2020.

60 C.A. Makridis and T. Wang, "Learning from Friends in a Pandemic: Social Networks and the Macroeconomic Response of Consumption," *SSRN,* May 17, 2020.

61 Associated Press, "U.S. Online Alcohol Sales Jump 243% during Coronavirus Pandemic," *MarketWatch,* April 2, 2020.

62 W. Oremus, "What Everyone's Getting Wrong about the Toilet Paper Shortage," *Medium Marker,* April 2, 2020.

63 Logistics Management Staff, "Parcel Experts Weigh In on FedEx and UPS So Far throughout the COVID-19 Pandemic," *Logistics Management,* June 8, 2020.

64 A. Palmer, "Amazon to Hire 100,000 More Workers and Give Raises to Current Staff to Deal with Coronavirus Demands," *CNBC,* March 16, 2020.

65 Anonymous, "US Oil Prices Turn Negative as Demand Dries Up," *BBC,* April 21, 2020.

66 A. Tappe, "Prices Are Tumbling at an Alarming Rate," *CNN,* May 12, 2020.

67 M. Wayland, "Worst Yet to Come as Coronavirus Takes Its Toll on Auto Sales," *CNBC,* April 1, 2020; P. LeBeau and N. Higgins-Dunn, "General Motors, Ford and Fiat Chrysler to Temporarily Close All US Factories Due to the Coronavirus," *CNBC,* March 18, 2020.

68 A. Villas-Boas, "Comcast, Charter, Verizon, and Dozens of Other Internet and Phone Providers Have Signed an FCC Pledge to 'Keep Americans Connected' Even If They Can't Pay during Disruptions Caused by Coronavirus," *Business Insider,* March 13, 2020.

69 Anonymous, "College Students: U-Haul Offers 30 Days Free Self-Storage amid Coronavirus Outbreak," *U-Haul,* March 12, 2020.

70 L. Rackl, "Demand for RVs Grows as Coronavirus Crisis Changes the Way We Travel. 'I Can See So Many People Doing It This Summer,'" *Chicago Tribune,* May 20, 2020.

71 Anonymous, "Considerations for Travelers—Coronavirus in the US," *CDC,* June 28, 2020.

72 J. Maze, "A Lot of Restaurants Are Already Permanently Closed," *Restaurant Business Magazine,* March 27, 2020.

73 Anonymous, "Small Business Impact Report," *CardFlight,* April 15, 2020; Anonymous, "Small Business Impact Report," *CardFlight,* May 13, 2020.

74 E. Luce, "Tata's Lessons for the Post-Covid World," *Financial Times,* May 1, 2020.

75 H. Goldman, "NYC to Close 40 Miles of Streets to Give Walkers More Space," *Bloomberg,* April 27, 2020.

76 E. Addley, "Eureka Moment? Law Firms Report Rush to Patent Ideas amid UK Lockdown," *The Guardian,* May 24, 2020.

77 C. Mims, "Reporting for Coronavirus Duty: Robots That Go Where Humans Fear to Tread," *Wall Street Journal,* April 4, 2020.

78 T.B. Lee, "The Pandemic Is Bringing Us Closer to Our Robot Takeout Future," *Ars Technica,* April 24, 2020.

79 D. Schneider and K. Harknett, "Essential and Vulnerable: Service Sector Workers and Paid Sick Leave," *University of California Shift Project,* April 2020.

80   E. Luce, "Tata's Lessons for the Post-Covid World," *Financial Times*, May 1, 2020.

81   E. Bernstein et al., "The Implications of Working without an Office," *Harvard Business Review*, July 15, 2020.

82   Ibid.

83   Bureau of Labor Statistics, U.S. Department of Labor, *Occupational Employment Statistics, Occupational Employment and Wages: 39-9011 Childcare Workers*, May 2017.

84   L. Hogan et al., "Holding On Until Help Comes: A Survey Reveals Child Care's Fight to Survive," *National Association for the Education of Young Children*, July 13, 2020.

85   N. Joseph, "Roll Call: The Importance of Teacher Attendance," *National Council on Teacher Quality*, June 2014.

86   G. Viglione, "How Scientific Conferences Will Survive the Coronavirus Shock," *Nature*, June 2, 2020.

87   L.B. Kahn, "The Long-Term Labor Market Consequences of Graduating from College in a Bad Economy," *Labour Economics* 2010; 17: 303–316.

88   E.C. Bianchi, "The Bright Side of Bad Times: The Affective Advantages of Entering the Workforce in a Recession," *Administrative Science Quarterly* 2013; 58: 587–623.

89   A. Grant, "Adam Grant on How Jobs, Bosses, and Firms May Improve after the Crisis," *The Economist*, June 1, 2020.

90   A. di Tura di Grasso, *Cronica Maggiore*, in "Plague Readings," A. Futrell (University of Arizona), 2002.

91   S. Correia et al., "Pandemics Depress the Economy, Public Health Interventions Do Not: Evidence from the 1918 Flu," *SSRN working paper*, June 11, 2020.

92   Ò. Jordà et al., "Longer-Run Economic Consequences of Pandemics," *Federal Reserve Bank of San Francisco Working Paper 2020-09*, June 2020.

93   Cathedral of Rochester, *Historia Roffensis*, in *The Black Death*, trans. and ed. R. Horrox, Manchester: Manchester University Press, 2013, p. 70.

94   E. Saez and G. Zucman, "Wealth Inequality in the United States since 1913: Evidence from Capitalized Income Tax Data," *Quarterly Journal of Economics* 2016; 131: 519–578.

95   T. McTague, "The Decline of the American World," *The Atlantic*, June 24, 2020.

96   V. Sacks and D. Murphey, "The Prevalence of Adverse Childhood Experiences, Nationally, by State, and by Race or Ethnicity," *Child Trends*, February 12, 2018; D.J. Bryant et al., "The Rise of Adverse Childhood Experiences during the COVID-19 Pandemic," *Psychological Trauma: Theory, Research, Practice, and Policy* 2020; 12: S193–S194.

97   M. Lin and E. Liu, "Does In Utero Exposure to Illness Matter? The 1918 Influenza Epidemic in Taiwan as a Natural Experiment," *Journal of Health Economics* 2014; 37: 152–163.

98   B. Mazumder et al., "Lingering Prenatal Effects of the 1918 Influenza Pandemic on Cardiovascular Disease," *Journal of Developmental Origins of Health and Disease* 2010; 1: 26–34.

99   D. Almond, "Is the 1918 Influenza Pandemic Over? Long-Term Effects of In Utero Influenza Exposure in the Post-1940 U.S. Population," *Journal of Political Economy* 2006; 114: 672–712.

100   R.E. Nelson, "Testing the Fetal Origins Hypothesis in a Developing Country: Evidence from the 1918 Influenza Pandemic," *Health Economics* 2010; 19: 1181–1192; J. Helgertz and T. Bengtsson, "The Long-Lasting Influenza: The Impact of Fetal Stress during the 1918 Influenza Pandemic on Socioeconomic Attainment and Health in Sweden, 1968–2012," *Demography* 2019; 56: 1389–1425.

101   L. Spinney, *Pale Rider: The Spanish Flu of 1918 and How It Changed the World*, New York: Public Affairs, 2017, p. 261.

102   V. Woolf, "On Being Ill," *The Criterion*, January 1926, p. 32.

103   Z. Stanska, "Plague in Art: 10 Paintings You Should Know in the Times of Coronavirus," *Daily Art Magazine*, March 9, 2020.

104   A. Swift, "In U.S., Belief in Creationist View of Humans at New Low," *Gallup*, May 22, 2017.

105   S. Neuman, "1 in 4 Americans Thinks the Sun Goes around the Earth, Survey Says," *NPR*, February 14, 2020.

106   E.C. Hughes, "Mistakes at Work," *Canadian Journal of Economics and Political Science* 1951; 17: 320–327.

107   D. Lazer et al., "The State of the Nation: A 50-State COVID-19 Survey," *Northeastern University*, April 20, 2020.

108   C. Funk, "Key Findings about Americans' Confidence in Science and Their View on Scientists' Role in Society," *Pew Research Center*, February 12, 2020.

109   K. Andersen, *Fantasyland: How America Went Haywire: A 500-Year History*, New York: Random House, 2017.

110 D. Lazer et al., "The State of the Nation: A 50-State COVID-19 Survey," *Northeastern University,* April 20, 2020.
111 US Department of Health and Human Services, "Dr. Anthony Fauci: 'Science Is Truth,'" *Learning Curve,* June 17, 2020.
112 M. Stevis-Gridneff, "The Rising Heroes of the Coronavirus Era? Nations' Top Scientists," *New York Times,* April 5, 2020.
113 S.K. Cohn Jr., *The Black Death Transformed: Disease and Culture in Early Renaissance Europe,* New York: Oxford University Press, 2002.

8. HOW PLAGUES END

1 G.C. Marshall, "Address of Welcome by the Honorable George C. Marshall," *Proceedings of the Fourth International Congress on Tropical Medicine and Malaria,* Washington, DC: Department of State, 1948, pp. 1–4.
2 A. Cockburn, *The Evolution and Eradication of Infectious Diseases,* Baltimore: Johns Hopkins University Press, 1963, p. 150.
3 R.G. Petersdorf, "The Doctor's Dilemma," *New England Journal of Medicine* 1978; 299: 628–634.
4 J. Lederberg, "Infectious Disease—A Threat to Global Health and Security," *JAMA* 1996; 275: 417–419.
5 F.M. Snowden, *Epidemics and Society: From the Black Death to the Present,* New Haven, CT: Yale University Press, 2019, p. 453.
6 Ibid., p. 458.
7 White House Office of Science and Technology Policy, "Fact Sheet: Addressing the Threat of Emerging Infectious Diseases," June 12, 1996.
8 US Department of Defense, *Addressing Emerging Infectious Disease Threats: A Strategic Plan for the Department of Defense,* Washington, DC: USGPO, 1998, p. 1.
9 CIA, "The Global Infectious Disease Threat and Its Implications for the United States," *NIE 99-17D,* January 2000.
10 K.E. Jones et al., "Global Trends in Emerging Infectious Diseases," *Nature* 2008; 451: 990–993.
11 L.A. Dux et al., "Measles Virus and Rinderpest Virus Divergence Dated to the Sixth Century BCE," *Science* 2020; 368: 1367–1370; M.J. Keeling and B.T. Grenfell, "Disease Extinction and Community Size: Modeling the Persistence of Measles," *Science* 1997; 275: 65–67.
12 C. Zimmer, "Isolated Tribe Gives Clues to the Origins of Syphilis," *Science* 2008; 319: 272; K.N. Harper et al., "On the Origin of the Treponematoses: A Phylogenetic Approach," *PloS Neglected Tropical Diseases* 2008; 2: e148.
13 J. Diamond, "The Germs That Transformed History," *Wall Street Journal,* May 22, 2020.
14 COVID-19 Response Team, "Severe Outcomes among Patients with Coronavirus Disease 2019 (COVID-19)—United States, February 12–March 16, 2020," *CDC Morbidity and Mortality Weekly Report* 2020; 69: 343–346; S. Richardson et al., "Presenting Characteristics, Comorbidities, and Outcomes among 5,700 Patients Hospitalized with COVID-19 in the New York City Area," *JAMA* 2020; 323: 2052–2059.
15 K. Modig and M. Ebeling, "Excess Mortality from COVID-19: Weekly Excess Death Rates by Age and Sex for Sweden," *medRxiv,* May 15, 2020.
16 J.R. Goldstein and R.D. Lee, "Demographic Perspectives on Mortality of COVID-19 and Other Epidemics," *NBER Working Paper 27043,* April 2020.
17 Ibid.
18 W.G. Lovell, "Disease and Depopulation in Early Colonial Guatemala," in *"Secret Judgments of God": Old World Disease in Colonial Spanish America,* ed. N.D. Cook and W.G. Lovell, Norman: University of Oklahoma Press, 1992, p. 61.
19 Korber et al., "Tracking Changes in SARS-CoV-2 Spike: Evidence That D614G Increases Infectivity of the COVID-19 Virus," *Cell* 2020; 182: 1–16, August 2020; Q. Li, et al., "The Impact of Mutations in SARS-CoV-2 Spike on Viral Infectivity and Antigenicity," *Cell* 2020; 182: 1–11, September 2020; H. Yao et al., "Patient-Derived Mutations Impact Pathogenicity of SARS-CoV-2," *medRxiv,* April 23, 2020; L. Zhang et al., "The D614G Mutation in the SARS-CoV-2 Spike Protein Reduces S1 Shedding and Increases Infectivity," *bioRxiv,* June 12, 2020.
20 S. Chen et al., "China's New Outbreak Shows Signs the Virus Could Be Changing," *Bloomberg News,* May 20, 2020.
21 L. Vijgen et al., "Complete Genomic Sequence of Human Coronavirus OC43: Molecular

Clock Analysis Suggests a Relatively Recent Zoonotic Coronavirus Transmission Event," *Journal of Virology* 2005; 79: 1595–1604.

22 M. Honigsbaum, *A History of the Great Influenza Pandemics: Death, Panic, and Hysteria, 1830–1920,* London: Bloomsbury, 2014.

23 A.J. Valleron et al., "Transmissibility and Geographic Spread of the 1889 Influenza Pandemic," *Proceedings of the National Academy of Sciences* 2010; 107: 8778–8781.

24 Anonymous, "The Influenza Pandemic," *The Lancet,* January 11, 1890, pp. 88–89.

25 G. Daugherty, "The Russian Flu of 1889: The Deadly Pandemic Few Americans Took Seriously," *History,* March 23, 2020.

26 A.J. Valleron et al., "Transmissibility and Geographic Spread of the 1889 Influenza Pandemic," *Proceedings of the National Academy of Sciences* 2010; 107: 8778–8781; J. Mulder and N. Masurel, "Pre-Epidemic Antibody against 1957 Strain of Asiatic Influenza in Serum of Older People Living in the Netherlands," *The Lancet* 1958; 1: 810–814.

27 Anonymous, "The Influenza Pandemic," *The Lancet,* January 11, 1890, pp. 88–89.

28 A.B. Docherty et al., "Features of 20,133 UK Patients in Hospital with COVID-19 Using the ISARC WHO Clinical Characterization Protocol: Prospective Observational Cohort Study," *British Medical Journal* 2020; 369: m1985; Y. Wu et al., "Nervous System Involvement after Infection with COVID-19 and Other Coronaviruses," *Brain, Behavior, and Immunity* 2020; 87: 18–22; S.H. Wong et al., "COVID-19 and the Digestive System," *Journal of Gastroenterology and Hepatology* 2020; 35: 744–748.

29 A.S. Monto et al., "Clinical Signs and Symptoms Predicting Influenza Infection," *Archives of Internal Medicine* 2000; 160: 3243–3247; J.H. Yang et al., "Predictive Symptoms and Signs of Laboratory-Confirmed Influenza," *Medicine* 2015; 94: 1–6.

30 L. Vijgen et al., "Complete Genomic Sequence of Human Coronavirus OC43: Molecular Clock Analysis Suggests a Relatively Recent Zoonotic Coronavirus Transmission Event," *Journal of Virology* 2005; 79: 1595–1604.

31 E.T. Ewing, "La Grippe or Russian Influenza: Mortality Statistics during the 1890 Epidemic in Indiana," *Influenza and Other Respiratory Diseases* 2019; 13: 279–287; D. Ramiro et al., "Age-Specific Excess Mortality Patterns and Transmissibility during the 1889–1890 Influenza Pandemic in Madrid, Spain," *Annals of Epidemiology* 2018; 28: 267–272.

32 J. Leng and D.R. Goldstein, "Impact of Aging on Viral Infections," *Microbes and Infection* 2010; 12: 1120–1124.

33 H. Rawson et al., "Deaths from Chickenpox in England and Wales 1995–7: Analysis of Routine Mortality Data," *BMJ* 2001; 323: 1091–1093; S. Chaves et al., "Loss of Vaccine-Induced Immunity to Varicella over Time," *New England Journal of Medicine* 2007; 356: 1121–1129.

34 S.K. Dunmire et al., "Primary Epstein-Barr Virus Infection," *Journal of Clinical Virology* 2018; 102: 84–92; S. Jayasooriya et al., "Early Virological and Immunological Events in Asymptomatic Epstein-Barr Virus Infection in African Children," *PLOS Pathogens* 2015; 11: e1004746; A. Ascherio and K.L. Munger, "Epstein–Barr Virus Infection and Multiple Sclerosis: A Review," *Journal of Neuroimmune Pharmacology* 2010; 5: 271–277.

35 T. Westergaard et al., "Birth Order, Sibship Size and Risk of Hodgkin's Disease in Children and Young Adults: A Population-Based Study of 31 Million Person-Years," *International Journal of Cancer* 1997; 72: 977–981; H. Hjalgrim et al., "Infectious Mononucleosis, Childhood Social Environment, and Risk of Hodgkin Lymphoma," *Cancer Research* 2007; 67: 2382–2388.

36 E.K. Karlsson et al., "Natural Selection and Infectious Disease in Human Populations," *Nature Reviews Genetics* 2014; 15: 379–393.; K.I. Bos et al., "Pre-Columbian Mycobacterial Genomes Reveal Seals as a Source of New World Human Tuberculosis," *Nature* 2014; 514: 494–497; B. Muhlemann et al., "Diverse Variola Virus (Smallpox) Strains Were Widespread in Northern Europe in the Viking Age," *Nature* 2020; 369: eaaw8977; S. Rasmussen et al., "Early Divergent Strains of *Yersinia Pestis* in Eurasia 5,000 Years Ago," *Cell* 2015; 163: 571–582.

37 M. Fumagalli et al., "Signatures of Environmental Genetic Adaptation Pinpoint Pathogens as the Main Selective Pressure through Human Evolution," *PLoS Genetics* 2011; 7: e1002355.

38 N.A. Christakis, *Blueprint: The Evolutionary Origin of a Good Society,* New York: Little, Brown, 2019.

39 J. Ostler, *The Plains Sioux and U.S. Colonialism from Lewis and Clark to Wounded Knee,* Cambridge: Cambridge University Press, 2004.

40 C.D. Dollar, "The High Plains Smallpox Epidemic of 1837–38," *Western Historical Quarterly* 1977; 8: 15–38.

41   R. Thornton, *American Indian Holocaust and Survival: A Population History since 1492,* Norman: University of Oklahoma Press, 1987.

42   J. Lindo et al., "A Time Transect of Exomes from a Native American Population before and after European Contact," *Nature Communications* 2016; 7: 1–11.

43   M. Price, "European Diseases Left a Genetic Mark on Native Americans," *Science,* November 15, 2016.

44   H. Laayouni et al., "Convergent Evolution in European and Roma Populations Reveals Pressure Exerted by Plague on Toll-Like Receptors," *Proceedings of the National Academy of Sciences* 2014; 111: 2668–2673.

45   A.C. Allison, "Protection Afforded by Sickle-Cell Trait against Subtertian Malarial Infection," *British Medical Journal* 1954; 1: 290–294; D.P. Kwiatkowski, "How Malaria Has Affected the Human Genome and What Human Genetics Can Teach Us about Malaria," *American Journal of Human Genetics* 2005; 77: 171–192; K.J. Pittman et al., "The Legacy of Past Pandemics: Common Human Mutations That Protect against Infectious Disease," *PLoS Pathogens* 2016; 12: e1005680.

46   I.C. Withrock et al., "Genetic Diseases Conferring Resistance to Infectious Diseases," *Genes and Diseases* 2015; 2: 247–254; A. Mowat, "Why Does Cystic Fibrosis Display the Prevalence and Distribution Observed in Human Populations?" *Current Pediatric Research* 2017; 21: 164–171; G.R. Cutting, "Cystic Fibrosis Genetics: From Molecular Understanding to Clinical Application," *Nature Reviews Genetics* 2015; 16: 45–56; E.M. Poolman et al., "Evaluating Candidate Agents of Selective Pressure for Cystic Fibrosis," *Journal of the Royal Society Interface* 2007; 4: 91–98.

47   D. Ellinghaus et al., "The ABO Blood Group Locus and a Chromosome 3 Gene Cluster Associate with SARS-CoV-2 Respiratory Failure in an Italian-Spanish Genome-Wide Association Analysis," *medRxiv,* June 2, 2020.

48   H. Zeberg and S. Pääbo, "The Major Genetic Risk Factor for Severe COVID-19 Is Inherited from Neandertals," *bioRxiv,* July 3, 2020.

49   A.M. Brandt, *No Magic Bullet: A Social History of Venereal Disease in the United States since 1880,* New York: Oxford University Press, 1985.

50   A.M. Brandt and A. Botelho, "Not a Perfect Storm—COVID-19 and the Importance of Language," *New England Journal of Medicine* 2020; 382: 1493–1495.

51   A. Wise, "White House Defends Trump's Use of Racist Term to Describe Coronavirus," *NPR,* June 22, 2020.

52   K. Gibson, "Survey Finds 38% of Beer-Drinking Americans Say They Won't Order a Corona," *CBS News,* March 1, 2020.

53   M. Honigsbaum, *A History of the Great Influenza Pandemics: Death, Panic, and Hysteria, 1830–1920,* London: Bloomsbury, 2014.

54   D.G. McNeil, "As States Rush to Reopen, Scientists Fear a Coronavirus Comeback," *New York Times,* May 11, 2020.

55   C. Ornstein and M. Hixenbaugh, "'All the Hospitals Are Full': In Houston, Overwhelmed ICUs Leave COVID-19 Patients Waiting in ERs," *ProPublica,* July 10, 2020; E. Platoff, "Gov. Greg Abbott Keeps Businesses Open Despite Surging Coronavirus Cases and Rising Deaths in Texas," *Texas Tribune,* June 25, 2020; A. Samuels, "Gov. Greg Abbott Warns If Spread of COVID-19 Doesn't Slow, 'The Next Step Would Have to Be a Lockdown,'" *Texas Tribune,* July 10, 2020.

56   S. Gottlieb et al., "National Coronavirus Response: A Road Map to Reopening," *AEI Report,* March 28, 2020.

57   G. Kolata, "How Pandemics End," *New York Times,* May 10, 2020.

58   E. Kübler-Ross, *On Death and Dying,* New York: Macmillan, 1969.

59   F.M. Snowden, *Epidemics and Society: From the Black Death to the Present,* New Haven, CT: Yale University Press, 2019; M.T. Osterholm and M. Olshaker, *Deadliest Enemy: Our War Against Killer Germs,* New York: Little, Brown, 2017; L. Garrett, *The Coming Plague: Newly Emerging Diseases in a World Out of Balance,* New York: Penguin, 1995; P.E. Farmer, *Infections and Inequalities: The Modern Plagues,* Berkeley: University of California Press, 1999.

60   D.M. Morens and A.S. Fauci, "The 1918 Influenza Pandemic: Insights for the 21st Century," *Journal of Infectious Diseases* 2007; 195: 1018–1028.

61   H. Sun et al., "Prevalent Eurasian Avian-Like H1N1 Swine Influenza Virus with 2009 Pandemic Viral Genes Facilitating Human Infection," *Proceedings of the National Academy of Sciences,* June 2020.

Index

Aachen, Germany, 149
absolute mortality, 195–197
ACE2 receptors: genetic variation in, 315; SARS-2 binding to, 26, 230; and sensitivity to infection, 183–184, 189; and vaccine development, 230, 235
ACT-UP (AIDS Coalition to Unleash Power), 245
adaptive immunity, 184
adjuvants, vaccine, 233–235
aerosolization, of virus, 194
African-Americans, 175, 195–197, 244
age: and impact of COVID-19 pandemic, 180–188; and susceptibility to SARS-2, 66–67. *See also* elderly people; young people
AIDS Coalition to Unleash Power (ACT-UP), 245
air travel, 12, 24, 43, 108–109
airborne transmission, 40
alcoholic beverages, 273–274
altruism: after disasters, 218; of clinical trial volunteers, 240–244; during epidemics, 209–211; NPI use as, 106, 216–217; psychological benefits of, 220
Amazon (company), 19, 274
American Indians, 197–198, 299–300, 306, 312–313
American Microbiology Society, 254
Americans with Disabilities Act, 187
Amoy Gardens housing complex (Hong Kong), 39–40
Andaman Islands, 225–226
Aneyoshi, Japan, 225
anger, 142
animals: ACE2 expression in, 235; bats, 3, 4, 20–21; camels, 59; chimpanzees, 255; domesticated, 299–300; fleas, 76–78; greeting behaviors of, 255; human pathogens from (*See* zoonotic pathogens); rats, 76, 82; in urban areas, 248
anosmia, 26
anti-Asian discrimination, 174–175
anti-elitism, 290–292
anti-Semitism, 179–180
antibiotics, 296
antibodies: duration of response involving, 236; testing for, 114–118, 201–202; treatments using, 210, 229–230
Apollo, xiii, xvi, 29, 324
Apple, 112, 260, 277
arousal, misattribution of, 213
arts, pandemics' influences on, 288–289
asymptomatic infection: and NPI use, 103–104, 108, 110, 117; rate of, 27, 46; with SARS-2 vs. other pathogens, 47–50; transmission by people with, 13, 29

Athens, plague of (430 BCE), 29, 141–142, 220, 261–262
attack rate: age and, 180–181, 182; defined, 60; ethnic and racial disparities in, 195; predicted future, 249; socioeconomic status and, 205
attitude changes, pandemic-related, 250–264
Australia, 92
Austro-Hungarian empire, 81–82
autoimmune reactions, 236
autonomous robotics, 276–277
Azar, Alex, 153, 194
azidothymidine (AZT), 245

bacteremia, 78
bats, 3, 4, 20–21
BCG vaccine, 184
Bedford, Trevor, 22, 23
Bellevue Hospital (New York City, N. Y.), 145–146, 222
Bellingham, Wash., 151–152
Bennett, William J., 155, 175
Bergensfjord (ship), 72–73
Biblical plagues, 83–84
big data technologies, 259–260
Bill and Melinda Gates Foundation, 297
biological end, of pandemic, 295–315
bioweapon, SARS-2 as, 160
bird flu, 69
birth rates, after disasters, 213
Black Death (1347), 29, 79; blaming others for, 80–81; border closings and, 107; economy after, 283, 285; generosity in, 221; genetic resistance to, 313–314; grief during, 139–140; long-term effects of, 294; plague as equalizer during, 179–180; symptoms of disease, 77–78; treatments used during, 86
blame, 80–81, 144–145, 171–177
bleach and bleaching agents, 162, 163–164
blood donations, 210, 230
blood type, COVID-19 severity and, 314–315
border closings, 10, 12, 81–82, 107–109, 176–177
Boston, Mass., 58, 146, 192
bovine coronavirus, 310
Brandt, Allan, 316, 321
Braskem America, 272
Brazil, 249, 288
Bridgeport Hospital, 222
Brigham and Women's Hospital, 266
bubonic plague, 76–83; arts and literature after, 289; blaming others for, 171–172; contact tracing in, 109–110; denial about, 157–158; fear about, 147–149;

bubonic plague (*cont.*)
and human genetic variation, 312. *See also specific pandemics, e.g.,* Black Death (1347)
Bukhara, Uzbekistan, 308
Bush, George W., 246, 298
business closure, 95, 275
butterfly effect, 31–33

California, 20, 75, 79, 90, 130, 264. *See also specific cities*
Camp Devens, 71
Camp Funston, 70
Camus, Albert, 34, 206, 295
Canada, 41, 208. *See also specific cities*
cardiologists, 268–269
CARES (Coronavirus Aid, Relief, and Economic Security) Act, 202, 271
case fatality rate (CFR): age and, 182–183; defined, 45–47; ethnic and racial disparities in, 195; of respiratory pandemics, 63, 64; and risk of death, 301
catastrophe compassion, 218
causal model, 199–200
cell phones, 161, 259–260
Centers for Disease Control and Prevention (CDC), xi, 48, 298; close contact defined by, 112; meatpacking plant outbreak analysis by, 193–194; modern plague case estimates from, 79; NPI guidelines from, 96, 103, 124, 153; pandemic preparedness documents from, 322; on SARS-2 outbreak in United States, 12–14; Spanish flu samples at, 69; test kits and testing by, 17, 114–116, 127, 150; travel advisories from, 18; trust in, 291
Central Intelligence Agency (CIA), 298
CFR. *See* case fatality rate
challenge trials, 242–244
Chandrasekaran, N., 276, 277
chemosignaling, 255
Chevy Chase House (Washington, D.C.), 34–35
Cheyne-Stokes respiration, 138
Chicago, Ill., 27–28, 152, 214, 245
chicken pox, 51, 311
child care, 119, 129, 207–208, 253, 279
children: impact of COVID-19 disease for, 180–186; impact of pandemic on, 287–288; independence of, 251–252; mortality of infectious diseases for, 180–182; school closure and safety of, 73–74, 118; severity of viral illnesses in adults vs., 311; transmission of SARS-2 by, 185
China, 92; 1957 influenza pandemic in, 64; 2003 SARS-1 outbreak in, 36–40, 42; age and case fatality in, 183; discrimination against Wuhan residents in, 200; global standing of, 286; and misinformation about SARS-2 origin, 161–162, 167–168, 174–175; NPI use in, xv, 112, 135; pandemics originating in, 174–175, 323; reemergence of SARS-2 in, 249; SARS-2 research from, 49–

50, 58, 307; testing in, 115; tracking in, xiv; travel restrictions on, 24. *See also specific cities*
Chinese Center for Disease Control, 5
Chinese Communist Party, 6, 7, 10, 42, 161
"Chinese virus," 317
chloroquine phosphate, 165
cholera, 204, 295, 312
Christians, blaming plague on, 172
chunyun migration, 7, 316
civil liberties, 110–112, 260
Clara Maass Medical Center, 223
climate change, 298–299, 323
clinical trials, 238, 240–244
Clorox, 164
close contact, defined, 112
closed-off management, 9–11
cold, common, 47, 229, 240, 307, 311
collective nonpharmaceutical interventions, 89, 107–126, 294, 323; border closings, 107–109; contact tracing, testing, and isolation, 109–114; school closures, 119–126
colleges and college students, 207–208, 257–258, 275, 281–282
Commerce Department, 271
common good, 116, 216–217
community transmission, 7; NPIs to prevent, 108, 116, 122; in United States, 16, 17; variant responsible for, 22–24
compassion, catastrophe, 218
complexity, dealing with, 292–293
Connecticut, 24, 173
connection, 141, 211–215
conspiracy theories, on SARS-2 origin, 159–162
consumption, after pandemics, 282–283
contact-reduction interventions, 89, 155
contact tracing: apps for, 259–260; for early US cases, 14, 16; in New York City, 127, 128; and nonpharmaceutical interventions, 109–114; relationship of testing and, 118; and social end of pandemic, 320–321
contactless payment methods, 277
control, 144, 159, 273
Cook County Hospital, 270–271
cooperation, 220; in drug development, 238–244; in vaccine development, 227–238
Copan, Honduras, xiv–xv, 25–26
coping mechanisms, 215–216
cordon sanitaire, 81–82
Coronavirus Aid, Relief, and Economic Security (CARES) Act, 202, 271
coronaviruses: characteristics of, 26; immunity to, 184, 200, 236; mutations in, 307, 310–311; symptoms associated with influenza vs., 309–310; types of, 20, 35, 47; vaccines for, 228, 230, 232, 233. *See also specific viruses, e.g.,* SARS-2 virus
COVID-19 (disease), xi; blood type and severity of, 314–315; CFR and IFR for, 45–46; diagnosis of, in New York City, 126–127; disparities in impact of, 180–200; first confirmed case of, 4; misinformation about

cures for, 162–168; narratives about seasonal flu and, 317–318; pathogen causing (*See* SARS-2 virus); risk of death from, 300–303; symptoms of, 26–29; and symptoms of Russian flu, 309–310
COVID-19 Mutual Aid USA, 207
COVID-19 pandemic: declaration of, 26; duration of, 249–250; effects of, xiii–xiv; excess deaths vs. diagnosed cases in, 74–75; lessons from, 323–324; lessons learned from, 322–324; long-term effects of, xvi; "perfect storm" analogy for, 316; social divisions and impact of, 180–200; social end of, 318–321; volunteerism in, 206–209. *See also specific topics*
COVID Challenge, 242
CovidSitters, 207–208
cowpox, 231
Cremona, Italy, 222
crises, love and connection during, 211–215
Crossroads Hospice, 222
crowded conditions, 192–194
Crown Heights neighborhood (Brooklyn, N. Y.), 138
cruise ships, 18–19. *See also specific ships*
culture, 223–227, 289–294
cumulative culture, 227
cumulative exposure, 117
Cuomo, Andrew, 127, 128, 130–133, 145–146
Cutter Laboratories, 236–237
cystic fibrosis, 314
cytokine storms, 28–29, 184
cytotoxic T cells, 231
Czech Republic, 105–106

Dahl, Ophelia, 113
dancing manias, 149
Dartmouth-Hitchcock Medical Center, 223
Darwin, Charles, 244, 255, 307
dating, 214–215
Davis, Calif., 65
de Blasio, Bill, 127, 129–132
death (generally): age and probability of, 180–182; alone, 137–139, 257, 320; health-care related, 267–269; impact of, on family, 217–218; risk of, 301; salience of, connection and, 212; untimely, 140. *See also* lethality; mortality rate
deaths, COVID-related: in China, 5; of health-care workers, 221–223; media coverage of, 204–205; in New York City, 24, 133–134; and reopenings, 319, 320; in United States, xiv, 15, 26, 248–249; worldwide, xiv, 26; of young people, 249
Defense of Production Act, 193
delivery companies, 274
denial, 151–159
Department of Defense, 298
Department of Health and Human Services, 115
Department of Labor, 94
DeSantis, Ron, 109, 176

Detroit, Mich., 190–191
dexamethasone, 239
Diamond Princess cruise ship, 18–19, 27, 46
diphtheria vaccine, 234
disability, COVID-related, 265
discrimination, 174–176, 200
dispersion, in R_0, 53–59
distress, 142
divorce, 213–214
domestic violence, 214
domesticated animals, 299–300
Dongmen Market, 42
Dow Jones Industrial Average, 95
Dowd, Patricia, 15–16, 46
droplet transmission, 40, 78, 103–105, 317
drug development, 87, 88, 238–246
Duh, Quan-Yang, 266

Ebola virus, 45, 51, 84, 139, 228
ecological release, 30–31
economic inequality, 178, 179
economic value of year of life, 304–305
economy: costs of NPIs in terms of, 94–98, 320; effect of COVID-19 pandemic on, 271–282; effect of plagues on, 136, 282–285; lessons from pandemic about, 323–324
Ecuador, 25
education, 279–282. *See also* schools
effective reproduction number (R_e), 50–53; dispersion of, 59; individual differences and, 55; nonpharmaceutical interventions and, 102, 105
efficacy testing, 241–242
Elbe River, 225
elderly people: COVID-related deaths of, 14–15, 302; impact of COVID-19 for, 183, 184, 186; mortality of respiratory infections for, 180, 182; volunteer services to help, 206–208
emergency medical technicians (EMTs), resuscitation rules for, 132, 188
emotional wellness, 142–143
empathy, 264
end of pandemic, 102, 295–324; biological end, 295–315; bubonic plague, 82–83; and characteristics of pathogen, 300–306; COVID-19, 31, 318–321; and eradication of infectious diseases, 295–300; and lessons learned from COVID-19, 322–324; social end, 315–322; via human genetic adaptation, 311–315; via pathogen mutation, 306–311
endemic disease, 51, 305, 312
England, 79, 308–309. *See also* United Kingdom
environment: COVID-19 related changes in, 247–249; reproduction number and, 52–53, 58–60
epidemic hysteria, 149
epidemics: denial in, 151–159; displays of generosity in, 209–211; exponential growth in, 97, 98; lethality calculations during, 45–46; psychological effects of, 137–159; solidarity in, 211. *See also* pandemics

Episcopal church, 217
Epstein-Barr virus, 311
equality, 178–180, 253
error rate, test, 117–118
essential workers: during bubonic plague, 80; at meatpacking facilities, 192–194; risk of infection for, 197; socioeconomic status of, 190; volunteer child care for, 207–208; worker protections for, 285–286
ethical corporate leadership, 282
ethical issues, with immunity-passport programs, 201–202
ethnicity, social divisions based on, 195–200
Europe, 276, 299; bubonic plague in, 78, 79; philanthropy in, 208; resistance to smallpox in, 312–313; river low-water marks in, 225; travel restrictions on, 24. *See also specific cities and countries*
ex-spouses, 214, 264
excess deaths, 74–75
expertise, 290–292
exponential growth, 97, 98
externalities, 190

face, touching of, 104
face masks, use of. *See* mask-wearing
FaceTime, 137
facial recognition software, 258, 260
false positive test results, 117–118
family members: impact of death on, 217–218; shunning of COVID-19 patients by, 203; time spent with, 251, 253, 257
Farmer, Paul, 113
Farr, William, 74, 302
Fauci, Anthony, 90, 165, 245, 248–249, 254, 292, 293, 321
fear, 143–151
Federal Communications Commission (FCC), 275
Federal Emergency Management Agency (FEMA), 132
Federal Trade Commission (FTC), 163
Finger, Anne, 145, 211
First Amendment rights, 264
5G cell phone towers, 161
flattening the curve, xv, 90–93, 101; costs associated with, 94; emotional wellness and, 143; exponential growth and, 97; with nonpharmaceutical interventions, 90, 96; and pharmaceutical interventions, 227–228, 239; testing and, 116
flipped classrooms, 281
Florida, 176
Floyd, George, 158
"flu," use of term, 318. *See also* influenza
fly theory of polio transmission, 145
Food and Drug Administration (FDA), 115, 163, 165
food banks, 191
Ford Motor Company, 272–274
Four Horsemen of the Apocalypse, 306

Fox News, 164, 165
Frankfurt, Germany, 263
French Hospital of Hanoi, 40
Frieden, Thomas, 116, 126, 319
Fujian Province, China, 12
funeral homes, 272

gardening, 252
gaseous plumes, 40
Gates, Bill, 237, 322
Gay Men's Collective, 175
gender-based social divisions, 143, 188–189
General Motors, 274
generosity, 206–246; cooperation in drug development, 238–244; cooperation in vaccine development, 227–238; in epidemics, 209–211; of health-care workers, 220–223; in HIV epidemic, 244–246; and love/ connection in crisis, 211–215; as part of culture, 223–227; physical distancing as form of, 215–220; volunteerism in COVID-19 pandemic, 206–209
Genesis II Church of Health and Healing, 162
genetic adaption, ending pandemics via, 311–315
genetic mapping: of 1918 influenza strain, 69–70; of SARS-1 virus, 44; of SARS-2 virus, 20–26, 126
genetic resistance, to animal pathogens, 298–300
genetically engineered animals, 235
genome sequencing, 20
Georgia (state), 48, 58, 106
GetUsPPE.org, 223
GlaxoSmithKline, 235
Google, 112, 260
Grand Princess cruise ship, 18, 21
greeting practices, 253–257
grief, 137–142
Grinnell, Iowa, 65
Guangdong Province, China, 12
Guangzhou Institute of Respiratory Diseases, 42
Guillain-Barré syndrome, 62, 236
gun sales, 204, 273
Guru-Nanda (company), 163

H1N1 influenza virus: 1918 Spanish flu pandemic, 69; 2009 pandemic, 57, 62, 84, 108, 116, 124. *See also* Spanish flu pandemic (1918)
H2N2 influenza virus, 63
H3N2 influenza virus, 64
H5N1 influenza virus, 64
HAART (highly active antiretroviral therapy), 245–246
Haitian earthquake (2010), 316
Halifax, Nova Scotia, 218
handwashing, 102. *See also* hygiene measures
Hangzhou Province, China, 12
Hanoi, Vietnam, 40–41
Hanover, N. H., 95
Harvard School of Public Health, 62
Hasidic Jewish population, 210, 230

health, COVID-19 mortality and, 188–189
health care, access to and responsibility for, 190, 251
health-care system: avoidance of, 94; COVID-related changes in, 265–271; effect of flattening the curve on, 91–93; and iatrogenic illnesses and injuries, 267–270; in New York City, 131–133
health-care workers: conserving masks for, 103; fear/shunning of, 145–147; generosity of, 209–210, 220–223; in New York City, 131–133; pandemic's effects on, 269–270; preventing infection of, 13; psychological benefit of cheering for, 144; SARS-1 transmission to, 36–38, 40–41, 44, 47–48; shortage of, 92–93; social media posts by, 151–153
health outcomes, ethnic and racial disparities in, 199–200
heart attacks, 269
heliotrope cyanosis, 68
helper T-cells, 230
herd immunity: in 1957 influenza pandemic, 67; defined, 56; end of pandemic due to, 305–306; and immunity passports, 203; and immunization, 243; NPI use and, 98–102; random testing to assess, 117; SARS-2 exposure after, 311; vaccine development and, 229
highly active antiretroviral therapy (HAART), 245–246
Hispanics, 195, 197
HIV. See human immunodeficiency virus
HKU1 coronavirus, 47
homeless people, 178, 192, 252
Homer, ix, 255
homophily, 198–199
homosexuals, HIV and, 154–155, 175
Hong Kong, 183
Hong Kong University, 37
hongi greeting, 256
Horace Mann school (New York City, N.Y.), 129
hormones, immune system and, 189
Hospital Universitario de Caracas, 222
hospitalization, risk of death after, 301
hospitals: economic impact of pandemic for, 267; silencing of health-care workers by, 151–153; triage policies of, 186–188
host factors, reproduction number and, 51–52, 55–58
Houston, Tex., 157, 319
Huanan Seafood Wholesale Market, 3–6, 9
Hubei Province, China, 7, 9, 14
human immunodeficiency virus (HIV): 1987 epidemic, 84, 154–155, 175, 222, 244–246, 289, 298; homophily and spread of, 198–199; misinformation about, 169; mismatch period for, 49, 50; vaccine development for, 229
Human Nature Lab, 260
humans: evolution of pathogens and, 297–298; genetic adaptation by, 311–315

Hunala app, xv, 260–261
Hurricane Hugo, 213
Hurricane Katrina, 186, 218, 316
hydroxychloroquine, 164–166
hygiene measures, 102, 125, 253–254

iatrogenic illnesses and injuries, 267–270
Iceland, 107, 108
Idaho, 36
identity, shared, 219
IFR (infection fatality rate), 45–46, 117
IgG, 114
IgM, 114
The Iliad (Homer), ix, xiii
Illinois, 16. *See also* Chicago, Ill.
immediate pandemic period, 250; child care in, 279; economy in, 271, 273; education in, 280, 287; government's role in, 294; health-care practice in, 265; NPI use in, 320–321; religious behavior in, 262
immune system: age and, 184; and COVID-19 symptoms, 27–29; of men vs. women, 189; in SARS-2 infection, 230–231; steroids' effects on, 239; and vaccines, 230–237
immunity: adaptive, 184; to coronaviruses, 200, 236; memory, 184, 231, 236, 240; social divisions based on, 200–203; and social network size, 57–58. *See also* herd immunity
immunity-passport programs, 201–202
immunization: in 1957 influenza pandemic, 67; proof of, 201; targeted populations for, 58, 67, 99–100; willingness to undergo, 243–244, 321. *See also* vaccines
in-person medical practice, 265–267
inactivated virus vaccines, 232
income, 143, 192–194
incubation period, 48–50
independent learning, 224
India, 79, 147–149. *See also specific cities*
Indian Health Service, 197
Indiana, 151–152
indigenous people, 306, 312–313. *See also* American Indians
indirect deaths, 75
individual learning, 224
individual nonpharmaceutical interventions, 88, 102–106
infection fatality rate (IFR), 45–46, 117
infectious diseases: age and impact of, 180–181; eradication of, 295–300
infectiousness, age and, 185
influenza, 60–67, 322; 1889 Russian flu pandemic, 307–309; 1957 pandemic, 62–67, 100, 134, 297; 1968 pandemic, 62, 297; 1976 swine flu pandemic, 62; symptoms of coronavirus infections vs., 309–310; types and subtypes of, 63–64; vaccines for, 236. *See also* seasonal flu; Spanish flu pandemic (1918)
innovation, during pandemic, 276–277
insecticides, 296–297

insurance reimbursement, for health care, 265–267
interest rates, 284–285
intermediate pandemic period, 249; child care in, 279; economy in, 271, 275, 276, 283; education in, 280; politics in, 294
International Boy Scout Jamboree, 65
intimacy, 214–215, 257
Invisible Hands, 206–207
Iran, 24, 25, 127, 222
Italy, 25, 79, 92, 124, 186, 221. *See also specific cities*

Jacob Javits Convention Center, 131
jail inmates, 192
Jalandhar, India, 248
Japan, 92, 124, 135, 225
Jenner, Edward, 231, 242
Jews: blaming plague on, 171–172, 179–180; blood donations by Hasidic population, 210, 230
Jinyintan Hospital, 5
job satisfaction, 278, 282
John of Ephesus, 136, 166–167
Johnson, Boris, 101, 249
Justinian, plague of (541), 29, 79, 80, 136, 166–167, 178

Kansas, 195
Kansas City, Mo., 195
Kent County, Mich., 195
Korea, 68. *See also* South Korea
Krumholz, Harlan, 94
"Kung flu," 317
Kwong Wah Hospital, 37

L-shaped curves, 181, 183, 310, 311
Lancet (journal), 6, 308
latent period, 48–50
learning: capacity for, 224–227, 253; individual/independent, 224; outdoor, 125; remote, 120, 251, 252, 257–258, 279–281; social, 224–226
lethality: of bubonic plague, 77, 78, 82; gender-based differences in, 188–189; and mask use, 104–105; metrics for, 301–304; with natural strategies, 100–101; of Spanish flu, 67–68; and vaccine/drug development, 87, 88; viral mutations that reduce, 306–312
Li, Wenliang, 5–7, 151
Liberty Loans Parade, 72, 97
libraries, repurposing of, 207
Life Care Center, 14–15, 19, 22
life expectancy, 300–304
literature, pandemics' influences on, 288–289
Liu, Jianlun, 37–38, 52
live attenuated virus vaccines, 231–232, 236–237
living conditions, impact of pandemic and, 192–194
location(s): naming pathogens after, 174; variation in mortality by, 64–65
lockdowns: closed-off management in China,

9–11; effect of, on virus, 11; flattening the curve with, 90–93; herd immunity strategies vs., 101; psychological effects of, 141; research done under, 185; seismic activity during, 247–248. *See also* shelter-in-place orders; stay-at-home orders
London, England, 308–309
loneliness, 142, 214
Los Angeles, Calif., 163
loss, 140–141
love, 211–215
Lowell, Mass., 173–174
Lunar New Year festival, 7
Lysol, 164

malaria, 298, 311, 314
Mamla, India, 147–148
manufacturing facilities, 237, 272–273
Marburg virus, 45
marriage, 212–213
Marseille, France, 157–158
mask-wearing, 102–106; as act of altruism, 210–211; anonymizing effect of, 257; exerting control through, 144; protests against, 150, 151; and social end of pandemic, 320; spill-over effects of policies on, 219; symbolic meaning of, 317
mass gatherings, 130, 150–151, 158–159
mass graves, 82
mass psychogenic illness (mass sociogenic illness), 149–151
Massachusetts, 113, 195, 264. *See also specific cities*
Match.com, 214–215
Mayans, 306
Mayo Clinic, 106, 210
measles, 51, 88, 99, 299
meatpacking plants, 192–194
media coverage, of pandemic, 204
medical costs, of NPIs, 94–98
medical errors, 267–268
medical licensure, 265
memory immunity, 184, 231, 236, 240
men: COVID-19 impact for, 188–189, 253; NPI use by, 254
mercury, 86
MERS. *See* Middle East respiratory syndrome
Metropole Hotel (Hong Kong), 37–41
Mexico, 108
Middle East respiratory syndrome (MERS), 47, 59, 84
Milan, Italy, 92, 172
military personnel, 70–71, 202–203
Military World Games, 161
Mindoro, Philippines, 295, 296
minority populations: blaming of, for bubonic plague, 171–172, 203–204; COVID-19 risk factors for, 198; impact of COVID-19 on, 195–200
misattribution of arousal, 213
misinformation, 151–170; about COVID-19 cures, 162–168; about SARS-2 origin, 159–162;

about Wuhan outbreak, 6; and denial, 151–159; and public education, 169–170; and scientists' use of preprint servers, 168–169
mismatch period, 48–50
Mississippi, 195
modern plague, 79–80
Moderna Therapeutics, 233
mortality rate: of chicken pox, 311; dexamethasone and, 239; NPI use and, 123–124; racial disparities in, 195–197; in United States, 60–62
motor type, mass psychogenic illness, 149
multigenerational families, 198
Mumbai, India, 79
music, making, 215–216
mutations: in coronaviruses, 307, 310–311; end of pandemic due to viral, 93, 306–311; in SARS-2 virus, 307; studying spread of SARS-2 using, 21–23
Mutual Aid Disaster Relief network, 207
mutual aid societies, 207

N-Ergetics, 162–163
namaste greeting, 256
Namiwake, Japan, 225
National Association for the Education of Young Children, 279
National Guard, 128, 191
National Health Commission, of China, 8, 9
National Institute of Allergy and Infectious Diseases (NIAID), xi
National Institutes of Health (NIH), 245
natural control strategies, 98–102
natural disasters, 212, 213, 225, 317. *See also specific events, e.g.,* Hurricane Katrina
natural experiments, 18–19
naturally-occurring pathogens, accidental release of, 160–161
Navajo Nation, 197
New Jersey, 75, 177
New Orleans, La., 186, 201
New Rochelle, N. Y., 127–128
New York City, N.Y., 119; 1889 Russian flu pandemic in, 309; 1918 Spanish flu pandemic in, 72–74; blood donations in, 210; disparities in impact of COVID-19 in, 189–190, 196, 197; domestic violence in, 214; mutual aid societies in, 207; NPI use in, 125–134; outdoor recreation spaces in, 276; PPE and supply shortages in, 92; resuscitation rules for EMTs in, 132, 188; SARS-2 outbreak in, 24
New York Philharmonic, 216
New York State, 24, 75, 90, 195, 196, 219, 264. *See also specific cities*
New York Times, 128, 141, 192
New York University Hospital, 188, 189
New York University Langone Health System, 152
New Zealand, 107, 108
Newport, R.I., 65
Newsom, Gavin, 20, 264
Nextstrain (platform), 24
Nicobar Islands, 225–226

NIH (National Institutes of Health), 245
NL63 coronavirus, 47
non-contact greetings, 256
nonessential items, consumption of, 273
nonpharmaceutical interventions (NPIs), xi, 85–136; closed-off management, 9–11; collective action, 107–126; and efficacy testing, 241–242; end of pandemic and, 305–306; fear and use of, 147; flattening the curve with, 90–93; and geographic variability in effects of pandemics, 134–136; herd immunity argument against, 98–102; in immediate pandemic period, 249, 320–321; individual activities, 102–106; medical, social, and economic costs of, 94–98, 284; in New York City, 126–134; pharmaceutical interventions vs., 85–89. *See also specific interventions, e.g.,* mask-wearing
North East London Foundation Trust, 222
Northwestern Memorial Hospital, 152
nosocomial infections, 268
novel coronavirus. *See* SARS-2
NPIs. *See* nonpharmaceutical interventions
nucleic acids fragments, vaccines using, 232–233
nursing homes, transmission in, 14–15

OC43 coronavirus, 47, 308, 310–311
Occupational Safety and Health Administration (OSHA), 223
online conferences, 282
OpenTable, 136
OSHA (Occupational Safety and Health Administration), 223
outdoor spaces, 125, 280
outsiders, blaming, for infection, 171–177
Oxford University, 239

pain relief, connection and, 211–212
pandemics, 34–84; declaring an end to (*See* end of pandemic); geographic variability in effects of, 134–136; lessons learned from previous, 83–84; "perfect storm" analogy for, 316–317; public education in, 169–170; ruinous effects of, xiii–xiv; and SARS-2 vs. SARS-1, 44–60; self-disclosure during, 219. *See also* epidemics; *specific pandemics*
panic-buying, 273
Partners in Health, 113
partnerships, 212–213
pathogen(s): characteristics of, 300–306; ending pandemics via mutations in, 93, 306–311; evolution of, 71–72, 297–298; and human societies, 76; naturally-occurring, 160–161; novel, 298–299; testing for, to contain epidemics, 114–118; zoonotic, 76, 161, 299. *See also specific viruses and bacteria by name*
Patient Zero, 13–17, 22, 23
PeaceHealth St. Joseph Medical Center, 151–152
peer-review process, 168

penicillin, 296
PEPFAR (President's Emergency Plan for AIDS Relief), 246
"perfect storm" analogy, for pandemics, 316–317
period of subclinical infectiousness, 49
person-to-person transmission, 6, 12–13, 16, 56
personal experience, of disease, 320
personal liberty, 106, 210–211, 260, 317
personal practices, COVID-related changes in, 250–264
personal protective equipment (PPE), xi; manufacturing of, 272–273; requests for, 151–152; and risks for health-care workers, 222–223; supplies of, 103, 132
personal responsibility, 251
Pew Charitable Trust, 95
pharmaceutical interventions, 85–89
phase 3 studies, 241–242
Philadelphia, Pa., 72, 74, 97, 196, 201, 256–257
physical distancing, 89; costs of, 94–96; in crowded conditions, 192; and droplet transmission, 40; failure to follow guidelines about, 150, 151; as form of generosity, 215–220; mask use to promote, 104; in New York City, 126; at religious services, 263; in schools, 125; and social end of pandemic, 321; strategic, 100; and testing/contact tracing, 118; voluntary adoption of, 135–136
Pittsburgh, Pa., 123–124
The Plague (Camus), 34, 206, 295
plague(s): Biblical, 83–84; COVID-19 vs. other, 306; history of, 29–31; psychological effects of, 136, 141; as social equalizer, 178–180. *See also* epidemics; pandemics
pneumonic plague, 78, 148
polio, 297; 1916 outbreak, 144–145, 211; age and impact of, 180–181; interest in eradicating, 321–322; safety of vaccine for, 236–237; shunning survivors of, 173–174
politics: COVID-related changes in, 289–294; framing of pandemics in, 174–176; and in-person religious services, 264; and misinformation in pandemics, 154–157; and NPI adoption, 96, 106, 116, 216, 317; party affiliation and public health behaviors, 292; and Wuhan outbreak of SARS-2 virus, 6, 7
pollution, 248
post-COVID syndrome, 29
post-pandemic period, 249; economy in, 273; education in, 280; government's role in, 294; health-care practice in, 265, 269; societal changes in, 282–289; working from home in, 277–278
PPE. *See* personal protective equipment
prenatal exposure, to virus, 182, 288
preprint servers, 168–169
President's Emergency Plan for AIDS Relief (PEPFAR), 246
Prince of Wales Hospital, 37–38, 221
prison inmates, 192
privacy, 257–261

proactive school closure, 122
protein fragments, vaccines using, 232
Providence Regional Medical Center, 13
psychological effects of epidemics, 137–159; and altruism, 220; for children, 287; denial, 151–159; effects of caring for patients, 221; and emotional wellness, 142–143; fear, 143–151; grief, 137–142; and NPI use, 94; and public education, 169–170; and social end, 315–322
public discourse, nuance in, 292–293
public education, 73, 96–97, 169–170
public health, 87, 110–112, 170, 292, 322
public information: in China, 20; communicating, xv, 154–155; from preprint servers, 168–169; SARS-1 case reporting as, 36; SARS-2 case reporting as, 5–7, 11; and teaching/learning in culture, 227
pure anxiety type, mass psychogenic illness, 149

Qualtrics, 278
quarantines, 73–74, 81, 127–128, 192. *See also* self-isolation
Quebec City, Que., 177

R_0 parameter, 51–55; and herd immunity, 99–100; for MERS vs. SARS-2 virus, 59; misinformation about, 169; of respiratory pandemics, 63, 64
rabies, 201
race-based social divisions, 195–200, 244
random samples, testing, 116–117
RaTG13 virus, 20
rats, 76, 82
R_e. *See* effective reproduction number
reactive control measures, 50, 129–130, 133–134
reactive school closure, 122, 129
Reagan, Ronald, 154, 155
real estate industry, 276
reality, as social construction, 159
recessions, 271, 282
recovered patients: antibody treatments from serum of, 210, 229–230; immunity passports for, 201–203
recreational vehicles (RVs), 275
recurrent outbreaks, 66
Red Cross, 97
Reformation, 294
religious sentiment, 80, 261–264
religious services, 153–154, 262–263
remdesivir, 238
remote learning, 120, 251, 252, 257–258, 279–281
reopenings, 101; in China, 10–11; denial during, 158–159; religious services after, 263–264; for restaurants, 275; for schools, 121, 125; and social end of pandemic, 319–320
repatriation, 273
reservoirs of infection, detecting, 116
residential segregation, 198
respiratory infections, 180–182, 322–323. *See also specific agents and diseases*

"responsibility" bonuses, 193
restaurants, 275
resuscitation, EMT rules about, 132, 188
riboviruses, 63
rinderpest, 299
risk of death, calculating, 301
RNA vaccines, 233
robotics, autonomous, 276–277
Roman Empire, collapse of, 227
Rudra, 29
Russia, 167–168, 308
Russian flu pandemic (1889), 307–310
RVs (recreational vehicles), 275

sadness, 142
safety, vaccine, 236–237
safety as a service businesses, 276
St. Louis, Mo., 123–124
St. Patrick's Day parade, 130
St. Peters (steamboat), 312–313
St. Petersburg, Russia, 308
St. Thomas Church (Hanover, N.H.), 263
Salem witch trials, 149–150
San Antonio Food Bank, 191
San Francisco, Calif., 113, 129, 207
Santa Clara, Calif., 15
saquinavir, 245
SARS (severe acute respiratory syndrome), xi,
  4–6
SARS-1 virus, xi, 84; 2003 pandemic caused by,
  35–60, 133, 183, 221; aerosolization of body
  fluids and spread of, 194; characteristics of
  SARS-2 vs., 44–60
SARS-2 testing, 17–18, 50; and caseloads, 248–
  249; misinformation about, 156–157; in
  New York City, 127, 128; of random
  samples, 116–117; RNA testing, 114, 115;
  and social end of pandemic, 320–321;
SARS-2 virus (SARS-CoV-2), xi, 3–33; butterfly
  effect and spread of, 31–33; characteristics
  of, 26–29, 44–60; disease caused by (*See*
  COVID-19 disease); ecological release of,
  30–31; genetic mapping of, 20–26; and
  history of plagues, 29–31; impact of, on
  human evolution, 312; lethality of, 45–47;
  mismatch period for, 47–50; mutations in,
  307; origin of, 4, 20–21, 159–162; reproduction number of, 52–60; SARS-1 vs.,
  44–60; sociological implications of name
  for, 317; tracking of, xiv; US spread of,
  12–20; USA/WA1/2020 variant, 12, 23;
  vaccine development for, 227–229, 232,
  233, 235; variants of, 249; Wuhan, China
  outbreak of, 3–12
Saudi Arabia, 59
Scarborough Grace Hospital, 41
scarlet fever, 88
sCFR (symptomatic case fatality rate), 45
school closure, 119–126; in 1918 Spanish flu
  pandemic, 73–74; and infectiousness of
  children, 185–186; in New York City, 129;

and remote learning, 120, 251, 252, 257–
  258, 279–281; social narratives about, 318;
  withdrawal of children prior to, 136
schools (generally): contact tracing in, 110–
  111; flipped classrooms in, 281; outdoor
  spaces at, 125, 280
science, culture's treatment of, 289–294
scientific community, xv; preprint server use by,
  168–169; respect for, 290–291; silencing of, by
  politicians, 154–157; Twitter use by, 23–24;
  vaccine and drug development in, 227–238
seasonal flu, 66; biology of virus responsible for,
  228; case fatality rate of, 46; narratives about
  COVID-19 and, xiv, 317–318; reproduction
  number for, 51, 59, 105; vaccine for, 236
seasonal forcing, 66
seasonality, of virus, 66, 100
Seattle, Wash., 14–23, 53, 90
Seattle Flu Study, 16–17, 22
seismic activity, during lockdown, 247–248
self-disclosure, 219
self-isolation, 50, 109–114, 320–321. *See also*
  quarantines
self-reflection, 264
self-reliance, 250–253
sepsis, 78
sexual activity, 215
sexually transmitted diseases, 318. *See also
  specific diseases, e.g.,* syphilis
shared identity, 219
shelter-in-place orders, 130–131, 134, 214, 273.
  See also lockdowns; stay-at-home orders
shingles vaccine, 235
ships: bubonic plague on, 80, 81; SARS-2 on, 18–
  19, 21, 27, 46, 176; Spanish flu on, 72–73
Shunde district, China, 36
sick leave, 190, 193, 277
sickle cell anemia, 314
silver, colloidal, 162–163
Singapore, 41, 52, 111, 115, 124, 135
Sinovac, 232
Sioux Falls, S.Dak., 193
small businesses, 275, 276
smallpox: contact tracing for, 110, 111; eradication of, 111, 297; and human genetic
  variation, 312–313; period of subclinical
  infectiousness, 49; R_0 parameter, 51; reason
  for forgetting epidemics of, 84; symptoms
  of, 204; vaccine for, 231
Smithfield meatpacking facility (Sioux Falls,
  S.Dak.), 193
smoking indoors, 250
Snohomish, Wash., 12–14
Snowden, Frank, 76, 298
social distancing, 89
social divisions, 171–205; age-based, 180–188; and
  blaming others for causing infection, 171–
  177; discrimination against Wuhan residents,
  200; and epicenters in New York City, 133;
  ethnic and racial, 195–200; gender-based,
  188–189; immune status as basis for, 200–203;

social divisions (*cont.*),
impact of COVID-19 pandemic and, 180–
200; lessons from pandemic about, 324;
living and working conditions as basis for,
192–194; plague as equalizer of, 178–180;
and psychological effects of pandemic, 143;
and school closures, 120; socioeconomic-
based, 189–191; violence/social unrest
caused by, 203–205
social justice, protests for, 264
social learning, 224–226
social media, 151–153, 167–168. *See also* Twitter
social mixing, 119, 122–123
social networks, 55–58, 261
social systems: butterfly effect in, 32–33; costs
of NPIs for, 94–98; end of a pandemic for,
315–322
social unrest, 158–159, 203–205
societal changes, 247–294; and changes in
natural environment, 247–249; and economic impact, 271–282; in expectation of
privacy, 257–261; in greeting practices, 253–
257; in health-care practice, 265–271; long-
term, 282–289; in personal attitudes/
practices, 250–264; and political and
cultural impact, 289–294; in religious senti-
ment/spirituality, 261–264; in self-reliance,
250–253; uncertainty about, 248–250
socioeconomic status, 87, 189–191, 285
Solidarity trial, 238
South Korea, 58, 92, 118, 135, 249, 262
South Pacific islands, 108
Southington, Conn., 173
Spaniards, blaming plague on, 172
Spanish flu pandemic (1918), 29, 67–75; age
and impact of, 180–181, 182; changes
catalyzed by, 250; COVID-19 pandemic vs.,
62; generosity of health-care workers in,
209–210; long-term effects of, 283, 284, 288,
289, 293; loss during, 140; memory of, 34–
35, 61; NPIs in, 86, 97, 103, 108, 119, 123–
124; origin of, 174; symptoms of, 204
spike proteins, 26, 230, 307
spillover effects, of policies, 219–220
spirituality, 261–264
spitting, public, 250
state borders, closing, 109
stay-at-home orders: costs associated with, 94;
domestic violence during, 214; and First
Amendment rights, 264; for flattening the
curve, 90–93, 107; misinformation about,
167; and school closures, 121; socioeconomic status and effects of, 190–192;
spillover effects of, 219–220. *See also* lockdowns; shelter-in-place orders
steroids, 239
stigma, for exposed persons, 202–203
Storm the NIH campaign, 245
Strasbourg, France, 171–172
substitute teaching, 279–280
Sun Labour Memorial Hospital, 36–37

super-spreaders, 37, 52–60, 137
Surat, India, 148
surveillance, 10, 257–260, 279
survivors, equality for, 178–179
Sweden, 101, 288, 302
swine flu, 62, 69
symptomatic case fatality rate (sCFR), 45
syphilis, 86, 109, 111, 201

T cells, 230, 231
Taiwan, 135, 288
Tata Consulting Services, 277
teaching: capacity for, 224–227; substitute, 279–280
technology, 324; for contact tracing, 112, 259–
261; in COVID-19 pandemic, 85–86, 276–
277; theology of, 263
telemedicine, 265–267
Tennessee, 158
testing: antibody, 114–118, 201–202; and NPIs,
114–118. *See also* SARS-2 testing
Texas, 263
thermal screening, 43
3M, 272–273
Thucydides, 141–142, 220, 261–262
toilet-paper shortages, 274
Torbat-e Heydarieh, Iran, 222
training, for health-care workers, 270–271
transmission: as goal of pathogen, 306–307;
interventions to reduce, 89, 155; rate of, for
SARS-2 virus, 22
travel: air, 12, 24, 43, 108–109; bubonic plague
and, 80, 81; and COVID-19 pandemic, 7,
127, 129; restrictions on, 24, 107–109, 212–
213 (*See also* border closings). *See also* ships
travel nurses, 146
triage policies, hospital, 186–188
Trump, Donald, 153, 155–157, 160, 162–166,
174, 248, 317
Tsimshian people, 313
tsunami stones, 225
tuberculosis, 88, 110–111, 184, 201, 312
Tulane Hospital, 222
Tuskegee syphilis study, 111
Twitter, xv, 23–24, 167–169
229E coronavirus, 47, 240
typhoid, 88, 110

U-Haul, 275
U-shaped curve, 180–182, 310
ultraviolet light, as treatment, 164
uncertainty, 170, 248–250, 324
unemployment, 94–95, 191, 271
United Kingdom, 100–101, 110–111, 161. *See
also specific cities*
United States: 1918 Spanish flu in, 67–71; 2003
SARS-1 pandemic in, 41; age and COVID-19
case fatality in, 183; beliefs about science
and politics in, 291–292; contact tracing in,
112–114; COVID-related deaths in, xiv, 18,
26, 75, 221–223; global standing of, 286–
287; gun sales in, 204; handwashing in,

253–254; health care policies in, 190;
health-care worker deaths in, 221–223;
hospital beds per capita, 92; mask protests
in, 106; mortality in, 60–61; pandemic preparedness in, 322; philanthropy, 208; proof
of immunizations for immigrants to, 201;
SARS-2 spread in, 12–20, 26; school closure
in, 124–125; social end of pandemic in,
319–321. *See also specific cities and states*
University of Pennsylvania Medical Center, 268
University of Washington, 257
urban areas, movement away from, 133, 252–253
urbanization, infectious disease and, 200–300
"us versus them" mentality. *See* social divisions
USA/WA1/2020 (WA1) SARS-2 variant, 12, 23
USS *Theodore Roosevelt,* 18, 192
Utah, 113

vaccine development, 227–238; approaches to,
230–233; clinical trials in, 241–244; and
convalescent serum treatments, 229–230;
and death rates, 87, 88; and duration of
pandemic, 249–250; misinformation about,
156; pandemic's effect on, 265; safety as
concern in, 236–238; for SARS-2 virus, 227–
229, 232, 233, 235; speed of, 228–229
vaccines: adjuvants of, 233–235; BCG, 184;
coronavirus, 228, 230, 232, 233; cross-
immunity from other, 184; diphtheria, 234;
and eradication of infectious diseases, 296;
and Guillain-Barré syndrome, 62, 236; and
herd immunity, 101–102, 305–306; and
immune system, 230–237; influenza, 236;
live attenuated virus, 231–232, 236–237;
from nucleic acid fragments, 232–233; from
protein fragments, 232; seasonal flu, 236;
shingles, 235; smallpox, 231
Venice, Italy, 81
ventilator shortages, 131–132, 186, 272–273
Vermont, 109, 113, 219
vertical transmission, 182
video chatting, dating via, 214–215
videoconferencing, 258–259
violence, 203–205
volunteerism, 206–209, 221, 240–244
vulnerability to infection, 100

W-shaped curve, 180–181
wages, 284–285
wai greeting, 256
war, marriage/divorce rates and, 213–214
Warren County High School (McMinnville,
Tenn.), 150
Warwick hotel (New York, N.Y.), 276
Washington State, 18, 20, 58
water, access to, 190–191

weather, 66, 83
Welch, H. Gilbert, 269
wet markets, 3–4
WHO. *See* World Health Organization
Whole Leaf Organics, 163
women: COVID-19 impact for, 188–189; NPI
use by, 254
worker protections, 285–286
working-age adults: economic impact of
pandemic affecting, 284–285; mortality
respiratory infections for, 181, 182
working conditions, crowded, 192–144
working from home, 19, 90, 189–190, 252, 274–
279
World Health Organization (WHO): in 2003
SARS-1 pandemic, 36, 37, 41, 42, 44;
COVID-19 case information from, 19;
declaration of pandemic by, 26; on misinformation, 169; NPI guidance from, 103;
pathogen naming conventions of, 174; in
Solidarity trial, 238
World War I, 71–72, 97, 209–210, 213, 283
World War II, 212, 213
worry, 142, 217–218
Wuhan, China: discrimination against residents
of, 200; disparities in COVID-19 impact in,
181–183, 188; hospital for COVID-19
patients in, 92; SARS-2 outbreak in, 3–12,
21; suppression of information from, 151;
travelers from, 12–13, 16
Wuhan Center for Disease Control and Prevention, 4, 5, 9
"Wuhan flu," 317
Wuhan Institute of Virology, 4, 160
Wuhan pneumonia, 174

X chromosome, 189

Yale New Haven Hospital, 186–187, 266
Yale University, xiv, xv, 62, 75, 104, 206
years of life lost (metric), 303–304
yellow fever epidemic (1793), 196–197, 256–257
Yersinia pestis, 76, 78–80, 314. *See also* bubonic
plague
young people: challenge trials involving, 242–243;
COVID-19 symptoms for, 16–17, 28; COVID-
related death for, 249, 302; impact of
COVID-19 for, 180–183, 287; transmission of
COVID-19 by, 185; volunteering by, 206–208

Zaandam cruise ship, 176
Zhejiang Medical University, 8
Zhong, Nanshan, 8–9, 42
zoonotic pathogens, 76, 161, 298–299
Zuckerman, James E., 270

About the Author

Nicholas A. Christakis is a physician and sociologist who explores the ancient origins and modern implications of human nature. He directs the Human Nature Lab at Yale University, where he is the Sterling Professor of Social and Natural Science in the Departments of Sociology, Medicine, Ecology and Evolutionary Biology, Statistics and Data Science, and Biomedical Engineering. He is the codirector of the Yale Institute for Network Science, the coauthor of *Connected*, and the author of *Blueprint*.